**CHILDREN AND HEALTH CARE: MORAL AND SOCIAL ISSUES**

# PHILOSOPHY AND MEDICINE

*Editors:*

H. TRISTRAM ENGELHARDT, JR.
*The Center for Ethics, Medicine and Public Issues*
*Baylor College of Medicine, Houston, Texas, U.S.A.*

STUART F. SPICKER
*University of Connecticut, School of Medicine,*
*Farmington, Connecticut, U.S.A.*

VOLUME 33

# CHILDREN AND HEALTH CARE

*Moral and Social Issues*

*Edited by*

LORETTA M. KOPELMAN

and

JOHN C. MOSKOP

*East Carolina University School of Medicine,*
*Greenville, North Carolina, U.S.A.*

KLUWER ACADEMIC PUBLISHERS

DORDRECHT / BOSTON / LONDON

Library of Congress Cataloging-in-Publication Data

Children and health care.

    (Philosophy and medicine; v. 33)
    Papers presented at a seminar series held at Greenville, N.C., in the fall of 1986. Sponsored by East Carolina University School of Medicine and other organizations.
    Includes index.
    1. Child health services–Social aspects–United States–Congresses. 2. Child health services–Government policy–United States–Congresses. 3. Decision-making in children–Congresses. 4. Health Behavior in children–Congresses. I. Kopelman, Loretta M. II. Moskop, John C., 1951– . III. East Carolina University. School of Medicine. IV. Series. [DNLM: 1. Attitude to Health–in infancy & childhood–congresses. 2. Child Advocacy–United States–congresses. 3. Child Health Services–United States–congresses. 4. Child Welfare–United States–congresses. 6. Health Policy–United States–congresses. W3 PH609 v.33 / WA 320 C5365 1986]
RJ102.C485   1989   362.1'9892'000973   88-13198
ISBN 1-55608-078-6

---

Published by Kluwer Academic Publishers,
P.O. Box 17, 3300 AA Dordrecht, The Netherlands

Kluwer Academic Publishers incorporates
the publishing programmes of
D. Reidel, Martinus Nijhoff, Dr W. Junk and MTP Press.

Sold and distributed in the U.S.A. and Canada
by Kluwer Academic Publishers,
101 Philip Drive, Norwell, MA 02061, U.S.A.

In all other countries sold and distributed
by Kluwer Academic Publishers Group,
P.O. Box 322, 3300 AA Dordrecht, The Netherlands.

All Rights Reserved
© 1989 by Kluwer Academic Publishers
No part of the material protected by this copyright notice may be reproduced or
utilized in any form or by any means, electronic or mechanical,
including photocopying, recording or by any information storage and
retrieval system, without written permission from the copyright owners.

Printed in the Netherlands

to my husband Arthur E. Kopelman,
a pediatrician,
and in memory of my mother
Gertrude M. Veitch Criden,
a nurse, LMK

to my daughter,
Megan Ruth Moskop, JCM

# TABLE OF CONTENTS

ACKNOWLEDGEMENTS     xi

INTRODUCTION     xiii

## SECTION I / CHILDREN'S HEALTH AS A SOCIAL AND POLITICAL ISSUE

INTRODUCTION     3

BARBARA STARFIELD / Child Health and Public Policy     7

STUART F. SPICKER / Comments on Barbara Starfield's 'Child Health and Public Policy'     23

ANN L. WILSON / Development of the U.S. Federal Role in Children's Health Care: A Critical Appraisal     27

TODD L. SAVITT / American Social and Political Thought and the Federal Role in Child Health Care     67

ROBERT J. LEVINE / Children as Research Subjects     73

LORETTA M. KOPELMAN / When is the Risk Minimal Enough for Children to be Research Subjects?     89

## SECTION II / CHILDREN, ILLNESS, AND DEATH

INTRODUCTION     103

ROSALIND EKMAN LADD / Death and Children's Literature: *Charlotte's Web* and the Dying Child     107

LORETTA M. KOPELMAN / Charlotte the Spider, Socrates, and The Problem of Evil     121

GARETH B. MATTHEWS / Children's Conceptions of Illness and Death    133

JOHN C. MOSKOP / Terminally Ill Children and Treatment Choices: A Reply to Gareth Matthews    147

## SECTION III / CHILDREN'S AND PARENTS' ROLES IN MEDICAL DECISIONMAKING

INTRODUCTION    155

ANGELA R. HOLDER / Children and Adolescents: Their Right to Decide about Their Own Health Care    161

ROBERT L. HOLMES / Children and Health Care Decisionmaking: A Reply to Angela Holder    173

DAN W. BROCK / Children's Competence for Health Care Decisionmaking    181

ROBERT L. HOLMES / Consent and Decisional Authority in Children's Health Care Decisionmaking: A Reply to Dan Brock    213

WILLIAM RUDDICK / Questions Parents Should Resist    221

H. TRISTRAM ENGELHARDT, JR. / Taking the Family Seriously: Beyond Best Interests    231

## SECTION IV / THE ROLE OF THE PEDIATRICIAN

INTRODUCTION    241

PETER C. ENGLISH / 'Not Miniature Men and Women': Abraham Jacobi's Vision of a New Medical Specialty a Century Ago    247

TODD L. SAVITT / The Development of Pediatrics as a Specialty    275

JOHN LADD / The Good Doctor and the Medical Care of Children    281

STUART F. SPICKER / Comments on John Ladd's 'The Good Doctor and the Medical Care of Children'    303

MYRON GENEL / Government by Case Anecdote or Case Advocacy: A Pediatrician's View    305

# TABLE OF CONTENTS

H. TRISTRAM ENGELHARDT, JR. / Advocacy: Some Reflections on an Ambiguous Term — 317

THOMAS G. IRONS / Loving the Chronically Ill Child: A Pediatrician's Perspective — 323

JOHN C. MOSKOP / Love and the Physician: A Reply to Thomas Irons — 331

NOTES ON CONTRIBUTORS — 337

INDEX — 339

## ACKNOWLEDGMENTS

The articles in this volume are based on papers presented at seminars on 'Children and Health Care: Moral and Social Issues' at East Carolina University School of Medicine in Greenville, North Carolina, in the fall of 1986. We wish to express our appreciation to those sponsoring the program: East Carolina University School of Medicine, The North Carolina Humanities Committee, The Duke Endowment, and Pitt County Memorial Hospital. To the faculty of the School of Medicine and especially our dean, William Laupus, M.D., we extend our special appreciation. His support for this program and the activities of the Department of Medical Humanities has been invaluable. We also wish to thank the now-retired chancellor of East Carolina University, John M. Howell, who introduced and supported this program. The Eastern Area Health Education Center of North Carolina, as always, was most cooperative. We would also like to thank the members and staff of the North Carolina Humanities Committee, the general editors of this series, H. Tristram Engelhardt, Jr., and Stuart F. Spicker, and Martin Scrivener of Kluwer Academic Publishers for their continuing encouragement, support, and friendship. Vicki Davenport Tyson, Claire Pittman, and Shirley Nett assisted us in preparing the symposium and subsequent volume. We thank them for all their hard work and dedication.

Loretta M. Kopelman, Ph. D.<br>
John C. Moskop, Ph.D.

# INTRODUCTION

Before a separate Department of Medical Humanities was formed, the editors of this volume were faculty members of the Department of Pediatrics at our medical school. Colleagues daily spoke of the moral and social problems of children's health care. Our offices were near the examining rooms where children had their bone-marrow procedures done. Since this is a painful test, we often heard them cry. The hospital floor where the sickest children stayed was also nearby. The physicians, nurses, and social workers believed that children's health care needs were not being met and that more could and should be done. Fewer resources are available for a child than for an adult with a comparable illness, they said. These experiences prompted us to prepare this volume and to ask whether children do get their fair share of the health care dollar.

Since the question "What kind of health care do we owe to our children?" is complex, responses should be rooted in many disciplines. These include philosophy, law, public policy and, of course, the health professions. Representing all of these disciplines, contributors to this volume reflect on moral and social issues in children's health care. The last hundred years have brought great changes in health care for children. The specialty of pediatrics developed during this period, and with it, a new group of advocates for children's health care. Women's suffrage gave a political boost to the recognition of children's special health needs. Also during this period legal rights were first granted to children independently of their parents. In recent years, too, the social and behavioral sciences have studied the development of competency and responsibility throughout the years of childhood, offering new insights about children's special needs and abilities. Most importantly, advances in the diagnosis, treatment, and prevention of disease have made good health care, like good education, housing, and food, almost indispensable to the proper development of a child's potential and opportunities. Since these advances have been the result of research, difficult questions arise about when children may be subjects in medical studies. Though government programs now subsidize health care for thousands of low income and chronically ill children, many still lack access to effective health care. All of these issues are explored in this volume.

*Loretta M. Kopelman and John C. Moskop (eds.)*
*Children and Health Care: Moral and Social Issues,* xiii–xiv.
© *1989 by Kluwer Academic Publishers.*

The volume is divided into four sections. In the first section, we examine children's needs for health care and the history of U.S. federal initiatives to provide for those needs. Children's own views of sickness and death are explored in the second section. In the third section, we raise questions about what role children should have in health care choices, and what the limits of parental authority should be. These questions are addressed from moral as well as legal standpoints. We reflect on the role of those directly caring for children in the final section, focusing on the pediatrician as representative of all pediatric health care workers. We focus on the role of pediatrics, but do not intend to slight other health professionals dedicated to children's care. We would also like to acknowledge pioneers in educating the public about child health and development, expecially Benjamin Spock, T. Berry Brazelton, and Haim Ginott.

Throughout the volume, we concentrate on the older child because, first, these children have not only important needs but also desires about their care. Second, the medical ethics literature on children's health care gives little attention to older children. Historians from a future age who survey this literature might conclude incorrectly that most of the moral or social problems in children's health care in 1970s and 1980s concerned sick newborn infants. We hope that this volume will help to correct that imbalance.

# SECTION I

# CHILDREN'S HEALTH AS A SOCIAL AND POLITICAL ISSUE

# INTRODUCTION

Chronically diseased or disabled children often have significantly fewer opportunities in life than their healthy counterparts. Thus, decisions such as how much money to spend on children's health care or whether to fund research projects on childhood diseases will have a direct effect on the lives of those children. Because prevention, treatment, and cure expand opportunities, we naturally want to provide them for our children, and perhaps we have some obligation to do so. The contributors to this section consider what it means to allocate an appropriate share of our resources, especially our health-care funds, to children. How do we decide when we have provided them proper access to the health care they need?

There are, of course, different arrangements made in different countries. Our authors have focused on their experiences in the United States. Their analyses raise questions about the United States' commitments to and implementation of equality of opportunity for children, child welfare, and fairness in the allocation of resources. Poor children may be disadvantaged once by poverty, disadvantaged again if they grow to adulthood unhealthy, and yet again in living with diminished potential that could have been avoided by good health care. If a society believes in the principle of equal opportunity and in the encouragement of each citizen's use of his or her potential, what implications does this have for provision of adequate health care for children? More generally, are our funding choices fulfilling our social or moral responsibilities to the children of the world?

Barbara Starfield, in 'Child Health and Public Policy,' marshals an array of empirical and philosophical arguments to urge better funding of children's health programs. First, she argues, the United States can afford to take better care of its children's health, and the need to do so is great. Second, in a society committed to equality of opportunity, we ought to do so. Finally, public and private utility would be maximized because we will need a healthy population of adults in the near future to support the growing percentage of elderly citizens predicted in the coming decades.

Starfield argues that children have been neglected in public policy, especially in regard to their health. She points out that one-fifth of our children live in poverty, and serious health problems are associated with poverty. For example, after the neonatal period but before one year of age, death rates among low-income children are double to triple the national average. She argues that as the health of our elderly has been vastly improved by better funding, so also would public commitment to better nutrition, screening, and health care be of great benefit to children. Citing evidence for the effectiveness of pediatric health care, Starfield concludes, "There is thus little question that access to health care can greatly improve health, particularly for those who are at greatest risk of poor health. That is the reason to be concerned about recent public policy that has apparently diminished access to care. ... If we favor equality of opportunity, if we believe that health care should be distributed according to need, or if we believe that it is important to maximize health in a generation that will have to support an increasing number of elderly people, the policy imperatives are clear" (p. 14, 17). She favors expanding Medicaid coverage and standardizing eligibility, maintaining food support programs, and collecting sound and important information on many levels to identify those who are most needy. Finally, Starfield invites us to reflect on what it would cost to provide good health care to children in relation to the amount of the world's current military spending.

In his commentary, Stuart Spicker highlights on Starfield's normative claims: first, that present funding programs are unfair to children; second, that children have great unmet needs; third, that there is much we ought to do to minimize children's risk of getting illnesses; fourth, that if we value equality of opportunity, we should try to achieve access to a basic standard of health care for all children; and fifth, that it will prove socially useful and cost-effective to have a healthier population of children. Spicker draws our attention to what such claims presume.

In 'Development of the U.S. Federal Role in Children's Health Care: A Critical Appraisal,' Ann Wilson reviews the history of federal provision for children's health care. This history, she suggests, is marked by three recurrent themes: the proper scope of federal authority over issues of children's health, the economic consequences of federal programs in this area, and the assertion on behalf of children of a human right to health care. Wilson begins her history in the late nineteenth century with the Progressives' concern for the health of the child victims of industrialization and mass immigration. These concerns led Progressives like Theodore Roosevelt to propose the establishment of a Children's Bureau to conduct research and provide education on children's health. Despite the vocal opposition of legislators who questioned the authority of the federal government in this area, and the need for federal assistance to promote child health (and who were accused of protecting the interests of those who exploited child labor), the Children's Bureau was established by Congress in 1912. During the next decade, this agency conducted studies on infant mortality; it also published and distributed educational pamphlets on maternal and child health.

In 1920, another proposal, sponsored by Senator Sheppard and Representative Towner, called for increased federal support for child welfare activities. Opponents of this bill ridiculed its strong support among recently enfranchised women. Though enacted in 1921, the Sheppard-Towner legislation remained controversial and was repealed before the end of the decade. Franklin D. Roosevelt's efforts to blunt the personal suffering of the Depression included provision for maternal and child health under the Social Security Act of 1935. An emergency federal maternal and infant care program was established for the wives and children of relocated servicemen during World War II. The years after World War II saw a gradual expansion of health programs for children, marked especially by the establishment of the Medicaid program in 1965 which supported health care for large numbers of low income children. In 1975, however, a reorganization of the Department of Health, Education and Welfare assigned many of the health and welfare programs of the Children's Bureau to other federal agencies, thus effectively diminishing the role of the Children's Bureau as an advocate for the welfare of children. Finally, Wilson notes that the reductions in funding for maternal and child health care programs in the early 1980s have begun to have an adverse effect on infant mortality rates in the United States. She cites recent calls for a new federal initiative in assessing the health and well-being of the nation's children.

In his commentary on Wilson, Todd Savitt seeks to locate the controversies over U.S. federal support for children's health care within a broader social and political context. Savitt points out that the major opposition to the creation of a Children's Bureau came from those who adhered to the late nineteenth century laissez-faire system of government using a justification based on Social Darwinism. He also notes that as social developments gradually eroded support for this position, a new source of opposition to federal involvement in health care arose. This opposition, first expressed strongly in the debate over the Sheppard-Towner Act, came from the medical profession in its attempt to prevent interference with its control over medical practice. Savitt notes the resemblance of the Reagan administration's suspicion of big government to the views of its predecessors by one hundred years, the Social Darwinists.

Biomedical research is also a social and political issue in the sense that societies must decide how important it is to acquire knowledge in this area. Efforts to regulate and to support research are based in part on estimates of its value. Robert J. Levine, in 'Children as Research Subjects', discusses the general problem of when research involving children as subjects may be permitted and the solution to that problem proposed by the National Commission for the Protection of Human Subjects of Biomedical and Behavioral Research (the National Commission). Levine begins by criticizing Ramsey's view that adherence to the ethical principle of respect for persons requires "that we leave persons alone unless they consent to be touched" (p. 73). According to Ramsey, non-consenting research subjects (including all small children) are, therefore, wronged whether or not there is risk because they are touched without their consent. Responses to this position held by McCormick, Freedman, Ackerman, and Gaylin are discussed in order to indicate some of the major themes in the controversy.

Levine points out that the recommendations of the National Commission "reflected its conclusion that, because infants and very young children have no autonomy, there is no obligation to respond to it through the usual devices of informed consent" (p. 76). The Commission recognized the need to do research so that children would not become therapeutic orphans. Interpreting 'respect for persons' differently from Ramsey, it concluded that the child's capacity for self-determination, where it exists, should be respected, but that respect is also shown by protecting children from harm. The National Commission, Levine points out, "recommended that the authority accorded to children or their legally authorized representatives to accept risk be strictly limited; any proposal to exceed the threshold of 'minimal risk' requires special justification" (p. 78). Levine then discusses some of the specific recommendations offered by the National Commission.

In 'When is the Risk Minimal Enough for Children to be Research Subjects?' Loretta M. Kopelman agrees with the general position adopted by the National Commission, expressed in the federal guidelines, and defended by Robert Levine. According to this view, research involving children may be permitted if it holds out the prospect of direct benefit to them, or if it is not too risky. Kopelman, however, focuses on a central notion of this policy, minimal risk, arguing that its definition is flawed. She holds that without further clarification about what this means, the widespread agreement about when children may be used as research subjects may be illusory; empirical evidence suggests that people tailor their notion of minimal risk to fit preconceived ideas about what research should be done. She shows why standards for estimating risk are needed, how minimal risk is a pivotal notion in the guidelines, and why the current definition of it is flawed.

BARBARA STARFIELD

# CHILD HEALTH AND PUBLIC POLICY

At one time in our history, children were very important. Were it not for the labor they contributed, many families would not have risen above penury. To a considerable extent, we owe the success of our national growth over the first century of our nation to the children. We "rewarded" them with a system of public education that, with all of its problems, is still a model for most of the world [4].

But, unfortunately, attention shifted away from children just when medical progress was entering a new revolution. Public policy, and particularly that aspect of it concerned with health, has neglected children. As a society, we rationalize the neglect by considering children "healthy". Only the awareness of teenage sexuality (with consequences for early childbearing) and drug misuse (with consequences for crime) have begun to shake our notion of carefree, disease-free youth. Yet still we fail to think of these phenomena as manifestations of neglect of children.

Consider the following:
– Poverty rates among children have been rising, and about one in every five children lives in poverty. One-third of all children born in this country in 1980 will spend some part of their lives living in poverty ([38], p. 44).
– In 1985, 49 states defined 'need' in the AFDC program for a family of three with children at an amount less than the poverty level. Eighteen states defined 'need' as less than 50% of the poverty level [8]. Income levels for Medicaid eligibility are on average less than 44% of the federal poverty level [11].
– While the number of children in poverty grew by 18% between 1980 and 1982, Medicaid eligibility stayed constant, or in some places, declined [9]. In 1985, Medicaid reached fewer than half of poor and near poor families as compared to 65% in 1969 [8].
– Even though ill health is more frequent among poor children, many serious health problems cut across social class. One in twenty high school seniors – not including the 20% who have already dropped out of school – smoke marijuana more than 20 times a month, or use cocaine at least once a month. One in five smokes cigarettes daily. Three in eight have five or more

*Loretta M. Kopelman and John C. Moskop (eds.)*
*Children and Health Care: Moral and Social Issues, 7–21.*
© *1989 by Kluwer Academic Publishers.*

alcoholic drinks in a row within each two-week period ([18], p. 46). One in ten newborns in urban, middle class families has an umbilical cord blood lead level that is elevated [5].

– About one-third of children born into two-parent families will spend at least some of their childhood in a family where their parents separate [13]. As children in families with divorce are subject to a wide variety of stresses, increased rates of parental separation are a cause for concern about children.

It is not, of course, that our society has failed to make great strides that promote health, social strides as well as scientific ones. Consider what has happened to the poverty rates of senior citizens: marked declines since 1970. Life expectancy has increased remarkably at older ages; between 1968 and 1980, life expectancy increased 4 years at age 70. Suicide, the ultimate effect of alienation, is much less common among the elderly than it was two decades ago. In the same period of time, there have been large increases in the participation of our elderly in the democratic process, as evidenced by voting rates. Our changing demographic profile assures that this will continue to be the case; advocacy for the elderly will continue at high levels. Their needs will receive attention, as they should.

Hardly the same can be said for children. Poverty rates in childhood have increased. Life expectancy at age 15 has increased only one year compared to the four years at age 70. Suicide rates have increased in childhood and adolescence, particularly among white males and females [25].

It seems only fair that society now consider a major public policy initiative for children as well. What are the problems of children and youth? Are there clues as to what kinds of interventions would be useful in dealing with the problems? What is the policy imperative, at the local level, the state level, and the national level?

## I. THE POLICY CHALLENGE

We know that the average child spends about five days bedridden in a year. About one child in twenty is reported to be in fair or poor health and about one in twenty-five is limited in his or her activities because of some persistent health problem. We also know that the prevalence of problems is related to family income, with low-income children experiencing a much greater frequency of problems, both acute and chronic, than middle-income children, who in turn experience more than children in higher-income families [22].

Table I shows the relative frequency of some *specific* problems of childhood in low-income children as compared with other children. The frequency of low (LBW) birthweight is double, the frequency of teenage births is triple, and the frequency of delayed immunizations is triple. Asthma is more common, but we are not quite sure how much more common because there is no standard way of diagnosing it, and reporting by parents is notoriously subject to biases according to the type of medical care that the child has received. Bacterial meningitis is twice as common in low-income children, rheumatic fever is more than twice as common, lead poisoning is three times as common.

TABLE I

Approximate relative frequency of selected health problems in low-income children

| | |
|---|---|
| LBW | double |
| Teenage births | triple[a] |
| Delayed immunizations | triple |
| Asthma | higher |
| Bacterial meningitis | double |
| Rheumatic fever | double-triple |
| Lead poisoning | triple |

[a] Estimated on basis of black/white differential.
Source: Abstracted from [33].

TABLE II

Approximate relative severity of selected problems in low-income children

| | |
|---|---|
| Neonatal mortality | 1.5 times |
| Postneonatal mortality | double-triple |
| Child deaths | |
|    due to accidents | double-triple |
|    disease related | triple-quadruple |
| Complications of appendicitis | double-triple |
| Diabetic ketoacidosis | double |
| Complications of bacterial meningitis | double-triple |
| % with conditions limiting school activity | double-triple |
| Lost school days | 40% more |
| Severely impaired vision | double-triple |
| Severe iron deficiency anemia | double |

Source: Abstracted from [33].

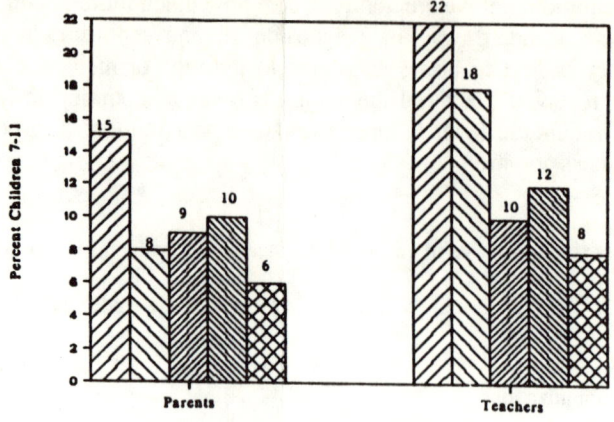

Fig. 1. Source: [41].

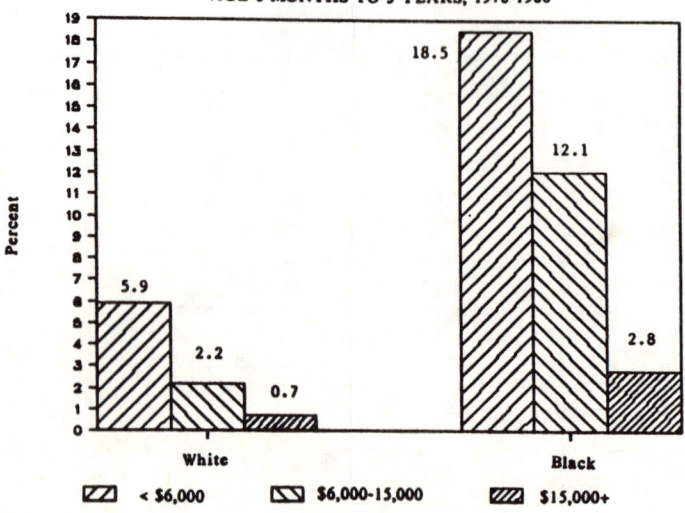

Fig. 2. Source: [3].

Table II shows the relative severity of problems when the problems occur among low-income compared with other children. Death rates are higher among low-income children, and the differential becomes greater with age during childhood. In the first year of life after the neonatal period, death rates are double to triple those of other children; after the first year, death rates due to diseases are triple to quadruple among low-income children. Low-income children are two to three times more likely to have complications from appendicitis, twice as likely to experience ketoacidosis if they are diabetic, two to three times more likely to have complications if they contract bacterial meningitis, two to three times more likely to have a condition that limits their school activities, 40% more likely to be absent from school with their health conditions, two to three times more likely to have severely impaired functional vision, and twice as likely to have severe iron deficiency anemia.

Figure 1 [41] shows that the reporting of increased problems among low-income children is not limited to the parents. Teachers also report higher frequencies of health problems among lower-income children, even though they do not know the income of the children in their classes. Thus, the consistency of the findings from a wide variety of sources – vital statistics data, medical facilities, parents, and teachers – makes it quite evident that the lower the family income, the more likely the child is to have a health condition, and when the child has a health condition, to have a more severe form of it.

Figure 2 [3] provides information on blood lead levels in U.S. children. Blood lead levels are higher in black children than in white children, but within each racial group, blood lead levels are higher the lower the family income. Please note, however, that the frequency of elevated blood lead levels, in this survey defined as 30 micrograms per deciliter, is not insubstantial among higher-income children – 1 to 2%. Since the time these data were analyzed, the Centers for Disease Control has lowered the cutoff point to 25 micrograms – so that an even greater proportion of children (9%) is in the danger zone ([21], p. 10).

In fact, very recent data indicate that there is no threshold level for the adverse effects of blood lead (from 5 to 35 micrograms per deciliter) on height, weight, and chest circumference. Elevated blood levels, even if minimal, are associated with impaired growth [28].

Mental health problems are of substantial frequency in the population of children. Most epidemiological studies, including those using well standardized instruments, indicate that about 15% of children have a mental health problem that requires at least some professional attention [15].

TABLE III

Benefits of food programs

Marked decrease in anemia during the 1970's
WIC evaluations show:
– Increased gestational age
– Less preterm birth
– Increased iron intake by child participants
– Increased verbal ability
– Increased digit memory
– Increased birth weight (small magnitude)

Sources: [27, 39].

Who are the children at high risk of health problems? How can they be identified so that resources can be targeted at them?

Poor children are certainly at higher risk, as Tables I and II demonstrated. Children using illicit drugs are at risk of adverse outcome as are children who smoke and children who drink heavily. Children born with low birth weight are also a risk group; they are more likely to have neurologic damage, congenital anomalies, and greater frequency of illness than other children [20]. Children exposed to environmental toxins such as lead are also at high risk. Those with cord blood lead levels of 10 micrograms or more per deciliter are at double the risk of a congenital anomaly and/or a 6–8 point deficit on intelligence testing than their peers without this exposure. Children in families with divorce are at double to triple the risk of school suspension, expulsion, or school behavior problems, and 3 to 5 times the risk of a mental health consultation [24, 40].

Another group of high risk children can be identified by their persistently high use of services. About one in eight children is consistently in the upper third of the distribution of utilization in his age-sex grouping over periods at least as long as six years. Over half of these children – about 53% – are found to experience a wide variety of types of illnesses: repetitive acute illnesses, chronic illnesses, injuries, and problems of a non-medical nature such as visual, hearing, speech, orthopedic, or psychosocial [35].

Contrary to popular belief, acute self-limited conditions of childhood are not necessarily so innocuous. In fact, children with many acute illnesses are very likely to be children who also experience illness of other sorts – injuries, chronic medical problems, chronic non-medical problems, and psychosocial problems. About 30% of children and youth are at risk of having a combination of many types of morbidity in a several year period; many of these children will need extensive services and will contract subsquent illnesses or disabilities as well [36].

Does access to adequate health services ameliorate the damage to children imposed by these risks? In fact, there is accumulating evidence that medical care, when targeted at those who need it, has a major beneficial impact. Much of the improvement in access to care is a direct result of public policy commitments that were made in the 1960s and early 1970s, in the form of Medicaid and other public programs.

In the early 1960s, before the entitlement programs such as Medicaid were instituted, poor children were hospitalized less frequently than non-poor children, but when they were hospitalized, they stayed much longer – evidence that they were much sicker. After Medicaid, hospitalization rates of poor children increased to levels surpassing those of non-poor children, and their lengths of stay were reduced, becoming more similar to those of their more fortunate peers [34].

Before Medicaid, a much greater proportion of poor children had not seen a doctor at all in the previous two years than was the case for other children, and this was true for both black and white children. After Medicaid, the rates became much more similar, although there is still a difference. Low income *sick* children are still at a disadvantage, even though this disadvantage is much less than it was 25 years ago [34].

Trends in infant mortality rates closely followed public policy as well as clinical advances. It is a truism, although not often remembered in funding decisions, that the most important scientific advances are of no use unless they are transformed into services delivered to those who need them. In the case of infant mortality, there have been many scientific advances, but mortality declined sharply only following programs to enhance the financing and delivery of care to pregnant women and their offspring. The sharp downturn in rates of postneonatal mortality just after 1965 (when Medicaid was instituted) is a case in point. During the same period, the disparity in postneonatal mortality rates between the poor and the non-poor and between blacks and whites narrowed. The translation of technical knowledge into family planning services and into neonatal intensive care was associated with a decline in neonatal mortality after a decade of absolute stagnation, and the availability of legal abortions in an increasing number of states between 1968 and 1973 was associated with another downturn in the rates. Unfortunately, decreases in coverage by Medicaid in the most recent few years has been accompanied in time by another stagnation in that component of infant mortality that is most responsive to the benefits of access to physician services: postneonatal mortality. From 1982 to the most recent year for which data are available (1984), postneonatal mortality did not decline at all.

Vaccines against many childhood diseases have been responsible for the near eradication of these diseases. Less well recognized, however, is the importance of public support for immunization programs and vaccine availability. The importance of this support is demonstrated by the increase in measles attack rates in the period immediately following reductions in federal vaccine expenditures, and a decrease in attack rates immediately following a resumption of public expenditures [6].

Table III shows the benefits of a public commitment to adequate nutrition for children, in the form of school lunch programs and the WIC (Women, Infants and Children) Program. The prevalence of anemia, particularly severe anemia, fell during the 1970s. WIC evaluations show that infants whose mothers participated in WIC were born after longer gestations and with less preterm births, and that children who participated have increased iron intake, increased verbal ability, and increased digit memory. Mean birthweight is also increased, although all evaluations show the magnitude of this effect to be rather small.

The situation with regard to birthweight is of interest. Mean birthweight has increased only very little over the past two decades; it does not seem to have been very responsive to improvements in access to services. However, there is evidence that women who are in programs where their prenatal care is part of ongoing care – such as is the case in enrollees of HMOs – do have infants of higher birthweight than women at similar social disadvantage who receive their prenatal care in other types of settings [23, 26, 30].

Public commitment, in the form of legislation to require screening for congenital metabolic conditions, was required to convert an efficacious procedure into one that would be applied when it was needed and in a timely fashion [12]. Public support for the organization and operation of comprehensive care programs in deprived areas was associated with a decline in rates of first attacks of rheumatic fever [14]. Improved access to services such as occurred after the institution of the Medicaid program was also associated with more rapid receipt of care for bacterial meningitis, diabetic crises, asthma, and appendicitis [33].

There is thus little question that access to health care can greatly improve health, particularly for those who are at greatest risk of poor health. That is the reason to be concerned about recent public policy that has apparently diminished access to care. By 1984, there were 35 million Americans without health insurance – a 22% increase since 1979. Although children under 18

CHILD HEALTH AND PUBLIC POLICY 15

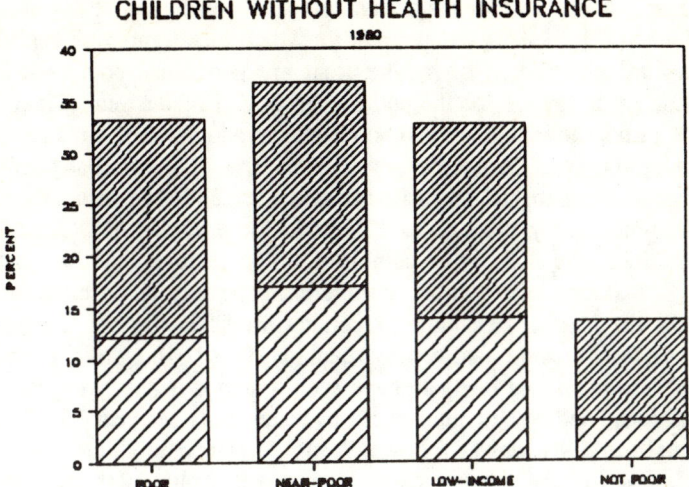

Fig. 3. Source: [7].

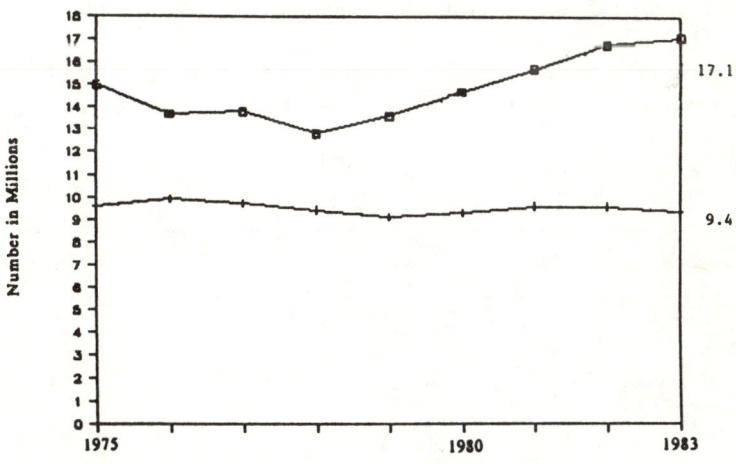

Fig. 4. Sources: ([16], p. 104; [37]).

comprise only about 25% of the population, they constituted nearly 40% of the uninsured [29]. In 1980, the last year for which such figures are available, over one-third of children in families that are poor, near-poor, and low-income had no insurance for the entire year or part of the year (Figure 3). About 4% of non-poor children had no insurance; another 10% had insurance only part of the year. In contrast, well over 10% of poor, near-poor and working-class children had no insurance and over 20% had insurance only part of the time [7]. By 1985, no state in the Union provided AFDC and food stamp benefit levels that, when combined, lifted recipient families out of poverty [8]. Although the number of children in poverty has risen since the mid 1970s, the number of children receiving Medicaid has not (Figure 4). Children in low-income families, particularly those who do not receive Medicaid, are much less likely to have seen a doctor in the past year than other children. The average number of physician visits per year for children under age 18 in 1981 was as follows: 7.7 for low-income children with no Medicaid, 9.3 for low-income children with Medicaid, 10.0 for middle-income children, and 10.5 for high-income children [22]. That is, children who clearly need more services because of their increased health problems are at a marked disadvantage in receiving them if public policy does not provide the means to help them. A substantial proportion of children, particularly poor children who live in cities, are not fully immunized in the preschool period, at a time when they should be immunized (Table IV). Only about half of such children had received the immunizations they should have to protect them from preventable communicable diseases, and even fewer had received all of the immunizations recommended for protection.

TABLE IV

Percent of children 1–4 years old who are immunized, by area of residence

| Residence | Measles | Rubella | DPT | Polio | Mumps |
|---|---|---|---|---|---|
| Central City | | | | | |
| Poor | 48 | 53 | 49 | 44 | 41 |
| Non-poor | 61 | 60 | 61 | 54 | 52 |
| Rest of SMSA | | | | | |
| (Surburban) | 66 | 65 | 69 | 62 | 57 |
| Rural | 66 | 64 | 68 | 63 | 58 |

Source: [21], p. 78.

What are the lessons from these data? If we favor equality of opportunity, if we believe that health care should be distributed according to need, or if we believe that it is important to maximize health in a generation that will have to support an increasing number of elderly people, the policy imperatives are clear. It is important to maintain and expand Medicaid coverage and to standardize eligibility. Standards vary enormously from state to state, and even the standards set by each state are not reflected in funds actually disbursed.

Food support programs must be maintained.

Data to document changes in the population's health must be maintained; without them it is not possible to monitor the impact of public expenditures and assure that changes in mechanisms and levels of funding are not having detrimental effects.

It is also important to monitor and evaluate the impact of new organizational arrangements that are made for specific population groups. In this category are programs to assign Medicaid recipients to various new and unproven forms of organizations such as preferred provider organizations – PPOs, or the special HIOs – health insurance organizations, some of which are developing specifically for the purpose of contracting with public agencies to provide care for Medicaid populations. We have little experience with these types of organizations, and their development specifically to serve the poor is a form of social experimentation that could be considered unethical unless carefully monitored.

On the state and local levels, it is important to identify and serve those who are most needy; information to identify this group is essential. Geocoded information is needed for vital statistics, for immunization rates on entry to day care and schools, and for school absenteeism. The capability to obtain such data is widespread but far from exercised.

States and localities also have the capacity to identify frequent users of services, particularly if the source of funding is centralized, such as is the case for Medicaid. Audits of care may identify situations where the provision of services is inadequate in meeting the child's needs – in the same way that audits of mortality in hospitals are responsible for focusing attention on deficits in the quality of care provided there.

Surveys of the extent of potentially preventable hospitalizations will also illuminate situations where there has been inadequate care. Children should not have complications of appendicitis or meningitis, and they should not be hospitalized in status asthmaticus. Children who are known diabetics should be receiving care sufficient to prevent ketoacidosis. Iron deficiency anemia should not be occurring, nor should delayed diagnosis of phenylketonuria or hypothyroidism.

In a larger sense, children's health is a byproduct of policy on a grander scale than that limited to health. In this world in which we live, 40,000 children die and another 40,000 are irreversibly maimed every day as a result of preventable morbidity – one child every other second is so affected. In each *second*, $25,000 is spent on the military – $5,000 of this in the United States alone. Consider these sums in terms of health benefits. The equivalent of:

– 1 hour's worth of the world's current military spending eradicated smallpox in the world;
– 3 hours' worth supports WHO's annual budget;
– 1/2 day's worth would fully immunize all the world's children;
– 4 days' worth would control malaria worldwide;
– 6 months' worth would eradicate hunger in the world in the next twenty years [32].

What could the $5,000 we spend in the United States each second do?

– The cost of 3 Army AH64 helicopters could have restored 90 million dollars cut from the community health centers and migrant health centers in 1982.
– The cost of 7 XM1 tanks would restore the 17 million dollars cut from the budget of the National Health Service Corps.
– The cost of 2.5 F15A airplanes would make available 75 million dollars cut from the budget in Los Angeles which resulted in the closing of 8 county clinics, reduced services 10% in 3 county hospitals, and eliminated 400,000 walk-in visits at 32 other clinics [31]. Lurie and her colleagues provided convincing evidence of the harm done to health from these actions [19].
– For each 226 MX missile (which costs $110 million dollars), poverty would be eliminated in over 100,000 female-headed families for a year [9].

We must invest in our children, if for no other reason than that we need them. Soon we shall need them as much as they were needed earlier in our country's history. In the 75 years between 1950 and 2025 the ratio of workers to retirees will be reduced more than half: from 4.9 to 2.4 [1]. There are other, equally persuasive reasons for investing in child health. A society committed to equality of opportunity ought to make it possible for children to compete equally. Children with reversible conditions, or conditions that can be ameliorated, cannot compete as equals with healthy children if they are denied the services that are required to overcome their disadvantage. Our greatest gains in reducing childhood poverty and improving child health occurred at a time when there was real growth in expenditures for children. There could be no stronger case for a national commitment and public policy for children.

*The Johns Hopkins University*
*Baltimore, Maryland*

BIBLIOGRAPHY

1. Ackerman, F.: 1982, *Reaganomics: Rhetoric vs Reality*, Southland Press, Boston.
2. Annest, J. L. and Mahaffey, K.: 1984, 'Blood Levels for Persons 6 Months – 74 Years', in National Center for Health Statistics, *United States, 1976–80, Vital and Health Statistics*, Series 11, No. 233, U.S. Public Health Service, Washington. DHHS Pub. No. (PHS) 84–1683.
3. Annest, J., O'Connell, D., Roberts, J., and Murphy, R. 'Blood Lead Levels From the Second National Health and Nutrition Examination Study, 1976–1980', in *Childhood Lead Poisoning Prevention and Control and Public Health Approaches to an Environmental Disease*, Maternal and Child Health Section, Office of Health Services and Environmental Quality, Department of Health and Human Resources, New Orleans.
4. Axinn, J. and Stern, M.: 1985, 'Age and Dependency: Children and the Aged in American Social Policy', *Milbank Memorial Fund Quarterly/Health and Society* **63**, 648–670.
5. Bellinger, D., Nedleman, H., Leviton, A., Waternaux, C., Rabinowitz, M., and Nichols, M.: 1984, 'Early Sensory-motor Development and Prenatal Exposure to Lead', *Neurobehavioral Toxicology* **6**, 387–402.
6. Blendon, R. and Rogers, D.: 1983, 'Cutting Medical Care Costs', *JAMA* **250**, 1880–1885.
7. Butler, J., Winter, W., Singer, J., and Wenger, M.: 1985, 'Medical Care Use and Expenditure Among Children and Youth in the United States: Analysis of a National Probability Sample', *Pediatrics* **76**, 495–507.
8. Children's Defense Fund: 1986, *A Children's Defense Budget*, Washington, D.C.
9. Children's Defense Fund: 1984, *American Children in Poverty*, Washington, D.C.
10. Children's Defense Fund: 1984, *CDF Reports*, Vol. 6, #4 (June), Washington, D.C.
11. Children's Defense Fund: 1985, Testimony before the Subcommittee on Taxation and Debt Management of The Senate Finance Committee regarding S.367 The Child Health Incentive Plan, Sept. 16.
12. Committee for the Study of Inborn Errors of Metabolism: 1975, *Genetic Screening: Programs, Principles, and Research*, National Academy of Sciences, Washington, D.C.

13. Furstenberg, F., Jr., Nord, C., Peterson, J. and Zill, N.: 1985, 'The Life Course of Children of Divorce: Marital Disruption and Parental Contact', *American Sociological Review* **48**, 656–668.
14. Gordis, L.: 1973, 'Effectiveness of Comprehensive Care Programs in Preventing Rheumatic Fever', *New England Journal of Medicine* **289**, 331–335.
15. Hankin, J. and Starfield, B.: 1986, 'Epidemiologic Perspectives on Psychosocial Problems in Children', in N. Krasnegor, J. Arasteh, and M. Cataldo (eds.), *Child Health Behavior*, John Wiley & Sons, New York, pp. 70–93.
16. 'Health Care Financing Trends', *Health Care Financing Review* 6:2 (Winter 1984), 97–105.
17. Hughes, D., Johnson, K., Simons, J. and Rosenbaum, S.: 1968, *Maternal and Child Health Data Book*, Children's Defense Fund, Washington, D.C.
18. Johnston, L., O'Malley, P. and Bachman, J.: 1985, *Use of Licit and Illicit Drugs by America's High School Students 1975–1984*, U.S. Department of Health and Human Services, National Institute of Drug Abuse, Rockville, Maryland. DHHS Publication No. (ADM) 85-1394.
19. Lurie, N., Ward, N., Shapiro, M., Gallego, C., Vaghaiwalla, R., and Brook, R.: 1986, 'Termination of Medical Benefits', *New England Journal of Medicine*, **314**, 1266–1268.
20. McCormick, M.: 1985, 'The Contribution of Low Birth Weight to Infant Morbidity & Childhood Mortality', *New England Journal of Medicine* **312**, 82–90.
21. National Center for Health Statistics: 1982, *Health, United States, 1982*. U.S. Public Health Service, Washington. DHHS Pub. No. (PHS) 83-1232.
22. Newacheck, P. and Halfon, N.: 1986, 'Access to Ambulatory Care Services for Economically Disadvantaged Children', *Pediatrics* **78**, 813–819.
23. Papiernik, E., Bouyer, J., Dreyfus, J., Collin, D., Winisdorffer, G., Guegen, S., Lecomte, M., and Lazar, P.: 1985, 'Prevention of Preterm Births: A Perinatal Study in Haguenau, France', *Pediatrics* **76**, 154–158.
24. Peterson, J. and Zill, N.: 1986, 'Marital Disruption, Parent-Child Relationships and Behavioral Problems in Children', *Journal of Marriage and Family* **48**, 295–307.
25. Preston, S.: 1984, 'Children and the Elderly: Divergent Paths for America's Dependents', *Demography* **21**, 435–457.
26. Quick, J., Greenlick, M., and Roghmann, K.: 1981, 'Prenatal Care and Pregnancy Outcome in an HMO and General Population: A Multivariate Cohort Analysis', *American Journal of Public Health*, **71**, 381–390.
27. Rush, D.: 1986, *The National WIC Evaluation, Vol 1: Summary*, Office of Analyses and Evaluation, Food and Nutrition Services, Department of Agriculture, Alexandria, Va., Contract No. 53-3198-9-87.
28. Schwartz, J., Angle, G., and Pitcher, H.: 1986, 'Relationship Between Childhood Blood Lead Levels and Stature', *Pediatrics* **77**, 281–288.
29. Schwartz, K.: 1985, *Who Doesn't Have Health Insurance, and What Is To Be Done?*, Urban Institute, Washington, D.C.
30. Shapiro, S., Jacobziner, H., Densen, P., and Weiner, L.: 1960, 'Further Observations on Prematurity and Perinatal Morbidity in a General Population and in the Population of a Prepaid Group Practice Medical Care Plan',

*American Journal of Public Health* **50**, 1304–1317.
31. Sidel, V.: 1986, 'Current Toll of the Final Epidemic: Death and Illness from the Weapons Race', *International Journal of Mental Health*, **15**, 40–55.
32. Sidel, V.: 1985, 'Destruction before Detonation. The Impact of the Arms Race on Health and Health Care', *Lancet* **II**, 1287–1289.
33. Starfield, B.: 1985, *Effectiveness of Medical Care: Validating Clinical Wisdom*, The John Hopkins University Press, Baltimore.
34. Starfield, B.: 1982, Family Income, Ill Health, and Medical Care of Children', *Journal of Public Health Policy* **3**, 244–259.
35. Starfield, B., Hankin, J., Steinwachs, D., Horn, S., Benson, P., Katz, H., and Gabriel, A.: 1985, 'Utilization and Morbidity: Random or Tandem?' *Pediatrics* **75**, 241–247.
36. Starfield, B., Katz, H., Gabriel, A., Livingston, G., Benson, P., Hankin, J., Horn, S., and Steinwachs, D.: 1984, 'Morbidity in Childhood: A Longitudinal View', *New England Journal of Medicine* **310**, 824–829.
37. U.S. Bureau of the Census: 1985, *Current Population Reports. Characteristics of the Population Below the Poverty Level: 1983*, Series P-60, No. 147, U.S. Government Printing Office, Washington, D.C.
38. U.S. House of Representatives, Committee on Ways and Means: 1985, *Children In Poverty*. WMCP: 99-8, U.S. Government Printing Office, Washington, D.C.
39. Zee, P., DeLeon, M., Roberson, P., and Chen C.-H.: 1985, 'Nutritional Improvement of Poor Urban Preschool Children: A 1983–1977 Comparison', *JAMA* **253**, 3269–3272.
40. Zill, N.: 1988, 'Behaviors, Achievement and Health Problems Among Children and Step Families: Findings from a National Survey of Child Health', in E. Mavis Hetherington, J. Arasteh (eds.), *The Impact of Divorce, Single Parenting and Step-Parenting*, Lawrence Erlbaum Associate, Hillsdale, New Jersey, 1988.
41. Zill, N.: 1980, *The State of American Children According to Their Parents and Teachers. National Survey of Children*, Foundation for Child Development, Washington, D.C.

STUART F. SPICKER

## COMMENTS ON BARBARA STARFIELD'S 'CHILD HEALTH AND PUBLIC POLICY'

In her review of the health status and plight of children in the United States during the past few years, Barbara Starfield has focused on our nation's allocation of material resources to the care of sick children. She understands care to be the access to and the ministration of medical treatment, and the prevention of various physical and mental maladies regarding which children are at serious risk. Hence Starfield has tallied essential facts that reflect the extent of our nation's commitment to the care of sick children. She draws conclusions from her data like: "... children who clearly need more services because of their increased health problems are at a marked disadvantage in receiving them if public policy does not provide the means to help them" ([9], p. 6); furthermore, "It seems only fair that society now consider a major public policy initiative for children" ([9], p. 8).

Although Starfield's presentation is generally straightforwardly empirical, it should be obvious that assertions like those noticed above are normative: they reflect Starfield's value judgments (1) that present financial allocations at the macro level of public policy are *unfair*, (2) that something (at least in the form of material resources and public funding) is seriously lacking with respect to the maintenance of the health of our nation's children – and perhaps elsewhere in the world as well, (3) that we as a people are morally responsible not only for the damage or "morbidity" that our children undergo, but for the high risk to poor health which they suffer as well, (4) that adherence to the value of equality of opportunity for children should encourage access to health care for all children, (5) that it will be useful and cost-effective for society to have a healthier population of children.

In general, I have no quarrel with the data which Dr. Starfield proffers describing the state of affairs with respect to what has been and what is being publicly allocated to the health maintenance of our nation's children, though at times her expression of the facts is not in accord with the full range of studies which always conclude with some global statistic – e.g., her claim that by 1984, "there were 35 million Americans without health insurance ..." ([9], p. 14). Though estimates vary, the correct figure is probably somewhere between 20 and 30 million persons[1]; moreover, a difference of 5 to 10 million here is very significant, since it represents a range of from 9 to 12% of the

23

population being without health insurance. But such discrepancies do not in themselves undercut Dr. Starfield's conclusion that "it is important to maintain and expand Medicaid coverage and to standardize eligibility" ([9], p. 7). Hence our task is not so much to challenge her normative conclusions as it is to discover her reasons for making them so forcefully. Let me explain.

Given, for example, the extensive resources being provided to the health care of the aged segment of our nation – data which is constantly being expressed in the media along with data which reflect the demographic momentum of the more aged segments of our population – can an argument be made that children should have at least as much consideration in the allocation of health care resources as the elderly? I think this is a reasonable question for Dr. Starfield. Answering this question would require a comparison of claims for health resources on behalf of different segments of the population.

It seems to me that those who wish to defend the nation's defense budget can counter Dr. Starfield's initially seductive (but regrettably unpersuasive) comparison of military and health spending by the following set of counter-arguments: We have children of our own, and we want to reduce their risk of poor health as they grow older, and we therefore would like to see more of our nation's material resources transferred to their health maintenance, but you do not appreciate the fact that the defense budget is actually a very complex series of budgets. For example, the entire computer revolution which made for important progress in the delivery of health care had its roots in research sustained by the defense budget. Furthermore, they might argue: we have for years observed billions of taxpayer dollars directed to medical research, and while pleased with past results, we are not inclined to fund research at the same high level indefinitely. We have other needs to meet. Moreover, and right to the point, why focus on the defense budget in the desire to reallocate resources to the young; couldn't one as easily (and perhaps more persuasively) look to cut other budgets? Finally, aren't you thinking too abstractly – being reminded of the philosopher G.F. Hegel – when you conclude by suggesting that monies spent by the world's governments on defense over four days could be used to control malaria world wide? How could such formulations possibly capture the complex and concrete political state of affairs which public policy can only reflect in the total dollars allocated to various purposes and goals?

In short, many of us believe that we are not only morally responsible for reducing the risk to poor health of our children, but that we are also morally responsible for all the lives of the nation's citizenry – the old old, the young old, those of middle years, and the young and neonates as well. We all have our *cause célèbre*; everyone who wishes and works to do good could use and expend additional material and human resources. Perhaps, in the end, we all shall have to make the best of it in circumstances best described as unfortunate but not justly judged unfair. That is, we may recognize that not all needs can be met.

*University of Connecticut Health Center*
*School of Medicine*
*Farmington, Connecticut*

NOTE

[1] Accurate information on the number of U.S. citizens without either private or public health insurance is difficult to ascertain. For example, from 1950–1976, some 25–50% of the increase in physician service expenditures was attributed to the *growth* in insurance coverage [3]. According to the U.S. Census Bureau, in 1983, 15% of U.S. citizens were uninsured [2]. Other sources claim that "In 1983, about 29.2 million people – 12.6 percent of the population – had no health insurance" ([5], p. 59). Mary O'Neal Mundinger's and Uwe E. Reinhardt's ([8], pp. 20–28) estimates differ only slightly. Mundinger remarks that "Thirty-five million people, or 15 percent of the population, are without [private or public] health insurance today – an increase of 10 million, or 25 percent, since 1977" ([6], p. 4).

Wilensky and Walden remark that the problem of the uninsured is often overstated, for data tend to fail to note that (1) many uninsured persons are uninsured only on a temporary basis, i.e., less than a year, (2) a large proportion of uninsured are relatively healthy younger persons between the ages of 6 and 24, (3) less than 1% are over 65 years, and many belong to the middle class or above and are presumably uninsured by choice. (4) Even the difference between the proportions of uninsured and insured who rated their health as poor was so small that it may well have been explained by normal sampling error [10]. Moreover, (5) although increases in coverage have been accompanied by even higher charges for insurance, this has not actually depressed service utilization, since these charges are not close enough in time to the use of services for consumers to link the two [7]. Again, (6) since over three-fifths of the population's insurance coverage is employer provided, the percentage of uninsured is necessarily quite volatile and fluctuates with employment trends.

Aday and Anderson are closer to the truth, perhaps, in their remark that in 1984 only 9 percent of the total population was uninsured ([1], p. 1335). In any event, individual citizens and other interest groups will have to determine what is to be done with respect to these uninsured citizens, especially those who truly do not have the ability to pay for their own health care or health insurance.

## BIBLIOGRAPHY

1. Aday, L. and Anderson, R.: 1984, 'The National Profile of Access to Medical Care: Where Do We Stand?' *American Journal of Public Health* **74**, 1331–1339.
2. Census Bureau: 1985, *Survey of Income and Program Participation*, U.S. Government Printing Office, Washington, D.C.
3. Dyckman, Z. Y.: 1978, *A Study of Physicians' Fees*, U.S. Government Printing Office, Washington, D.C.
4. Halper, T.: 1984, 'Aging Policy in the Eighties: Second Thoughts on a Strategy That Has Worked', in S.F. Spicker and S. Ingman (eds.), *Vitalizing Long-term Care*, Springer Publ. Co., New York, pp. 3–13.
5. Iglehart, J. K.: 1985, 'Medical Care of the Poor – A Growing Problem', *New England Journal of Medicine* **313**, 59–63.
6. Mundinger, M. O.: 1985, 'Health Service Funding Cuts and the Declining Health of the Poor', *New England Journal of Medicine* **313**, 44–47.
7. Pauley, M. V. and Langwell, K. M.: 1982, *Research on Competition in the Market for Health Services: Problems and Prospects*, Applied Management Systems, Silver Springs, Maryland.
8. Reinhardt, U. E.: 1985, 'Economics, Ethics, and the American Health Care System', *The New Physician* **34**, 9, 20–28, 42.
9. Starfield, B.: 1989, 'Child Health and Public Policy', in this volume, pp. 7–21.
10. Wilensky, G. R. and Walden, D. C.: 1981, *Minorities, Poverty, and the Uninsured*, Department of Health and Human Services, Hyattsville, Maryland.

ANN L. WILSON

# DEVELOPMENT OF THE U.S. FEDERAL ROLE IN CHILDREN'S HEALTH CARE: A CRITICAL APPRAISAL

> We the People of the United States, in Order to form a more perfect Union, establish Justice, insure domestic Tranquility, provide for the common defense, promote the general Welfare, and secure the Blessings of Liberty to ourselves and our Posterity, do ordain and establish this Constitution for the United States of America.

In this paper I will examine the history of one way in which the federal government of the United States attempts "to promote the general Welfare" and "secure the Blessings of Liberty to ourselves and our Posterity", namely, through the provision of health care for children. My account will begin at the turn of the century when the idea of a Children's Bureau to gather data on the status of the nation's children was first conceived. I will argue that, since that time, three themes are dominant in Congressional and public debate about federal responsibility for children's health care. They are: what is the *authority* of government to create legislation and policy affecting children's health; what are the *economics* of supporting health care for children; and whether health care for children should be a basic *human right* compelling its provision as a humane act of government. These are issues elected officials have faced and attempted to resolve as they have debated, voted on, and funded legislation for children's health care.[1]

## I. THE PROGRESSIVE ERA AND CHILDREN'S HEALTH CARE

At the turn of the century, over half (60.3%) of the nation's population was rural, living in communities with populations of 2,500 or less [67]. While this is more than twice today's proportion of rural population [68], the late 1800's was in fact a time of growth for urban areas. Their population tripled during the thirty years prior to 1900. Urban growth was spawned by post-Civil War industrialization and immigration. New problems emerged for the nation as its social fabric changed [41, 39].

With the rise of big business, natural resources and human labor were exploited and it was recognized that a continuation of the government's hands off or laissez-faire approach to problems would not resolve these social ills. Those who sought to make government more responsive to these problems became known as the Progressives. They strived "to secure the real and not the nominal rule of the people" [60]. The Progressives achieved many democratic innovations, including the recall of elected officials, the primary system for elections, women's suffrage, and the popular election of United States Senators. Among the social reforms most stridently sought by the Progressives was the regulation of child labor.

The 1870 census reported that among children aged 10 to 15 years, 765,000 were gainfully employed. By 1900, 1,750,000 children were employed with 40% working in industrial settings and 60% in agriculture [67]. Sprago, in his 1906 classic book *The Bitter Cry of Children*, wrote:

The industrial revival in the South from the stagnation consequent upon the Civil War has been attended by the growth of a system of child slavery ([63], p. 149).

Child labor, however, was not unique to the southern states. Mills, mines, quarries, canneries, and other factory work employed the energies of young children throughout the country. Those who employed children treated them without respect for their physical limitations or for how labor was impeding their education, health, or emotional development. Rather, the economic peril of immigrant and other recently urbanized families created a population of cheap labor for those managing the industries of the time, and children were blatantly exploited.

During these same years, there was a growing recognition of how the health care needs of children differed from those of adults. Pediatrics emerged as a medical specialty during the latter part of the 1800's. The first children's hospital in the country was established in 1855 in Philadelphia and the first professorship of pediatrics was held at Columbia University by Abraham Jacobi in 1860. The American Medical Association recognized a section on Diseases of Children in 1879, and in 1888 the American Pediatric Society was formed [74].

Lesser, in reviewing the history of children's health care, observed that in the late 1800s a dynamic view of child health and preventive care was promoted by the understanding of bacteriology and communicable disease. This new knowledge also fostered a growing awareness of how social conditions affect the health of children [49]. The Progressives publicized this new knowledge as they attempted to improve the plight of the laboring class. Theodore Roosevelt, in an essay "Conservation of Womanhood and Childhood" published in 1911, wrote:

To permit women and little children to live and work in wretchedly lighted, badly ventilated rooms, polluted by overcrowding, both from the materials and workers, so that the air is contaminated, furnishes new recruits to the ranks of tuberculosis victims ([59], p. 1015).

Among the data gathered to testify to the evils of child labor were findings suggesting that the growth of children employed in factories was different from that of unemployed children from affluent families. Stated Sprago:

It is the consensus of opinion among those having the best opportunities for careful observation that physical deterioration quickly follows a child's employment in a factory or workshop ([63], p. 175).

Furthermore, he observed:

The moral ills resulting from child labor are numerous and far reaching. When children become wage earners and are thrown into constant association with adult workers they develop prematurely an adult consciousness and view of life ([63], p. 181).

Not only were associations made between ill physical and mental health and poor working conditions, but also between mortality and poverty. In 1910, Newsholme made his classic observation:

Infant mortality is the most sensitive index we possess of social welfare. If babies were well born and well cared for, their mortality would be negligible. The infant death-rate measures the intelligence, health, and right living of fathers and mothers, the standards of morals and sanitation of communities and governments, the efficiency of physicians, nurses, health officers and educators [55].

Sprago, in 1906, made the similar observation that "poverty and death are grim companions" ([63], p. 5) and presented data showing that between 45% and 85% of early childhood deaths were related to "bad conditions or poverty."

A recognition of the bacteriological causes of illness and the greater vulnerability of the working classes to diseases and death led community efforts to improve the health of the poor. In the late 1800s, dirty and unsafe milk came to be known as a threat to infant health. Dr. Henry L. Coit founded the movement for Certified Milk and emphasized the need for high standards in dairy sanitation. Nathan Straus, a New York philanthropist, recognized that pasteurization was the only process that could ensure safe milk. Beginning in 1893, he established numerous milk stations for the poor in New York City where mothers were also taught how to prepare infant feedings and given other instructions about child care. This service reflected the emerging recognition that effective preventive health services for children must include education for parents.

City involvement with child health care grew in the early 1900s. In New York City, Lillian Wald, founder of the Henry Street Settlement, suggested to the City Health Commissioner that a nurse be assigned to the schools to provide health education for children and home visits to parents [4]. In 1908, the first Bureau of Child Hygiene was established in New York City with Josephine Baker, M.D., as its first chief. This bureau employed nurses who visited tenement homes of newborn babies to help mothers better care for them and prevent diarrheal disease. It also provided services to school children, supervised midwives, and regulated children's institutions and boarding homes [2].

The social circumstances of these times and their accompanying grim health problems reinforced the requests of the Progressives for greater governmental involvement in monitoring the general welfare of the nation's citizens. They pursued the formation of a Children's Bureau within the administrative branch of government and initiated Congressional debate about Federal responsibility for the lives of the nation's children.

## II. THE CREATION OF THE CHILDREN'S BUREAU

Though first introduced in 1906, not until 1912 was the legislation creating the Children's Bureau passed by Congress and signed into law by President Taft. Julia Lathrop, its first chief, was the first woman to hold such an administrative position in the federal government. Miss Lathrop had been a close professional associate of Jane Addams at Hull House, the well-known Chicago settlement house.

Four months after its creation by Congress, Miss Lathrop addressed the Biennial Meeting of the General Federation of Women's Clubs and reviewed for this national audience how the Children's Bureau came to be. Speaking with assurance, she described how the Children's Bureau was an idea developed by Lillian Wald, a nurse who founded the Henry Street Settlement in New York City. She credited Florence Kelley, secretary of the National

Consumers League, an organization opposed to child labor, with outlining the matters that were to be investigated by the Bureau. In describing the value of the Bureau entrusted to her direction, Miss Lathrop observed:

> The bureau was first urged by women who have lived long in settlements and who by that experience have learned to know as well as any persons in this country certain aspects of dumb misery which they desired through some governmental agency to make articulate and intelligible ([47], p. 318).

The Children's Bureau was conceived to gather data, not to provide services. It was to investigate conditions and disseminate educational materials to citizens, not to directly involve itself with the lives of families. Today this seems like a benign request to make of Congress. In the early part of the twentieth century, however, it represented a new role for the federal government. Specifically, the Children's Bureau, as originally proposed, "was to investigate all matters pertaining to the welfare of children." The specific issues to be investigated were:

> infant mortality, the birth rate, physical degeneracy, orphanages, juvenile delinquency and juvenile courts, desertion and illegitimacy, dangerous occupations, accidents, and diseases of children of the working classes, employment, legislation affecting children in the several states and territories, and such other facts having a bearing upon the health, efficiency, character, and training of children.

The road leading to the enactment of the bill creating the Children's Bureau was not smooth.

In January 1909, President Roosevelt had convened a national conference on "Care of Dependent Children," which later became known as the First White House Conference on Children. A similar conference was sponsored by the White House once a decade for the next 60 years. Among the recommendations of this first conference was the request that Roosevelt send a special message to Congress urging its passage of the legislation that would create the Children's Bureau. Roosevelt complied. In his statement to Congress he observed that the data-collecting activity of the Children's Bureau would be important since:

> ... in the absence of such information as should be supplied by the Federal Government many abuses have gone unchecked; for public sentiment with its great corrective power, can only be aroused by full knowledge of the facts ([6], p. 2363).

During Roosevelt's tenure in office, Congress did not take action on the proposed Children's Bureau, yet the former President continued to speak out on its behalf. In 1911, he wrote:

The chief reason for the failure of the bill thus far has been that it is solely in the interest of the conservation of human resources, and does not directly touch any material interest. It is devoid of features which would win selfish and self-interested advocacy and the persons who desire to continue to exploit the labor of children are naturally against it ([59], p. 1014).

While Roosevelt and the other advocates of this bill believed the federal government should play an active role in monitoring the welfare of the nation's children, some members of Congress opposed this activity. Senator Joseph Bailey of Texas was particularly concerned about how a Children's Bureau "contemplates the establishment of a control through the agencies of government over the rearing of children" ([7], p. 189). Furthermore, he stated:

We have for a 100 years or more left these matters concerning children to the proper authorities, which are mothers, fathers, and guardians ... We have produced the greatest race of men and women that ever blessed any land in the history of the world and as for my part I am not willing to invoke the Government's aid in this greatest of all work ([7], p. 188).

Senator Bailey feared that once a new governmental initiative were begun, it would be difficult to limit its authority and activity:

Men who are familiar with the course of legislation understand perfectly that these matters come first in the shape of requests for statistics and they are invariably followed by legislation. Unless the Congress of the United States intends to predicate some legislation upon these statistics, then it has no right to appropriate the public money for their collection ([7], p. 188).

Concerns about Congressional authority were also raised regarding how a Children's Bureau would interfere with state rights. New Hampshire's Senator Jacob Gallinger commented:

We do not need to have the General Government throw its protecting arms about us in every little matter that affects the welfare of the people of our several communities and our several States. We will paralyze the initiative and the work of the States if the General Government is going to do everything that the people of this country think ought to be done to promote their welfare and their happiness ([10], p. 1523).

In responding to such attacks on the constitutional authority of such legislation, members of Congress were quick to cite the Preamble clause of the

United States Constitution and to note that a Children's Bureau would "promote the general welfare." Precedents for such authority were also frequently cited by members of Congress who observed that authority was granted to Congress to expend considerably more money than requested for the Children's Bureau to investigate plant industries, to control disease of cotton, truck, and forage crops, to investigate and improve tobacco and its methods of production, to control diseases of orchards and other fruit, and to eradicate Southern cattle ticks. Stated Senator William Borah of Idaho:

If we have the constitutional power to gather statistics and facts with reference to diseases of hogs and cattle and sheep, it seems to me we have the constitutional power to gather the data with reference to diseases of children. The constitution was not made for the hogs alone, but also for men ([8], p. 704).

In the House, Representative Andrew Peters of Massachusetts declared the nation's need for information and compared American reluctance to extend federal authority in this area to efforts that were currently taken in Europe. He noted:

Not only England, but Germany as well with an eye to its future citizens has made a very complete and careful study of the facts relating to child life. Our own students of these problems have been compelled to turn over most of their information to foreign sources and to do the best they could to apply them properly to American conditions ([12], p. 4218).

Through these debates there was the Progressive recognition that the times had changed and that the Government needed to respond to the social concerns that evolved from the industrialization of the nation. Senator Gilbert Hitchcock of Nebraska responded to the remarks of Senator Bailey that a Children's Bureau would not have even been considered "50 years ago." He stated:

Fifty years ago we were not confronted by the problems which now confront all States in dealing with these questions; 50 years ago there was, practically speaking, no child-labor problem in the United States; 50 years ago this country was not confronted by the serious evils which have grown up in the great industrial centers, in the great mining regions, in the overgrown cities of the United States; and the only reason why such legislation as this was not considered 50 years ago was that the conditions of those days were radically different from the conditions of the present day ([9], p. 1251).

Prior to this time, private philanthropy had responded to the welfare needs of communities. The Children's Bureau represented a rethinking of relationships between public and private agencies concerned with the welfare of the nation's citizens. The enormity of the demands of philanthropic efforts was well illustrated in the comments made by Nathan Straus, whose personal efforts created infant milk depots throughout New York City. He is reported to have addressed the Alderman of New York City in 1909 by saying:

I have done as much as one man and one purse can do to save the lives of children in the city. Now I must put the work up to the City ([6], p. 420).

Congressional proponents observed that information collected by the Children's Bureau would benefit private charities in giving them a sound data base, which could help guide them in focusing their efforts on existing needs. Not all private agencies, however, agreed with this rationale. Notably, John D. Lindsay, President of the New York Society for the Prevention of Cruelty of Children, opposed the creation of the Bureau as he felt it duplicated the work of the Census Bureau, Bureau of Labor, and Bureau of Education. Dr. William O. Stillman, President of the American Humane Association, also declared opposition to the bill in a letter read before the Senate on the grounds that he did not approve of the attempt to centralize such activity in a department of the federal government [10]. These major philanthropic organizations opposed the bill and the way in which it would enable federal authorities to enter into areas of current private concern.

In contrast, many Senators acknowledged the widespread support for it among women who were urging its passage. In criticizing these supporters, Senator Bailey stated:

I am willing outside of this Chamber to give the good women all the credit they could crave, but I believe and I do not hesitate to declare it that the more a woman knows about the things she ought to understand the less she knows about how we ought to perform our duty here ([10], p. 1534).

Senator James Reed of Missouri had a different, frankly economic, perception of the lobbying forces affecting the passage of the bill:

Why is this bill opposed? Why is it that men contend that our right to expend the public revenues is comprehensive enough to reach the ox in his stall but not broad enough to embrace the child in its cradle? Why is this opposition coming from Senators who represent States that have cotton mills in them? Is it because pale-faced children are trooping every morning into the sweatshops of labor? Is it because there are unspeakable conditions in those mills? Is it because men are putting the guinea stamp upon the souls of young children? Or do they really fear they are treading upon the Constitution? ([11], p. 1517).

Representative Henry Barnhart of Indiana also described this economic factor that he believed to be motivating those who opposed this legislation. In floor debate, he stated:

The truth is, the bill is not wanted by some, not because it is in violation of any constitutional provision, but because if enacted it will become a big obstacle in the way of the use and abuse of child labor and women labor in many of the States ([12], p. 4220).

Claiming a Children's Bureau was worthy for its humanitarian mission, he further commented:

It is declared that a million young American mothers are annually bringing children into the world with no qualification for the proper care of their health except love and devotion. The Nation and the States have bureaus of scientific research which disseminate information on all topics of animal and industrial welfare, but scarcely a line is printed for free distribution which would prepare young parents for such care of children as would make of them sturdy and successful citizens ([12], p. 4440).

After six years of consideration, the House voted 177 to 17 [12] and the Senate 54 to 20 [11] to pass this legislation. While opposition was vocal, the large margin of support reflects general acceptance of this new activity of the federal government.

## III. EARLY WORK OF THE CHILDREN'S BUREAU

Congress appropriated $29,400 for the initial year of the Children's Bureau's work "to investigate and report upon all matters pertaining to the welfare of children among all classes of people." It is noteworthy that of the original issues delineated for investigation by the Children's Bureau, illegitimacy, juvenile delinquency, and physical degeneracy were excluded from the bill that eventually became law.

The initial work of the Children's Bureau focused on investigating the incidence of infant mortality, but a major methodological problem immediately hampered this effort: In 1910 the census data available on mortality were based on approximately one-half of the nation's population. While these data were sketchy, Miss Lathrop noted:

There was still greater ignorance as to the number of children born. For not a single state, not a single city has complete registration of births ... To know anything about the death-rate, we must register the children born as well as the children who die. To know how to stop the loss of 200,000 infants yearly, we must know first why they die, and when and where. We must register their births and deaths as the essential element of intelligent life-saving ([47], pp. 324, 328).

Miss Lathrop challenged the General Federation of Women's Clubs to assist the Children's Bureau's effort to register the births of America's newest citizens. The Federation responded and played an important role in achieving this end. By 1915, ten states and the District of Columbia had become "birth registration areas," but not until 1933 did all states register births.

Recognizing limitations in conducting national studies of infant mortality, the Bureau examined mortality in specific communities. The findings of a Manchester, New Hampshire, study were reported to Congress in 1917 in an attempt to secure funds for the continuation of this work [13]. In Manchester 86% of fathers were employed by textile mills, and the overall infant mortality rate was 165 per 1,000 live births. The Children's Bureau study

TABLE I

Maternal and infant mortality rates

|  | Maternal mortality rates | | | Infant mortality rates | | |
|---|---|---|---|---|---|---|
|  | (Rates per 10,000 live births) | | | (Rates per 1,000 live births) | | |
|  | Total | White | Other | Total | White | Other |
| 1915[a] | 60.8 | 60.1 | 105.6 | 99.9 | 98.6 | 181.2 |
| 1921 | 68.2 | 64.4 | 107.7 | 75.6 | 72.5 | 108.5 |
| 1929 | 69.5 | 63.1 | 119.9 | 67.6 | 63.2 | 102.2 |
| 1935 | 58.2 | 53.1 | 94.6 | 55.7 | 51.9 | 83.2 |
| 1945 | 20.7 | 17.2 | 45.5 | 38.3 | 35.6 | 57.0 |
| 1955 | 4.7 | 3.3 | 13.0 | 26.4 | 23.6 | 42.8 |
| 1965 | 3.2 | 2.1 | 8.4 | 24.7 | 21.5 | 40.3 |
| 1975 | 1.3 | 0.9 | 3.0 | 16.1 | 14.2 | 24.2 |
| 1982 | 0.8 | 0.6 | 1.6 | 11.5 | 10.1 | 17.3 |

Data from U.S. Census Bureau [67, 68].

[a] For this time period, incomplete registration greatly limited accuracy of statistics. Early Children's Bureau reports (approximately 1913) indicate an infant mortality rate of 124 per 1,000 live births. Massachusetts data for 1905–1909 indicate an infant mortality rate of 134 per 1,000 live births.

linked family income with mortality showing that 25% of all babies whose families had yearly incomes of less than $450 died, while mortality was 6% for babies in families with an income of over $1,050. Living conditions and the employment of mothers were also shown to be related to infant mortality. (See Table I for Census Bureau data on maternal and infant mortality, 1915–1982.)

In addition to investigations of mortality, the early work of the Bureau included the writing and distribution of pamphlets educating the public about maternal and child care. Its first pamphlet, *Prenatal Care*, was immediately acclaimed by the popular journalism of the time as a major contribution to society. Noted *The Outlook*:

In this pamphlet the Government has made a beginning of doing for children of the country what it has done superbly for the country's crops and herds ... Through the Children's Bureau the Government is now undertaking to act as a sort of expert home-counselor or consulting mother ([57], p. 60).

A government publication printed a typical letter from a woman requesting *Prenatal Care*. The letter is revealing of the life circumstances for mothers of the times. It read:

Dear Miss Lathrop:
I should like very much all the publications on the care of myself, who am now pregnant, also on the care of a baby. I live sixty-five miles from a Dr ... I am 37 years old and I am so worried and filled with perfect horror at the prospects ahead. So many of my neighbors die at giving birth to their children. I have a baby 11 months old now in my keeping, whose mother died. When I reached their cabin last November it was 22 below zero, and I had to ride 7 miles horse back. She was nearly dead when I got there, and died after giving birth to a 14 pound boy ... Will you please send me all the information for the care of myself before and after and at the time of delivery. I am far from a doctor, and we have not means, only what we get on this rented ranch ... ([70], p. 26).

The Bureau next published *Infant Care*, later to become the most widely distributed publication of the Government Printing Office [70]. Other Children's Bureau publications included pamphlets on baby saving campaigns, infant mortality, birth registration, child labor legislation, and mothers' pensions in the U.S., Denmark, and New England. Florence Kelley, in an article in a 1915 weekly Progressive publication *The Survey*, described the frustration of the Bureau, which was flooded with requests for these publications but was limited in its response to them due to its "niggardly initial appropriation and meager staff" ([44], p. 632).

## IV. THE SHEPPARD-TOWNER BILL

Just as Senator Bailey had predicted seven years earlier, the data gathering activity of the Children's Bureau led to the proposal of legislation to respond to problems identified. In her 1917 annual report, Miss Lathrop, Chief of the Children's Bureau, published a plan for "public protection" of maternity and infancy. Her plan included the use of public health nurses for instruction and service and the development of adequate confinement care and accessible hospital facilities for mothers and children [69]. The first woman to serve in Congress, Jeanette Rankin of Montana, responded to this plan and introduced a bill in 1918 for "instruction in the hygiene of maternity and infancy and for proper care of maternity and infancy in rural districts." Though Representative Rankin's bill died through lack of "public interest" [1], in 1920, Senator Morris Sheppard of Texas and Representative Horace Towner of Iowa introduced a similar bill "for the public protection of maternity and infancy and providing a method of cooperation between the government of the U.S. and the several states."

The bill proposed a five-year cooperative federal-state effort, the first of its kind for health care; its passage established a precedent for future similarly funded programs. Included in the bill were requirements that states have child hygiene or welfare bureaus in order to receive matching funds from the federal government to support efforts to improve health care for mothers and children. Specifically, these efforts, as described by Senator Sheppard, would consist of:

> ... distributing information and instruction in the hygiene of maternity and infancy through bulletins, public-health nurses, consultation centers, lectures, and other suitable methods. Wherever necessary, and especially in remote areas, medical and nursing care for mothers and infants may be provided, in so far as available funds will permit ([14], p. 417).

The proposed legislation became known as the Sheppard-Towner Bill. Some veteran members of Congress believed it aroused more intense debate than any other issue during their tenure in office. While many perceived Sheppard-Towner provisions for safe childbearing as a humane protection of a basic human right, some expressed great dismay and fear about what such proposed governmental activity represented.

During the debate on Sheppard-Towner, members of Congress heard repeatedly that 250,000 infants and 20,000 mothers died annually. Data were also presented describing how more women (ages 15 to 45) died from causes incidental to childbearing than any other cause except turberculosis [14] and that this accounted for over 10% of all female deaths in this age group [22]. Furthermore, the Chief of Obstetrics at John Hopkins testified that 75% of these maternal deaths could be prevented [17].

Perhaps the most graphic illustration of the extent of mortality was presented by Dr. Josephine Baker, Chief of the Child Hygiene Bureau in New York City. She testified that the number of mothers who died in childbirth during the 18 months of World War I almost equalled the number of soldiers lost and killed in battle. In further explaining this observation, she noted "for every soldier killed, a mother died in childbirth, and for every soldier killed, six babies died at childbirth, and all because the social and the economical conditions are so poor" ([17], p. 3143). Dr. Baker then described how the educational efforts of her bureau "decreased the baby death rate from 144 to 85 per thousand births" ([17], p. 3143) and predicted that if the Sheppard-Towner Bill were enacted, it could annually save the lives of 15,000 mothers and 100,000 babies.

Women's magazines presented these facts to their readers and decried the nation's failure to attend to the health needs of its citizens. An article published in the *Ladies Home Journal* entitled "Safe Motherhood" (1920) stated:

Criminal negligence must be charged against the nation for this fearful wastage of lives, neglect of the elementary safeguards which it is as much the duty of the nation to throw about motherhood and infancy as it is its duty to protect our national security by armed force ([35], p. 42).

*Good Housekeeping* published a three-part series of articles in a campaign of the support for the Sheppard-Towner Maternity Bill. The 1920 readers were presented with testimony given to the Children's Bureau such as this from a Montana mother whose last baby died:

We intended to go to the hospital again for my next baby, but the terrible expense of my last baby got us into debt and then I couldn't get away in time because all the autos in the neighborhood were being used for sheep shearing ([50], p. 20).

Writing for the *Ladies Home Companion*, Representative Towner reported that there were 14 other countries in the world where it was "safer to be a mother" ([65], p. 4).

Clear recognition was given to the fact that advocates for mothers and babies were needed if this situation were to change. Echoing the comments made by Theodore Roosevelt twenty years prior to the passage of the Sheppard-Towner Act, an editorial in *Nation* commented:

Babies needless to say have little direct influence in politics and until recently even that subtle indirect influence – so highly spoken of in presuffrage days – seemed to produce small effects ([78], p. 724).

Florence Kelley observed in 1920 that within a month's time a $33 million bill was introduced in Congress and signed into law by President Wilson to increase the pay of postal employees. Yet, the $4 million "babies' bill" was not moving. In analyzing the differences, she described postal workers as voting: "But babies have no votes, no organization" ([45], p. 401).

The ratification of the Nineteenth Amendment in 1920 granted women suffrage and they played an active role in promoting the Sheppard-Towner Bill. In this effort to present to his colleagues in Congress the wide support that existed for this bill, Senator Sheppard cited the names of over 40 women's organizations that had contacted him. He also read approximately 20 letters of endorsement of the bill he received from governors.

Not all members of Congress were impressed by such advocacy. Senator Charles Thomas of Colorado was among those who made this view known to his colleagues. In expressing his belief that the Sheppard-Towner Bill provided inappropriate federal involvement in a state function, he also commented that federal employees were becoming as thick as "lice in Egypt" ([14], p. 420). In belittling the support of the bill by the women who now voted, he claimed that its endorsement by governors was meaningless and commented:

What is a poor governor going to do when the representatives of 30 women's associations come to his office, gather around him with pleas and with tears, with flattery and with threats, and with suggestions regarding his ability and the need for reelecting him – what is he going to do? In nine hundred and ninety-nine cases out of one thousand he will not only write a letter to the proponent of the measure, but he will publish the fact of his adhesion as broadly as possible. I have no doubt I might do it myself if I were the governor of my State and wanted to be reelected. That sort of support to me, however, means absolutely nothing ([14], p. 421).

Although while nine years earlier Missouri's Senator Reed had endorsed the Children's Bureau, he now made acrid attacks against this proposed legislation. He ridiculed the Children's Bureau and its employees, many of whom were unmarried women:

When we employ female celibates to instruct mothers how to raise babies they have brought into the earth, do we not indulge in a rare bit of irony? ... I cast no reflection on unmarried ladies. Perhaps some of them are too good to have husbands. But any woman who is too refined to have a husband should not undertake the care of another woman's baby when that other woman wants to take care of it herself ([18], p. 8759).

Senator Reed furthermore minimized the educational activities proposed by this legislation, stating:

Ever since Eve first hugged Cain to her breast women have known how to feed a baby, what to feed a baby, and when to feed a baby. The mother of today has sense enough to know in general what her baby needs. When she is in doubt she resorts to the assistance of her husband, the counsel of some good old mother, and the advice of the family doctor. It is now proposed to turn the control of the mothers of the land over to a few single ladies holding Government jobs at Washington. I question whether one out of ten of these delightful reformers could make a bowl of buttermilk gruel that would not give a baby the colic in five minutes (laughter). We would better reverse the proposition and provide for a committee of mothers to take charge of the old maids and teach them how to acquire a husband and have babies of their own (laughter) ([18], p. 8764).

These were among the most derogatory comments made about women and their emerging involvement in advocating legislation and developing public policy. Clearly, there were those in Congress who believed:

... we have gotten along pretty well in the good old way, the people attending to their own business the best they can, the government attending to its business the best it can ([14], p. 423).

While the bill continued to be debated, its enormous support from citizen organizations throughout the country was recognized. The *Literary Digest* reported an article appearing in the *Times*, claiming that every important religious denomination endorsed what became dubbed the "Better Baby Bill" [64]. Two groups of citizens, however, expressed opposition to the bill: the National Association Opposed to Woman Suffrage [16] and certain physicians. Senator Kenyon summarized the concerns expressed by the latter group who claimed the bill was "socialistic," "looked to birth control," and that "its purpose was to develop free love and other kindred absurdities" ([17], p. 3142). He reported, as well, on the oppositional comments made by physicians from Massachusetts, who argued against the bill saying "control by the individual is democracy, control by the state is autocracy and in other words socialism" ([16], p. 252).

Both of these groups' fears reflected, in part, concerns about foreign influences, which they anticipated would spread if Congress assumed a role in supporting health-related activities for mothers and children. Senator Reed was particularly provoked by activities of the Children's Bureau, which had included an examination of how other nations provided for the welfare of children. At a 1919 Children's Bureau conference, a Japanese official participated and Senator Reed commented, "Think of importing a Jap to tell an American mother how to take care of her baby" ([18], p. 8763). Representative Caleb Layton from Delaware, a physician, commented as well on how the Sheppard-Towner Bill could enable

... disseminating and approving all sorts of hectic ideas germinated in the deranged minds of the peoples of distracted foreign lands concerning birth control, the use of contraceptives, sex hygiene, endowment of motherhood, wages for mothers, State support of children, false economics, the economic independence of mothers from husbands ([21], p. 7929).

In spite of these concerns raised, mortality statistics continued to be used to support the Sheppard-Towner legislation. Some members of Congress, however, questioned if current mortality rates were reason for concern. Among this group was Representative Layton, who in commenting on Sheppard-Towner observed that there was no demand for its passage "by reason of any unusual mortality either in expectant mothers or in the newborn children" ([21], p. 7927).

These comments were illuminated by his adherence to a belief that only the fittest should survive. In response to those who spoke about the need to respond to the alarming statistics, Representative Layton replied:

Is it desirable to have an overwhelmingly large population in America? ... is there not some virtue, some real reason in nature's law – the survival of the fittest? ... Has the father and the mother of today so degenerated that their own Government must turn itself into an almshouse for the preservation of their children ([20], p. 7146)?

Economic concerns may have mingled with Representative Layton's philosophic beliefs about natural selection. The Sheppard-Towner Bill was, he said, "unnecessary, an unexcusable expense and plainly socialistic." He made clear to Congress that his physician colleagues were "opposed to this measure and its ultimate purpose, the nationalization of medicine" ([21], p. 7929).

Representative Layton believed it would be unproductive to let "ill-trained women" rather than physicians provide education and care. He claimed lack of economic resources and not "lack of knowledge nor care on the part of the physicians of the land" was largely responsible for the mortality among mothers and their babies ([21], p. 7928). He argued:

National prosperity is a prerequisite to national health at all times and under all circumstances. What we should be working for, straining every energy for is a revival of business, a nation wide revival of employment. A people employed is a happy people, because they are a healthful people ([21], p. 7929).

These views differed significantly from those who held that attention and funding should be focused directly on the educational and care needs of mothers and their children. Many believed this direct approach would have lasting benefits to society. For example, Representative Clarence Lea from California wrote:

Every child that we save is not only an economic asset to the Nation but we conserve the highest purposes of humanity in protecting and promoting the welfare of the child. That child is her mother's most sublime experience, the father's greatest inspiration and ultimately the source of the Nation's renewal ([22], p. 7988).

Senator Joseph France of Maryland agreed and took on skeptic Senator Joseph Frelinghuysen of New Jersey in a debate about expenditures and investments. Senator Frelinghuysen stated:

I feel that under my constitutional oath of office I have no right to vote for any bill, however meritorious, when I know that the condition of the Public Treasury will not permit the expenditure of money for legislation of this character. We are told at the present time that there is a deficit of $1,200,000,000 ([15], p. 454).

Senator France queried:

Knowing that the Senator is a banker, I want to ask him if he makes no distinction between an expenditure and an investment ([15], p. 454).

Senator Frelinghuysen replied:

Mr. President, I note the statement of the Senator. I am not a banker ([15], p. 454).

The issue then became one of priorities and how funds are allocated. Nevada's Senator Key Pittman rose in defense of the bill and stated:

Mr. Frelinghuysen will probably feel that he is obeying his oath of office when he votes for an appropriation of $400,000,000 for the Army; and yet he thinks that he would be violating his oath of office if he voted for the little sum of $4,000,000 to assist in saving 23,000 women and 200,000 children who die every year ([15], p. 455).

In the end the Sheppard-Towner Bill passed in the Senate by a vote of 63 to 7 [19] and in the House by a vote of 279 to 39 [22]. While Senator Sheppard had described the bill as having "no partisan authorship or motive" ([17], p. 3145), it did have opponents who succeeded in slowing its passage. Frances Keyes, a senator's wife, wrote a monthly column for *Good Housekeeping* on her experiences in Washington, D.C., and had followed closely the long floor debate on this bill. In describing her observations, she wrote:

If the smallest women's club in the most insignificant village conducted its business in such a way, it would be an object of scorn to every man in the place ([46], p. 32).

Indeed, the passage of the bill was acclaimed as a great success for women. While some members of Congress believed that the death rates reflected "normal mortality" ([14], p. 423), the testimony showed great need for governmental intervention to prevent these needless maternal and infant deaths. Once again, voting records showed that those who opposed the Sheppard-Towner Bill also voted against suffrage. While the views of the minority were vocally represented, support for the legislation was strong. Passage of the Sheppard-Towner Bill by a wide margin suggested Congressional acceptance of authority to allocate general revenues "to promote the general welfare" by providing health education and care. This act of Congress also reflected its recognition of a responsibility to respond to the basic human needs it addressed.

## IV. THE CONTINUING CONTROVERSY OVER SHEPPARD-TOWNER

The Sheppard-Towner Act continued, however, to be a source of controversy. The constitutionality of the Act was questioned before the U.S. Supreme Court in the cases of *Massachusetts v. Mellon* and *Frothingham v. Mellon et al.* These cases contended that the appropriations were not for national but for local purposes of the states and constituted an effective means of inducing the states to yield a portion of their sovereign rights [38]. The court upheld the Act, but Massachusetts was among three states that never participated in the program.

In 1921, Julia Lathrop resigned from her position. Grace Abbott assumed the position of Chief of the Children's Bureau. It was during Miss Abbott's tenure that the Sheppard-Towner Act was administered. The programs that

# MOTHERS' DAY

## SAVE THE MOTHERS

| IN OTHER COUNTRIES AND IN OURS | DEATH RATE OF MOTHERS PER THOUSAND BIRTHS |
|---|---|
| Denmark | 2.4 |
| The Netherlands | 2.4 |
| Sweden | 2.5 |
| Italy | 3.0 |
| Norway | 3.0 |
| Uruguay | 3.4 |
| Japan | 3.8 |
| England and Wales | 3.9 |
| Hungary | 4.0 |
| Finland | 4.4 |
| Germany | 4.9 |
| Australia | 5.0 |
| New Zealand | 5.1 |
| Spain | 5.2 |
| Ireland | 5.5 |
| Switzerland | 5.5 |
| France | 5.7 |
| Scotland | 6.2 |
| **UNITED STATES** | **6.8** |

18,000 of our mothers died in childbirth in 1921. Most of them could have been saved. Why weren't they?

Taken from:
Mother and Child, IV, 220, February 1923.

Fig. 1.

developed in the states varied, depending on the needs of their geographic areas. However, four general activities were common to many of them. These included the promotion of birth registration; cooperation between health authorities and physicians, nurses, dentists, and nutrition workers; establishment of maternity centers; and educational classes for mothers, midwives, and household assistants or mothers' helpers. Figure 1 presents the effects of this activity in a cartoon published in 1923.

In its ongoing crusade of support for the Sheppard-Towner Act, *Good Housekeeping* published an article, "Making America Safe for Mothers," and acclaimed the progress made by the Act [40]. In describing activities sponsored by Sheppard-Towner funds, it reviewed how "Neighborhood Institutes" were held in Florida and nurses came and talked to neighborhood groups of women about prenatal care and "confinement." In Georgia, South Carolina, Arkansas, and Maryland, a "health truck" traveled, holding clinics in remote areas. Most of the states instituted mothers' classes and what became known as "Little Mothers' Classes" for young girls who were taught about infant care.

In 1927, the five-year Act was due to expire, and attempts were made to renew it for another five years. This action once again brought great debate to the floors of Congress. At this time, legislators examined not only the merits of the Act but also its accomplishments during its five-year history. The issue of states' rights was again lustily discussed. In the mid-twenties, however, fears were also expressed about how federal support for maternal and child health care resembled communistic activities.

Such concerns were stated by the board of managers of Woman Patriot Publishing Company. The *Woman Patriot* had been the official publication of the National Association Opposed to Woman Suffrage. Delaware's Senator Thomas Bayard submitted its views regarding Sheppard-Towner to the *Congressional Record*:

... this legislation is an integral part and direct result of a comprehensive communist legislative program ... to socialize and nationalize the care, control, and support of American children ([24], p. 12919).

The Catholic Church also opposed continuation of the Sheppard-Towner legislation and mounted pressure for its repeal [3, 77]. The most apparent opposition to extending the Sheppard-Towner Act, however, came from the American Medical Association (AMA).

In May 1922, the AMA stated that "the Sheppard-Towner law is an imported socialistic scheme unsuited to our form of government." This view was submitted to and published in the *Congressional Record* ([23], p. 1113). The AMA's ongoing opposition to this form of legislation was well represented in Congress and revealed the fears its members had about limiting their professional autonomy:

Maternal and infant health work cannot be separated from health work generally. If the government maintains supervision over maternal and infant health in the States, it must ultimately gain control over all other health activities ([25], p. 1116).

Indeed, physicians' concern motivated New York's Senator Royal Copeland to say:

The doctors are very fearful of what they call "state medicine" ... in the medicine profession every man is an authority unto himself. The standards are not fixed. He (the physician) dreads the time which he fears may come conferring power to the State to give directions as to the treatment which shall be accorded disease and the method by which that treatment shall be applied ([26], p. 839).

The AMA also challenged reports tying improved infant mortality with Sheppard-Towner activities. Furthermore, the AMA described how mortality rates improved in states that did not cooperate with the federal program and observed that international data were not comparable to those collected in the United States [26]. The AMA did not, however, comment on the fact that activity supported by the Sheppard-Towner legislation was concurrent with continuing efforts to improve the registration of births and deaths. Perhaps better record-keeping was affecting the nature of data reported.

Recognizing the difficulty of using mortality statistics, the American Public Health Association in praising the Sheppard-Towner Act commented that its success:

... is not to be measured as yet in lowered infant and maternal mortality rates, but by the attitude of the states in desiring to continue the program in its full scope after the federal support has been withdrawn ([66], p. 257).

With all said and done, Congress in 1927 reached a compromise to extend the provisions of the Act until June 30, 1929, and then voted for it to be repealed. In response to this action, *Good Housekeeping* published an article, "Play-Days on Capitol Hill," and begged its readers: "Vote these Senators out, and vote in somebody who is for you, and not against you!" ([34], p. 27).

Though the Act was extended for two more years, the mounting opposition and the severe economic stress of the times led to failure in attempts to revive it in 1930 and 1932. Members of Congress lamented the growth of federal agencies and the authority they assumed. In remembering the legislation that created the Children's Bureau in 1912, the comment was made that "it was never conceived that it would develop into its present swollen proportions" ([26], p. 822). Referring to Sheppard-Towner, the observation was made that "like many similar measures passed by Congress under the guise of a temporary need, they become permanent when once the door has been opened" ([27], p. 6281).

Supporters, however, charged that self-serving economic interests were the source of most of the opposition to legislation. Senator T.H. Caraway of Arkansas in 1930 said it

... comes from sources that hope to profit by the defeat of the legislation; that is, they are professional men who feel that the field of their activity is being invaded and their chance of making a profit out of human suffering is being limited ([26], p. 817).

Indeed, the AMA continued to oppose this type of legislation, going so far as to declare in 1930 that it was "unsound, wasteful, unproductive and promoted communism" ([26], p. 829). The same day these comments were added to the *Congressional Record*, Senator James Davis of Pennsylvania, the former Secretary of Labor, inserted into the *Record* official data demonstrating improved survival rates for mothers and infants during the years of Sheppard-Towner. He reported:

Countless lonely mothers and children isolated in rural districts have been reached by the maternity and infancy nurse, riding on horseback where an automobile could not take her ... more than 122,000 babies born in 1929 survived infancy, who would have died if the conditions of 20 years ago had prevailed. There were 76 infant deaths for every 1,000 live births in 1921; 68 in 1929 ... ([26], pp. 832–833).

While the AMA had fought the Sheppard-Towner Bill, physician opposition to it was not universal. There were physicians who shared the views of Senator Davis and supported this federal involvement in health care. For example, Senator Henry Hatfield, a West Virginia physician, expressed his dismay with his professional organization and commented:

I am a member of the American Medical Association, as I am a member of my State association, but I do not know why they object, and I can not for the life of me understand why any doctor would object to this class of legislation ([26], p. 949).

On the floor of the Senate, a letter from Dr. Charles Mayo was also read saying:

Federal aid to States for maternal and child hygiene activities is justifiable and advisable in reducing excessive mortality rates among mothers and infants ([26], p. 939).

Ironically, this controversy over the official position of the American Medical Association helped form the American Academy of Pediatrics. The Pediatric section of the American Medical Association favored the continuation of the Sheppard-Towner Act, but it was not permitted to publish its views independently. In protest, in 1930, these members broke away from the American Medical Association and formed the American Academy of Pediatrics with its motto, "For the Welfare of Children."

The authority exercised by Congress in considering and passing the Sheppard-Towner legislation was debated during the seven years the law was in force. The ultimate evaluation of this authority was the law's favorable examination by the Supreme Court. While the attempts of Congress to protect the rights of mothers and babies to a safe childbirth were upheld, citizen groups argued that Sheppard-Towner infringed on their rights and the expression of these views played a role in its repeal. Important, as well, was the severity of the general economic conditions of the nation, which were leading to the Great Depression. These factors together contributed to the unwillingness of Congress to continue funding what had originally been described as a time-limited activity.

## V. THE DEPRESSION YEARS AND SOCIAL SECURITY FOR CHILDREN

After Sheppard-Towner, the states attempted to continue the promotion of maternal and child health, but the deepening economic depression limited such efforts. This was noted in the "Twentieth Annual Report of the Chief of the Children's Bureau to the Secretary of Labor." In this report, Grace Abbott observed that in 1930 the rate of puerperal septicemia causing 36% of maternal deaths was the lowest ever reported, but she noted that these deaths were largely preventable and encouraged future studies of maternal mortality [5].

In describing how between 1921 and 1929 (the years Sheppard-Towner funds were available) there had been a statistically downward trend in infant mortality, she observed "popular education has proved its value and should be continued" ([5], p. 167). The closing statement of the report, however, revealed the concerns of these times:

Child welfare workers everywhere look to the coming year with much anxiety. It will take great effort to maintain the standards of service for children which were slowly developed during the years before the depression ... Neglect of the health, education, and general welfare of children will be permanently costly to the children and to the future of the country ([5], p. 168).

During the early years of the Depression, the Children's Bureau recognized that state budgets for health programs for mothers and children were shrinking but that the needs for assistance were growing. The need for a federal response to the growing social problems was acutely recognized. Abbott sent a plan for children's health and welfare in 1934 to Edwin Witte, the Executive Director of the Committee on Economic Security, created by Franklin Roosevelt. The President gave this committee the mission of making recommendations for legislation to provide safeguards against the misfortunes of the era. It acknowledged the importance of support for children in its final report.

In hopes of appeasing its anticipated foes, attempts were made to frame these measures for children in a way that distinguished them from the Sheppard-Towner Act [77]. Specifically, the new proposed assistance to the states was drafted to be less of a partnership between federal and state governments so as to give states greater latitude in developing their unique programs.

In spite of these efforts, the American Medical Association held a special delegate convention in 1935 and adopted a resolution vigorously condemning the maternal-child proposal. Mr. Witte, in analyzing the AMA's ultimate failure to interfere with this proposed measure, commented that they were at the time:

... far too alarmed about the possibility of (governmentally supported) health insurance to present any serious objection to the administration of the child and maternal health services through the Children's Bureau ([77], p. 167).

On January 4, 1935, President Roosevelt addressed a joint session of Congress. He stated:

I shall send to you, in a few days, definite recommendations (which) ... will cover the broad subjects of unemployment insurance and old-age insurance, of benefits for children, for mothers, for the handicapped, for maternity care, and for other aspects of dependency and illness where a beginning can now be made ([28], pp. 94–95).

The bill submitted to Congress became HR 7260 "to provide for the general welfare by establishing a system of federal old-age benefits and by enabling the several States to make more adequate provisions for aged persons, dependent and crippled children, maternal and child welfare, public health and the administration of their unemployment compensation laws to establish a Social Security Board; to raise revenue; and for other purposes." In short, this became known as the Social Security Bill.

Representative Harry Sauthoff of Wisconsin was among those who saw the value of Title V of the Social Security Bill, which provided for children's health care. He observed that Title V would be a:

... splendid forward step in the march of progress in social security (Applause). I want to keep on with that forward march just as long as we can possibly do so. As far as I am personally concerned, the 9,000,000 children who come under this beneficial legislation are more important than either the old-age people or the unemployed with the $4,880,000,000 work-relief bill. It now remains for us to make some substantial contribution to the future in securing not a temporary relief measure, but a definite, permanent, social security plan, and this is it ([29], p. 5553).

Representative Joseph Pfeifer from New York expressed the views of the AMA. Of note is the softer position of organized medicine to federal assistance to states. Unlike its opposition to Sheppard-Towner, the AMA views now expressed in Congress were focused solely on the administration of the Title V provisions. Representative Pfeifer described:

The medical profession recognizes the necessity under conditions of emergency for Federal aid in meeting basic needs of the indigent; the house of delegates of the American Medical Association deprecates, however, any provision whereby Federal subsidies for medical services are administered and controlled by a lay bureau. While the desirability of adequate medical service for crippled children and for the preservation of child and maternal health is beyond question, the house of delegates deplores and protests those sections of the bill which place in the Children's Bureau of the Department of Labor the responsibility for the administration of funds for these purposes ([32], p. 6050).

The Social Security Bill was not without its foes. Representative Charles Eaton from New Jersey, a physician member of Congress, made comments echoing those made a decade earlier by Representative Layton, also a physician. His comments went beyond those of the AMA as he critically observed:

This legislation does not provide adequate care for the aged, but it does lay a new and intolerable burden of taxation and control upon American industry without solving the problem of unemployment. It is simply one more step toward sovietizing our distinctive American institutions, devitalizing the self-reliance and enterprise of our people, and mortgaging our future by a debt so mountainous that we will be in grave danger of repudiation or inflation (Applause) ([30], p. 5583).

The broad support for this measure, however, was well revealed by Representative Thomas Jenkins from Ohio, a Republican member of the House Ways and Means Committee, who stated:

It is not legislation that belongs to any party. This is legislation that has sprung up out of a desire of the people of this country to have the Federal Government participate and help out the States in this grand and wonderful work ([31], p. 5680).

With few opposing votes, the Social Security Bill was passed in Congress and signed into law by President Roosevelt on August 14, 1935.

In 1934, Katharine Lenroot was appointed by President Roosevelt as the third Chief of the Children's Bureau. In addressing a session of the Health Officers' Section of the American Public Health Association in October, 1935, she reviewed with them the functions of the Title V legislation [48]. She described the intention of this title of the Social Security Act to extend and strengthen services to mothers and children in rural areas and those where economic distress was greatest. To support the need for this effort, she noted that in 1933 to 1934 the infant mortality rate increased one point from 57 to 58 in urban areas whereas it increased three points from 59 to 62 in rural areas.

The Crippled Children's services were a new activity for the Children's Bureau, and the funds were allocated to be used for the purpose of enabling each state to extend and improve, especially in rural areas, services for identifying crippled children. Specifically, Title V enabled provisions for medical, surgical, corrective, and other services and care and facilities for diagnosis, hospitalization, and after-care for children who were crippled or who were suffering from conditions that might lead to impairment. The provision enabling services for conditions *leading to impairment* was the basis for preventive programs reflected throughout the work of the Children's Bureau.

In contrast to the Sheppard-Towner Act, Title V had no time limit. The efforts to justify the extension of Sheppard-Towner were dogged by the recognition that funding for less than a decade could not, in a major way, demonstrate major effects on mortality statistics. Martha Eliot, M.D., who became Chief of the Bureau, observed that as a policy issue, in omitting any time limit, Congress adopted the position that the federal government

... has a responsibility to use its central taxing power to assist the States on a continuing basis in their efforts to improve the health and welfare of mothers and infants ([36], p. 137).

With the dire conditions of the Depression, fears of governmental interference in family life seemed to evaporate. In 1935, the needs of the citizenship were apparently perceived as so enormous that governmental intervention was mandated, not only for the elderly but also for children. Social Security had come to be perceived as a right, and the authority of the federal government to provide for it was now broadly accepted.

## VI. WORLD WAR II TO 1980

During the early 1940s, programs continued to be developed under Title V of the Social Security Act. With the United States' entry into World War II, a new situation faced those providing health services for mothers and children. Young wives of servicemen had accompanied their husbands to the military posts where they had been stationed prior to being sent overseas, and many of these women became pregnant without a true residence or source of medical care. The Children's Bureau was made aware of this problem and initially attempted to address the needs of this population with Title V funds, but soon realized that special appropriations were needed. On March 18, 1943, Congress unanimously approved the Emergency Maternity and Infant Care (EMIC) program which paid for maternity care and infant care for the babies' first year for dependents of every serviceman in the lowest four pay grades [69].

Interestingly, the Academy of Pediatrics, which had established itself as an advocate for children, initially opposed the EMIC program. The Children's Bureau's attempt to better inform the pediatricians of the enormous need for this program led the Academy to conduct its first survey on child health services and pediatric education. The findings of this study led the Academy to support Federal measures to provide care for children and convinced many of its members of the ability of public agencies to improve the health status of children [37].

The EMIC program did much to provide a humane service not only for 250,000 young wives and their babies, but also for the morale of service men.

From a policy perspective, this was the largest public medical care program ever entirely supported by general tax funds without a local match required. With the EMIC program, states made great progress in licensing and upgrading hospital maternity care, resulting in a shift away from home deliveries [69].

Between 1935 and 1963 Congress made few changes in the Title V program, but the overall level of appropriations for them was increased. The Crippled Children's program was expanded in the 1940s to treat children with rheumatic heart conditions and in the 1950s to cover children with congenital heart disease. Awareness of how perinatal complications affect the infant mortality rate was also growing, and many states began to use Title V funds to pay for infant intensive care.

White House conferences on children during these years reflected concerns of the times. The 1940 conference called for improvements in maternity care. By now, the death rate for mothers was approximately half of what it was in 1920 (38 vs. 80 per 10,000 live births). The 1950 conference brought a demand to ban public school racial segregation. At this conference, more attention was paid, as well, to the needs of the mentally retarded; subsequent to it, the National Association for Retarded Children was formed [74].

The need for child health research gained growing attention during these years. In the 1930s, elected officials paid little attention to this activity, and in 1949, a proposal for a National Child Research Act was defeated. In 1961, a national center for basic research in child development was established when the National Institute of Child Health and Human Development was created [61].

There were important medical advances affecting mothers and children during these times. In the 1930s, Vitamin D was added to milk to prevent a deficiency causing rickets. In the late 1950s, vaccines against polio, measles, and rubella were developed. Antibiotics made major contributions in reducing mortality from infection. By 1960, the infant mortality rate was 26 per 1,000 live births compared to its rate of 86 in 1920 [74].

In 1946, President Truman proposed that with the exception of activities related to child labor, the Children's Bureau be transferred from the Department of Labor and placed in the Federal Security Agency. Congress approved this transfer and in 1953 the Federal Security Agency was named the Department of Health, Education, and Welfare and the Children's Bureau began to reside in this new agency of the federal government.

The 1960s were a period of renewed interest in contemporary social problems. The White House Conference on Children and Youth in 1960 included the participation of youth for the first time, and much of its focus was on the context of childhood. Attention was paid to the effects of poverty, deprivation, denial of civil rights, and racial discrimination [74].

Changes were made during this decade in broadening the extent of services provided under Title V. With President Kennedy's personal interest in mental retardation, a panel on this topic was convened during his administration, which led to an increase in appropriations for special projects for mentally retarded children. Maternal and infant care projects were also authorized with the recognition that retardation could be reduced if complications of childbearing were minimized by better maternal care. New Congressional appropriation also enabled the development of Children and Youth (C & Y) projects to provide comprehensive health care services to preschool children.

President Johnson's "War on Poverty" during the 1960s led to growing federal expenditures for the disadvantaged. In 1965, Title XIX was added by Congress to the Social Security Act creating Medicaid, the important health insurance program which includes coverage for low-income children.

Amendments to the Title V program in 1967 authorized and mandated the development of three new projects: family planning, dental health, and intensive care. The inclusion of family planning as a mandatory part of the overall Title V program reflects changing policy views about this subject, which had always been considered very sensitive. The first time this subject was discussed at a federally supported meeting was in 1938 when the Children's Bureau allowed a conferee to bring up the matter at a conference on Better Care for Mothers and Babies.

In August, 1967, an administrative order dictated the reorganization of the Department of Health, Education, and Welfare. With this order, the health and welfare components of the Children's Bureau were split apart with its functions largely distributed to other agencies. With what became a dismantling of the Children's Bureau, Title V began to be administered by the Public Health Service. Dr. Eliot, who was then a former Chief of the Children's Bureau, commented on the meaning of this anticipated action when she received the Howland Award from the American Pediatric Society in 1967. She observed:

Is it in the public interest to have in the government an agency which is a spokesman, an advocate for children? Other population groups have their lobbies. Who will speak strongly enough for children if not a highly-placed government agency? If either the maternal and child health or child welfare activities are removed, the voice of the advocate for children will become dim or lost ([37], p. 572).

Though an act of Congress would be required for it to be abolished, the reorganization in the Department of Health, Education, and Welfare greatly decreased the stature and function of the Children's Bureau. No longer is there a single agency in the federal government serving as a "voice for children."

The years between World War II and the current administration were a time of growth for programs providing children's health care. Federal authority in this area was essentially unquestioned, and federal funding grew in many areas. Medicaid represented a de facto recognition of children's right to health care, and many services expanded to address the needs of those who lived in poverty. By the end of the 1970s, it was estimated that there were 260 federal programs for children administered in 20 different agencies. Many of these programs were health-related [54]. Ironically, while many developed with the availability of federal funds, their overall coordination at a federal level became less focused.

## VII. PRESIDENT REAGAN'S POLICIES AND INFANT MORTALITY IN THE 1980S

Approximately 80 years following the Progressives' attempt to make the federal government more responsive to the needs of the citizens, President Ronald Reagan in his inaugural address expressed his view that "government is not the solution to our problem. Government is the problem" ([33], p. 541). He urged the nation to rein in the powers of government, and this philosophy was well reflected in the Omnibus Reconciliation Act of 1981 passed by Congress in the first summer of President Reagan's administration. Included in the Act were sweeping changes in how federally funded social programs were to be administered. Previously independently funded programs were combined into "block grants," which gave to the states greater authority in deciding what activities should receive funding. The Reagan administration proposed the creation of block grants combining programs for children and for adults. Fearing that pressure from adult constituencies could rob children's programs' access to appropriations, the American Academy of Pediatrics and the National Maternal and Child Health Resource Center lobbied successfully for a separate block grant for Maternal and Child Health Care.

The Maternal and Child Health Care Block Grant combined eight previously independently funded health programs, including Title V of the Social Security Act, into one grant to the states. It is currently administered by the Office of Maternal and Child Health Care in the Public Health Service. Though administration of the block grant is a new mission for the Public Health Service, the greater challenge was responding to the cuts in appropriation. In 1982 and 1983, the Maternal and Child Health block grant was funded at a level 18% lower than the combination of the funding previously received by its eight programs. The 18% decrease did not truly represent the extent of curtailed funding. At this time, the inflation rate for medical care was approximately 10%, and the needs for publically funded services in the nation's communities were rapidly growing [68]. The Medicaid program also was cut in ways that eliminated services for many mothers and their children.

Early in the Reagan administration, the decision was made for the first time in 70 years not to hold the "once a decade" White House Conference on Children. At this time, the nation was also in the grip of a painful recession with over 11% unemployment reported in 1982 [68]. States were feeling increasing pressure to provide care for mothers and children while concurrently services were being eliminated due to federal budget cuts.

Probably nowhere was this situation more acutely felt than in Michigan. In January, 1983, Bailus Walker, Ph.D., Director of the Michigan Department of Public Health, testified before the House of Representatives subcommittee on Labor Standards [73]. He reported that Michigan's unemployment rate of 17% required that 15% of the population receive some form of public assistance and described how cutbacks in funds had forced the curtailment of maternal and child health services. Associated with this grave economic situation, Dr. Walker reported:

We are distressed by the fact that Michigan's infant mortality rate increased from 12.8 deaths per 1,000 live births in 1980 to 13.2 deaths in 1981. This increase represents a reversal of a three decade trend which saw our infant mortality rate cut by 50%. Some areas of the state have actually realized a 100% increase in one year ([73], p. 3).

News articles during this time reported that neighborhoods in Detroit had infant mortality rates that were approaching those reported in Third World countries [62].

As noted by Newsholme in 1910, the effects of social circumstances are well reflected in infant mortality rates. In 1979, *Healthy People*, the Surgeon General's Report on Health Promotion and Disease Prevention, published the goal for the nation to achieve an infant mortality rate of 9 per 1,000 live births by 1990 [72]. In her annual report, "Health United States – 1984," Secretary Margaret Heckler of the Department of Health and Human Services claimed adherence to this goal; however, she reported to the *American*

*Medical News*, "infant mortality statistics do show a declining rate of improvement" [52].

Indeed, the nation's overall progress in reducing infant mortality has slowed. The rate of improvement averaged approximately 3% per year between 1981 and 1983 [68]. This is the slowest decline in 18 years and about half the average annual decline during the 1970s. As noted on Table 1, the current infant mortality rate is approximately 10% of the rate recorded when data were first collected. However, a trend over time has persisted in the difference between the white and non-white rates. Over the course of the past 75 years, black infants have been shown to die at approximately twice the rate of white infants. In 1983, this gap between black and white infant mortality became the widest recorded in more than 40 years [42].

Perhaps the most significant change in health indicators is that observed in the postneonatal mortality rate. This is a measure of deaths occurring between the 29th day and the end of the first year of life and is sensitive to the environmental and social influences affecting infant survival [42]. The postneonatal mortality rate increased by 3% for all infants between 1982 and 1983 and rose 5% for blacks, representing the first reported increase for this group in 18 years [42]. Sobering as well are data showing how the current infant mortality rate compares with those reported by other nations. Data from 1984 reveal that the United States currently ranks eighteenth for infant mortality among 25 countries in the world with populations of more than 2.5 million [75].

Analyses of the United States' current infant mortality rate attribute our international standing to our failure to make progress in decreasing the incidence of low birth weight newborns. Between 1950 and 1982, there has been less than a 1% (7.6 to 6.8%) decrease in the percentage of births of low birthweight infants [68]. Caring for low birth weight infants is a highly expensive venture, and the application of costly technology has led to their improved survival. The Institute of Medicine of the National Academy of Sciences showed that in 1985 for every dollar spent for prenatal care for a high risk group of women, $3.38 could be saved in the total cost of caring for low birthweight infants ([43], p. 17). The Children's Defense Fund (a private children's advocacy organization), however, tells us that in 1983 for the third year in a row the percentage of pregnant women receiving late or no prenatal care increased nationwide [42].

Like the conditions at the turn of the century, cuts in services during the 1980s show the effects of poverty on children's health [51, 53, 62, 76]. Moreover, poverty among children is increasing. Table II shows this trend with over 20% of all U.S. children in 1984 living in poverty. The increasing rate of poverty for children must also be viewed in the context of what has happened to the family. In 1980, 18% of all births were to single women; 10% of white and 55% of black babies born to single mothers. Overall, 25% of all children live in single parent households with 19% of white children and 58% of black children living with one parent [68].

TABLE II
Children and poverty

|  | 1959 | 1969 | 1979 | 1983 | 1984 |
| --- | --- | --- | --- | --- | --- |
| All children less than 18 years | 27% | 14% | 16% | 22% | 21% |
| White children less than 18 years | 21% | 10% | 11% | 17% | 16% |
| Black children less than 18 years | 67% | 40% | 41% | 46% | 46% |
| Children in female headed households | 72% | 54% | 49% | 56% | 54% |

Data from U.S. Census Bureau [68].

Unlike the turn of the century, however, the federal government in 1983 took a very active position in intervening in health care to protect the rights of one group of children, handicapped infants. This was vividly displayed in what became known as the case of Baby Doe. A newborn infant with Down syndrome died following a parental decision, upheld by the courts, not to permit a life-saving surgery. In response to this well-publicized death, the Reagan administration, citing the Rehabilitation Act of 1973, ruled that notices be posted in all hospitals that included the statement "Failure to feed and care for infants may also violate the criminal and civil laws of your State." A toll-free Handicapped Infant Hotline was established so that any interested party could report suspected "abuse" of handicapped infants by health providers.

This action stirred enormous controversy and a group of child health care providers, including the American Academy of Pediatrics, filed suit against the Department of Health and Human Services. Within weeks of issuance of the regulation, Judge Gesell set aside the regulation requiring the posted statements in hospitals. In his ruling, the judge is reported to have stated that the government in invoking such a regulation:

... did not appear to give the slightest consideration to the advantages and disadvantages of relying on the wishes of the parents who, knowing the setting in which the child may be raised, in many ways are in the best position to evaluate the infant's best interests [58].

Legal and legislative efforts have continued to pursue the issue of how the rights of handicapped infants can best be protected. This case illustrates the Reagan administration's willingness to insert governmental authority into decisions made together by private physicians and their patients' families. Is this a departure from its philosophy regarding the use of federal authority?

Major concerns facing children's health care today include preventing the birth of costly low birth weight infants, improving infant mortality rates, and responding to the needs of children who are surviving with ongoing complex health care needs. Those who testify to Congress about these concerns make a familiar plea for federal assistance. Kenneth Osgood, M.D., representing the American Academy of Pediatrics in 1983 at a hearing before the Joint Economic Committee, stated:

I believe the President should assign major new responsibilities to the Children's Bureau with Department of Health and Human Services to gather data on the status of children in America, to prepare comprehensive reports annually to Congress on the status of children, how Federal programs are affecting that status and to coordinate issues within the Federal Government and the Nation dealing with child health, nutrition, education, and other related children's issues ([56], p. 9).

Recognizing the same need, Dr. Walker of Michigan urged Congress to create a new unit of the government with the familiar charge to "investigate and report on the conditions affecting the health and welfare of America's children, youth, and families" ([73], p. 12). In response, Senator Bentsen of Texas submitted a resolution that urged the President to reorganize the Children's Bureau, granting it administrative power to gather extensive data on the status of children and the impact of federal programs on them, and to submit to Congress an annual report on these findings [71]. To date, this has not occurred.

### VIII. ANALYSIS

Controversies about federal authority, economics, and human rights shape the history of federal involvement in children's health care. Times, of course, have changed, but many concerns remain the same. Today, we are not accommodating to new pressures of industrialization as in the Progressive era but rather face the pressures of technological advances that have brought unemployment and the social displacement of labor. Technology, however, has also led to the extension of life with babies born today having an expectation of living 20 years longer than did babies born in 1920 [68]. While as a nation we today recognize the heterogeneity of our citizenship, we struggle with accommodating the needs of families who frequently do not resemble the traditional two-parent structure more likely to have existed in the past. Unlike 85 years ago, currently we have a labyrinth of children's programs supported by the federal government. Yet, today, we have no one agency that represents children's needs.

The issue of what the appropriate role of federal authority should be in supporting children's health care persists. Children elicit powerful sentiment, but as was observed in the Senate in 1912:

We can not be blamed because the states failed to grant to the Federal Government the power to reach down and take every unfortunate little child by the hand and lead it to hope, opportunity and success ([10], p. 1527).

Sixty-nine years later, President Reagan reminded the country of how government is not the solution to problems but is often the problem itself.

As a nation, we have a history of intense respect for the autonomy of the family. In the 1920s, the notion of income standards for families brought shivers to spines of some members of Congress who expressed great fear that the Sheppard-Towner Act might lead to such "deranged" policy. Of interest today is the fact that every industrialized country except the United States provides a yearly allowance to all families with children [54].

In the early part of the twentieth century, the constitutional authority establishing a bureau "to investigate and report upon all matters pertaining to the welfare of children" was fervently argued. Today we debate the efficacy of cutting funds for programs and the grounds upon which a federal agency can investigate medical decisions made by parents and physicians that affect the care of their children and patients. Some see federal involvement as an infringement on the rights of states. Others argue that federal involvement in matters of national significance serves as a stimulus for state activity.

The improvement in maternal and child health since 1900 is clear and unmistakable. Unlike the earlier part of the century, childbirth is no longer a life-threatening event. The American Medical Association and those opposed to federal involvement claimed improvements in maternity care would have occurred in communities without a Sheppard-Towner Act. Its proponents argued that without a federal initiative, state and local action would have been longer in coming. Though it is difficult to identify cause and effect relationships, racial and socioeconomic disparities in mortality for children provide important evidence for the relation of social factors and children's health status [76]. In spite of such evidence, today one hears the same philosophy that was espoused in the 1920s by Dr. Lawton who in the House of Representatives claimed "national prosperity is the best health measure that we can possibly institute" ([21], p. 7929). Yet, can babies wait for prosperity to reach them?

In the early decades of the century, the United States' world ranking for infant and maternal mortality was deplored by those who sought federal attention to these problems. We hear similar concern expressed today, as our infant mortality rate is higher than the rate in seventeen other nations. Such data continue to attest to the fact that childhood in this country is a time of unnecessary medical vulnerability. Babies and young children who "go without" knowledgeable parents or material resources are more likely to succumb to illness or death.

As Theodore Roosevelt and Florence Kelley decried, "babies don't vote." Their needs must be expressed by those who care about their welfare and are willing to be advocates for their rights to health care. Indeed, resources are limited, and wisdom and foresight must accompany their expenditure. Yet, accompanying such thought, I believe, should be Senator France's recognition expressed in 1920 that appropriations made for children are not just expenditures but investments in the future.

*University of South Dakota*
*School of Medicine*
*Sioux Falls, South Dakota*

## NOTE

[1] To document these historical trends, several different references have been used as resources. To capture the sense of the views of elected federal legislators, the *Congressional Record* has been reviewed and excerpts from early debates are quoted. To capture the information available to the public, the *Reader's Guide to Periodical Literature* has served as a means of identifying how journalists reported on Congressional activity surrounding this concern and provided their editorial comments. Historical analyses have been consulted to gain insight into this history. Finally, in tracing the history of trends in this federal activity, maternal and infant mortality rates are provided to show how these health indicators are related to changing public policy.

## BIBLIOGRAPHY

1. 'American Women Urged to Vote for State Protection of Motherhood', *Current Opinion* (March 1920), 375.
2. Baker, S. J.: 1939, *Fighting for Life*, Macmillan, New York.
3. Berkelhamer, J. E., Noyes, E. J., and Chen, R. T.: 1982, 'Child Health Policy – an Overview of Federal Involvement', *Advances in Pediatrics* **29**, 211–228.
4. Buhler-Wilkerson, K.: 1985, 'Public Health Nursing: In Sickness or in Health?' *American Journal of Public Health* **75** (October), 1155–1176.
5. 'Child Hygiene', *American Journal of Public Health* **23** (February), 1933, 166–168.
6. Congressional Record: 1909 (February 15), 2363.
7. Congressional Record, The Senate: 1911 (December 11), 188–189.
8. Congressional Record, The Senate: 1912 (January 8), 704.
9. Congressional Record, The Senate: 1912 (January 24), 1251.
10. Congressional Record, The Senate: 1912 (January 30), 1523–1534.
11. Congressional Record, The Senate: 1912 (January 31), 1517–1579.
12. Congressional Record, House of Representatives: 1912 (April 2), 4218–4440.
13. Congressional Record, The Senate: 1917 (January 22), 1747–1748.
14. Congressional Record, The Senate: 1920 (December 16), 417–423.
15. Congressional Record, The Senate: 1920 (December 17), 454–455.
16. Congressional Record, The Senate: 1921 (April 27), 252.
17. Congressional Record, The Senate: 1921 (June 28), 3142–3145.
18. Congressional Record, The Senate: 1921 (June 29), 8759–8764.
19. Congressional Record, The Senate: 1921 (July 22), 4216.
20. Congressional Record, House of Representatives: 1921 (November 1), 7146.
21. Congressional Record, House of Representatives: 1921 (November 18), 7927–7929.
22. Congressional Record, House of Representatives: 1921 (November 19), 7988–8036.
23. Congressional Record, The Senate: 1926 (January 5), 1113.
24. Congressional Record, The Senate: 1926 (July 3), 12919–12950.
25. Congressional Record, The Senate: 1927 (January 5), 1116.
26. Congressional Record, The Senate: 1930 (December 16), 817–949.
27. Congressional Record, House of Representatives: 1931, (February 27), 6281.
28. Congressional Record, Joint Session: 1935 (January 4), 94–95.
29. Congressional Record, House of Representatives: 1935 (April 12), 5553.
30. Congressional Record, House of Representatives: 1935 (April 13), 5583.
31. Congressional Record, House of Representatives: 1935 (April 15), 5680.
32. Congressional Record, House of Representatives: 1935 (April 19), 6050.
33. Congressional Record, Joint Session: 1981 (January 20), 541.
34. Cranston, C.: 1927, 'Play-Days on Capitol Hill', *Good Housekeeping* **84** (May) 27, 282–288.
35. Dean, W. H.: 1920, 'Safe Motherhood', *The Ladies Home Journal* (December), 42–43.
36. Eliot, M. M.: 1960, 'Origins and Development of the Health Service', *Children* **7** (July-August), 135–149.

37. Eliot, M. M.: 1967, 'The United States Children's Bureau', *American Journal of Disabled Children* **114**, 565–573.
38. 'Experience Under the Federal Maternity Act, 1921–1929', *Monthly Labor Review* **29** (August 1929), 348–351.
39. Faulkner, H. U.: 1931, *The Quest for Social Justice*, The Macmillan Company, New York.
40. Glover, K.: 1926, 'Making America Safe for Mothers', *Good Housekeeping* (May), 98, 270–280.
41. Hofstadter, R.: 1955, *The Age of Reform*, Vintage Books, New York.
42. Hughes, D., Johnson, K., Simon, J., and Rosenbaum, S.: 1986, *Maternal and Child Health Data Book*, Children's Defense Fund, Washington, D.C.
43. Institute of Medicine: 1985, *Preventing Low Birth Weight*, National Academy Press, Washington, D.C.
44. Kelley, F.: 1916, 'Our Embassy to Childhood', *Survey* **33** (March 6), 632.
45. Kelley, F.: 1920, 'Why Let Children Die?' *Survey* **45** (December 25), 401.
46. Keyes, F. P.: 1921, 'Letters from a Senator's Wife', *Good Housekeeping* **73** (October), 31–32.
47. Lathrop, J. C.: 1912, 'The Children's Bureau', *American Journal of Sociology* **18**, 318–330.
48. Lenroot, K. F.: 1935, 'National Aspects of the Social Security Program as They Pertain to the Children's Bureau', *American Journal of Public Health* **25** (December), 1327–1333.
49. Lesser, A. J.: 1985, 'The Origin and Development of Maternal and Child Health Programs in the United States', *American Journal of Public Health* **75** (June), 590–598.
50. Martin, A.: 1920, 'We Couldn't Afford a Doctor', *Good Housekeeping* (April 20), 19–20, 133.
51. Miller, C.A.: 1985, 'Infant Mortality in the U.S.', *Scientific American* **253**, 31–37.
52. 'Minority Infant Death Rate Far From 1990 Goal', *American Medical News* (April 12, 1985), 37.
53. Mundinger, M. O.: 1985, 'Health Service Funding Cuts and the Declining Health of the Poor', *New England Journal of Medicine* **313**, 44–47.
54. Murphy, C.: 1982, 'Kids Today', *The Wilson Quarterly* **6** (Autumn), 61–82.
55. Newsholme, A.: 1910, Report by the Medical Officer on Infant and Child Mortality. Supplement to the 30th Annual Report of the Local Government Board, London.
56. Osgood, K.: 1983, Testimony before hearing on the Impact of Budgetary Cuts in the Maternal and Child Health Block Grant, Economic Goals and Intergovernmental Policy Subcommittee, Joint Economic Committee, November 17, 1–11.
57. 'Our New Mother – The Government', *The Outlook* **105** (September 13, 1913), 60–61.
58. Peark, R.: 1983, 'Judge Strikes Rule Requiring Care for Infants with Defects', *New York Times* (April 15), I, 1,3.
59. Roosevelt, T.: 1911, 'The Conservation of Womanhood and Childhood', *The Outlook* **100** (December 23), 1013–1019.

60. Roosevelt, T.: 1912, 'Who is a Progressive?' *The Outlook* **100** (April 13), 809–814.
61. Schmidt, W.: 1973, 'The Development of Health Services for Mothers and Children in the United States', *American Journal of Public Health* **63** (May), 419–427.
62. 'Spending Cuts Cited for Infant Death Rate', *American Medical News* (December 9, 1983), 1–2.
63. Sprago, J. S.: 1906, *The Bitter Cry of the Children*, Macmillan, New York.
64. 'The "Better Baby" Bill', *Literary Digest* **70** (July 16, 1921), 28.
65. Towner, H. M.: 1920, 'Mothers and Babies First', *Women's Home Companion* **47** (September), 4–5.
66. 'Two More Years for the Sheppard-Towner Law', *American Journal of Public Health* **17** (March 1927), 257.
67. United States Census Bureau: 1975, *Historical Statistics of the United States: Colonial Times to 1970*, Washington, D.C.
68. United States Census Bureau: 1986, *Statistical Abstract of the United States*, Washington, D.C.
69. United States Department of Health, Education, and Welfare: 1956, *Four Decades of Action for Children: A Short History of the Children's Bureau*, Publication 358, Washington, D.C.
70. United States Department of Health, Education, and Welfare: 1976, *Child Health in America*, Public Health Service, Washington, D.C.
71. United States Senate, Resolution 237, 98th Congress, First Session, October 3, 1983.
72. United States Surgeon General: 1979, *Healthy People*, The Surgeon General's Report on Health Promotion and Disease Prevention, United States Department of Health and Human Services, Washington, D.C.
73. Walker, B.: 1983, 'The Impact of Unemployment on the Health of Mothers and Children in Michigan: Recommendations for the Nation', Testimony presented to a hearing of the Subcommittee on Labor Standards, Committee on Education and Labor, Washington, D.C., January 31.
74. Waserman, M.: 1976, 'An Overview of Child Health Care in America', *Children Today* (May/June), 24–29, 44.
75. Wegman, M. E.: 1986, 'Annual Summary of Vital Statistics – 1985', *Pediatrics* **78**, 983–994.
76. Wise, P. H., Kotelchuck, M., Wilson, M. L., and Mills, M.: 1985, 'Racial and Socioeconomic Disparities in Childhood Mortality in Boston', *New England Journal of Medicine* **313** (August 8), 360–366.
77. Witte, E. E.: 1962, *The Development of Social Security Act*, University of Wisconsin Press, Madison.
78. 'Women and Children First', *The Nation* **111** (December 22, 1920), 724.

TODD L. SAVITT

## AMERICAN SOCIAL AND POLITICAL THOUGHT
## AND THE FEDERAL ROLE IN CHILD HEALTH CARE

The twentieth-century history of federal involvement in children's health affairs has followed an interesting pattern: from little concern to deep commitment and then to reluctant participation. Ann Wilson has shown how between 1906 and 1912 Theodore Roosevelt's administration prodded Congress to establish a Children's Bureau to determine children's and parents' medical and other needs; how Congress backed away from that commitment in the 1920s; how, during the Depression of the 1930s, President Franklin Roosevelt and Congress joined together to pass social legislation to help those in need, including children; and how the current Reagan administration, despite such actions as the "Baby Doe" regulations, has sought to reduce federal involvement in health care financing, Congress' wishes notwithstanding [5]. Though the three presidents mentioned – Theodore and Franklin Roosevelt, and Ronald Reagan – engineered these changes in government policy, they did not act in a vacuum. The general mood of the country, economic conditions, and political philosophy all played a role. To some extent, the medical profession also influenced public policy in this area.

Let us first go back to the late nineteenth century, before Theodore Roosevelt, to see what attitudes prevailed to keep government out of children's health problems. This post-Civil War era was one of rapid economic growth, especially for business and industry. Andrew Carnegie, John D. Rockefeller, Jay Gould, J. P. Morgan, and Leland Stanford are familiar names from the period. The federal government interfered little in individual citizen's business affairs, adhering to the prevailing *laissez-faire* philosophy of the time. The pursuit of individual self-interest was best for the country, according to this law of political economy, and government could promote that goal best by permitting free and unfettered competition among members of society. One historian has characterized this practical interpretation of Charles Darwin's then-recently-promulgated theory of evolution in the following manner:

All men, the theory read, applied themselves in the search for wealth and found rewards according to their ability. A few, the highest types of their race, discovered more effective ways to combine land, labor, and capital, and drew society upward as the rest reorganized behind their leaders. The large majority, possessing no more than

*Loretta M. Kopelman and John C. Moskop (eds.)*
*Children and Health Care: Moral and Social Issues, 67–71.*
© *1989 by Kluwer Academic Publishers.*

ordinary talent, divided a fund that was fixed by the requirements of the dearer resources, land and capital. The weakest simply disappeared. Meanwhile, government maintained order, conducted some public services at minimum cost, and above all did nothing to disrupt the laws of free competition. The society that abided by these principles slowly and steadily progressed, enjoying an ever superior utilization of its resources and an ever improving race winnowed by competition ([4], p. 135).

But this *laissez-faire* system did not always operate according to plan. Corporations sometimes abused their power and hurt individual citizens or, as they got more powerful, stifled free and fair competition. Local, state, and eventually even national government found it necessary to interfere in railroad, labor, and immigration issues, ostensibly in order to preserve *laissez-faire*. Changing social and economic conditions in the late nineteenth century forced legislators, judges, and politicians to reassess their "Social Darwinian" view of American political economy. They and an ever larger group of vocal citizens chose to abandon *laissez-faire*, even in social matters. They saw what rapid industrialization and the sudden large influx of immigrants from Eastern and Southern Europe brought to burgeoning cities: overpopulation, poverty, crime, and an overload on local social services. The community, previously able to extend a helping hand to those in need, found it difficult, if not impossible, to care properly for all those wanting assistance. The greatest sufferers, yet with the smallest voices, were children. The plight of these children left to fend for themselves while parents worked, or forced to work at dangerous, physically demanding jobs, or sent to school underfed and in ill health eventually attracted widespread attention.

To end human suffering and to preserve a national resource – the nation's future adults – child welfare groups began looking beyond overtaxed local sources of support to the federal government. The "Progressive Era" of early twentieth-century America, with its positive vision of improving society through involved government and concerned citizens, thus opposed the Social Darwinism of the earlier "Gilded Age." According to one historian of the period, a central focus of these Progressives was the child, the one who would fulfill the Progressives' vision of an ideal society, if properly nurtured and educated ([4], p. 169). Here then was one source of support for the Children's Bureau legislation of 1906–1912, pushed by a leading Progressive of the era, President Theodore Roosevelt.

Most of the objections to the Children's Bureau and later to the Sheppard-Towner Act arose out of the "survival-of-the-fittest" philosophy of limited government the Progressives were trying to overthrow. Social Darwinism had served the era of big business well, but times were changing and the arguments of a passing era, as Wilson described them from Congressional debates on these bills, were no longer persuasive.

Medicine remained rather silent during the Children's Bureau discussions of the first decade of the century, but came to life quite quickly when Sheppard-Towner was introduced in the next two decades. Physicians were just coming into their own in the early twentieth century as they began to make use of recent discoveries like germ theory, X-rays, antisepsis, and laboratory tests. They were also just becoming politically organized after a major internal reform of the American Medical Association in 1901 ([2], pp. 318–320). So even if they had perceived the Children's Bureau as a threat to the free enterprise system in medicine in 1906, they were not ready to raise a united voice against it. They were not yet a "big business." Thus the opposition to the Children's Bureau came from those still preaching the Social Gospel of survival-of-the-fittest and of *laissez-faire* government (i.e., those who stood to lose from new legislation).

What complicates the story by the time of Sheppard-Towner in 1921 is the voice of medicine, the AMA, now grown quite strong. The AMA had recently reversed its stand on federal involvement in health care in the form of compulsory health insurance [1]. Briefly stated, the AMA, in 1916, had looked favorably on the idea of compulsory health insurance to protect the American labor force. It seemed like a good idea to many physicians who saw such contract practice as guaranteeing a basic income (competition for patients was quite keen and physician income generally low). Furthermore, most European countries had already adopted social and medical insurance plans and the wave seemed inevitably headed across the Atlantic. But the sudden revulsion to things German (health insurance was reputedly of German origin) after American entry into World War I in April 1917, the growing fear that physicians would actually lose money under compulsory health insurance, and the lessening sense of inevitability as more and more people came out against health insurance plans, caused the AMA as a body and leading individual American physicians to strongly oppose any form of state involvement in medical practice. (The Russian Revolution of 1917 did not help either.)

Sheppard-Towner thus came along at a sensitive time for American physicians who were still feeling relieved at their close call with compulsory health insurance. Linking Sheppard-Towner and compulsory health insurance, the editor of one state medical journal wrote: "This bill is a menace and represents another piece of destructive legislation sponsored by endocrine perverts, derailed menopausics and a lot of other men and women who have been bitten by that fatal parasite, the upliftus putrifaciens ... all of whom are working overtime to devise means to destroy the country" (quoted in [1], p. 107). If there was any question that the strength of the Progressive Movement had waned by the 1920s, the persistent opposition of organized medicine to Sheppard-Towner throughout the decade, ending in its repeal in 1929, confirmed it.

Just a few years later, in 1935, however, the mood of the country was quite different from the boom times of the 1920s. FDR's Social Security legislation passed Congress in 1935, with Title V included. Though issues of government involvement in individual family affairs and of state rights to control local health matters certainly received discussion, the economic hard times dictated a change in national social policy. So children were included in Social Security and the federal government remained a presence in child health care. Organized medicine (the AMA) steadfastly and successfully opposed any inclusion of health insurance in Social Security legislation beyond that already named in Title V. (It also opposed the Medicare and Medicaid plans some thirty years later for many of the same reasons, but accepted these programs after their passage.)

Once again in 1980 the mood of America seemed to shift in its view of federal involvement in local matters. After years of providing funds through various programs, the now politically and economically conservative federal government made an effort to return power and money to the states, including, as Wilson points out, support for child health. As with Social Darwinism, the Reagan administration's "trickle down" theory assumed that the economic success of those at the top of the socio-economic scale would eventually improve the condition of all Americans through employment and private philanthropy. Big government would stay out of the American public's lives as much as possible. The "Baby Doe" regulations, however, kept a government presence in the family household, the hospital, and the doctor's office.

Wilson [5] and Starfield [3] question whether the Reagan administration is applying a false economy to its child health appropriations. Wilson seems, in both the tone and the content of her paper, to have answered that question in the affirmative, showing how federal money does make a difference in the state of health of American children, despite the huge expense and the possible loss of state rights.

*East Carolina University School of Medicine*
*Greenville, North Carolina*

## BIBLIOGRAPHY

1. Numbers, R. L.: 1978, *Almost Persuaded: American Physicians and Compulsory Health Insurance, 1912–1920*, Johns Hopkins University Press, Baltimore.
2. Rothstein, W. G.: 1972, *American Physicians in the Nineteenth Century, From Sects to Science*, Johns Hopkins University Press, Baltimore.
3. Starfield, B.: 1989, 'Child Health and Public Policy', in this volume, pp. 7–22.
4. Wiebe, R. H.: 1967, *The Search for Order, 1877–1920*, Hill and Wang, New York.
5. Wilson, A.: 1989, 'The U.S. Federal Role in Child Health Care: A Historical Perspective', in this volume, pp. 27–66.

ROBERT J. LEVINE

# CHILDREN AS RESEARCH SUBJECTS[1]

In 1975 the National Commission for the Protection of Human Subjects of Biomedical and Behavioral Research (the National Commission) began its studies of the ethics and regulation of research involving children as subjects. In its report, *Research Involving Children*, published in 1977, the National Commission presented its recommendations for the ethical conduct of research involving children [22]. These recommendations formed the basis of federal regulations that were proposed in 1978 [21] and promulgated in final form in 1983 [19].

The fundamental question presented to the National Commission was this: Should we do research involving children as subjects? If we do, then we shall be doing research on individuals without their informed consent. If we do not, then we shall be depriving children as a class of persons of the benefits of research. In this paper I shall first discuss these two issues. Then I shall discuss the major features of our federal policy regarding research involving children.

## I. INFORMED CONSENT

The central problem presented by proposals to do research involving children as subjects is that children, as a class of persons, lack the legal capacity to consent. In addition, many of them, particularly the younger ones, are incapable of sufficient comprehension to meet the high standards of consent to research developed in such documents as the Nuremburg Code. The Declaration of Helsinki reflects an awareness of this problem by calling for consent by the parent, guardian, or legally authorized representative; such consent is commonly referred to as proxy consent. Much of the debate about the ethics and law of research involving children has focused on the nature of the procedures for which a proxy may consent.

An extreme position in this debate is presented by Ramsey, who bases his argument on a strict interpretation of the ethical principle of respect for persons, i.e., that we leave persons alone unless they consent to be touched [15]. Consequently, he argues that the use of a non-consenting subject (e.g., a child) is wrong whether or not there is risk, simply because it involves an "unconsented touching." As he puts it, one can be wronged without being harmed. "Wrongful touching" is rectified only when it is for the good of the individual, because then the person is treated as an end as well as a means. Hence, proxy consent may be given for non-consenting subjects only when the research activity includes therapeutic interventions related to the subject's own recovery. Ramsey acknowledges that, in some cases, there may be powerful moral reasons for involving children in research having no therapeutic components; however, "it is better to leave (this) research imperative in incorrigible conflict with the principle that protects the individual human person from being used for research purposes without his expressed or correctly construed consent." He continues that it would be immoral either to do or not to do the research, but he maintains that one should "sin bravely" in the face of this dilemma by sinning on the side of avoiding wrong or harm rather than attempting to promote welfare. Recently, Ramsey has clarified his position on this issue [16]. His use of the term "touching" is literal; if the child is not touched physically, he or she is not wronged. In this interpretation, infants and very young children have no privacy rights to be respected.

When the principle of respect for persons is interpreted strictly, the unconsented-to touching of a competent adult is wrong even if it benefits that person. In that case, why should potential benefit justify such touching for a child? McCormick proposes that the validity of such interventions rests on the presumption that the child, if capable, would consent to therapy [10]. This presumption, in turn, derives from a person's obligation to seek therapy, an obligation that people possess simply as human beings. Because people have an obligation to seek their own well-being, we presume that they *would* consent if they *could* and thus presume also that proxy consent for therapeutic interventions will not violate respect for them as persons.

By analogy, people have other obligations, as members of a moral community, to which one would presume their consent; McCormick calls this "their correctly construed consent." One such obligation is to contribute to the general welfare when to do so requires little or no sacrifice. Hence, McCormick concludes that non-consenting subjects may be used in research not directly related to their own benefit so long as the research fulfills an important social need and involves *no discernible risk*. In McCormick's view, respecting persons includes recognizing that they are members of a moral community with its attendant obligations. (For a discussion of the philosophical bases for the argument that children bear obligations to a moral community, particularly the community of similarly situated children, see

Pence [14].)

Ramsey counters this argument by claiming that children are not adults with a full range of duties and obligations [15]. Therefore, they have no obligation to contribute to the general welfare and respect for them requires that they be protected from harm and from unconsented touching. Ramsey further insists that there is an important difference between "no discernible risk" and "discernibly no risk"; the latter standard presupposes empirical evidence of safety while the former does not.

Freedman bases his argument on the same premises as Ramsey's, i.e., a child is not a moral being in the same sense as an adult [4]. However, his analysis yields the same conclusion as McCormick's. Precisely because children are not autonomous, they have no right to be left alone. Instead, they have a right to custody, i.e., to be taken care of. Thus, the only relevant moral issue is the risk involved in research; the child must be protected from harm. Therefore, Freedman agrees with McCormick that children may be used in research unrelated to their therapy, provided it presents to them no discernible risk.

Ackerman argues that we tend to fool ourselves with procedures designed to show respect for the child's very limited autonomy [1]. He claims that the child tends to follow "the course of action that is recommended overtly or covertly by the adults who are responsible for the child's well-being." He further contends that, in general, this is as it ought to be. "Once we recognize our duty to guide the child and his inclination to be guided the task becomes that of guiding him in ways which will achieve his well-being and contribute to his becoming the right kind of person."

Gaylin tells the story of a man who acted in accord with Ackerman's position [5]. After directing his 10-year-old son to cooperate with venipuncture for research purposes, he explained that his direction arose from his perceived moral obligation to teach his child that there are certain things one does to serve the interests of others, even if it does cause a bit of pain: "This is my child. I was less concerned with the research involved than with the kind of boy that I was raising. I'll be damned if I was going to allow my child, because of some idiotic concept of children's rights, to assume that he was entitled to be a selfish, narcissistic bastard."

In developing his concept of variable competence, Gaylin goes even further [5]. In considering research procedures presenting low risk to the child, he claims that if a parent has no sense of obligation to the community, it would be good for the child as well as the community for the state to instruct the parent as well as the child on the topic of social responsibility. He uses as his example the collection of urine for epidemiologic research purposes. "Refusal of permission for such an experimental involvement would be trivial, arbitrary and ungenerous." Therefore, he would place extreme limits on the authority of an individual to refuse to cooperate and be "extremely generous" in according to the child independent authority to choose to participate.

These are some samples of the major themes in the controversy over the role of consent in research involving children.[2] As we shall see, the recommendations of the National Commission reflected its conclusion that, because infants and very young children have no autonomy, there is no obligation to respond to it through the usual devices of informed consent. Rather, respect for infants and very small children requires that we protect them from harm. "Minimal risk" was identified as a threshold standard in that research presenting more than minimal risk requires special justifications; "no discernible risk" was rejected as too restrictive. The National Commission also recommended procedures for respecting the developing autonomy of older children and adolescents.

## II. THERAPEUTIC ORPHANS

As a consequence of the uncertainties about the ethical propriety of and legal authority to do research on children, there has been a great reluctance in the United States to do studies to determine the safety and efficacy of drugs in children. As a result, as Shirkey observed in 1968, "Infants and children are becoming the therapeutic orphans of our expanding pharmacopoeia" [17]. Since 1962, nearly all new drugs have been required by the Food and Drug Administration (FDA) to carry on their label one of the familiar "orphaning" clauses: e.g., "not to be used in children," "... is not recommended for use in infants and young children, since few studies have been carried out in this group ...," "clinical studies have been insufficient to establish any recommendations for use in infants and children," "... should not be given to children ..." By 1975, over 80% of all drugs prescribed for children bore such orphaning clauses on their labels [12].

The therapeutic orphan phenomenon is not limited to children. Very similar conditions obtain in the use of drugs in pregnant women [12, 13] and in young women generally [7]. Pregnant women are usually excluded from drug trials and from many other types of research because the fetus is seen by some as a non-consenting subject who might be peculiarly vulnerable to the effects of drugs ([8], Chapter 13). Women who are biologically capable of becoming pregnant are commonly excluded from many types of research, including drug trials. Given suitable plans for contraception, it would be reasonably safe to include them in most studies [7]; however, many investigators do not wish to assume the burden of discussing plans for contraception with prospective subjects ([8], Chapter 5). Moreover, many investigators fear the potential legal consequences of a failure to prevent pregnancy during a drug study.

In passing, I wish to point out that the term "therapeutic orphan" has picked up a new meaning. Most current literature that refers to therapeutic orphans concerns "orphan drugs," viz., drugs that if developed would be useful only in the treatment of relatively uncommon diseases. Because the potential market value for these drugs is small, most drug companies do not wish to invest the huge sums of money required to secure FDA approval for marketing a new drug in the United States. In this paper, I use the term "therapeutic orphan" exclusively in the sense intended originally by Shirkey.

The therapeutic orphan phenomenon represents a serious injustice. If we consider the availability of drugs proved safe and effective through the devices of modern clinical pharmacology and clinical trials a benefit, then it is unjust to deprive classes of persons, e.g., children and pregnant women, of this benefit. This injustice is compounded as follows. If we were to do Phase II and III clinical trials in children as we now do in adults, the first administration of various drugs would be done under conditions much more controlled and much more carefully monitored than is customary in the practice of medicine. It is likely that adverse drug reactions that are peculiar to children would be detected much earlier than they are now; consequently, either we could discontinue administration of the drugs to children or we could issue appropriate warnings to physicians who are using the drugs. The prevailing practice in the United States is to ignore the orphaning clauses on the package labels [9]. Consequently, we have a tendency to distribute unsystematically the unknown risks of drugs in children and pregnant women, thus maximizing the frequency of their occurrence and minimizing the probability of their early detection. Parenthetically, it should be noted that most drugs proved safe and effective in adults do not produce unexpected adverse reactions in children; however, when they do, the numbers of harmed children tend to be much higher than they would be if the drugs had been studied systematically before they were introduced into the practice of medicine. Historical examples include the development of phocomelia with thalidomide administration to pregnant women and the development of "grey sickness" in infants treated with chloramphenicol. In addition, we have insufficient knowledge of the proper doses of many drugs to be used in children [2].

The recommendations of the National Commission and the ensuing regulations should go far to reduce the magnitude of the problem associated with the therapeutic orphan phenomenon. In particular, their recommendation that risks "presented by an intervention that holds out the prospect of direct benefit for the individual subject" may be considered differently from the risks presented by procedures designed to serve solely the interests of research should facilitate the ethical conduct of clinical trials in children. It is also noteworthy that the FDA has announced its intention to propose regulations that will require that new drugs with major therapeutic utility in children be tested in children as a condition of approval of such drugs for marketing [3].

## III. PUBLIC POLICY

The National Commission concluded that children, because they have limited capacities to consent, are vulnerable or disadvantaged in ways that are morally relevant to their involvement as subjects of research. Therefore, the National Commission interpreted the principle of justice as requiring that we facilitate activities that are designed to yield direct benefit to the children-subjects and that we encourage research designed to develop knowledge that will be of benefit to children in general. However, we should generally refrain from involving children in research that is irrelevant to their conditions as individuals or at least as a class of persons. The principle of respect for persons was interpreted as requiring that we show respect for a child's capacity for self-determination to the extent that it exists. Although they are legally incapable of consent, many can register knowledgeable agreements (assents) or deliberate objections, terms that I shall discuss shortly. To the extent that the capacity for self-determination is limited, respect is shown by protection from harm. Accordingly, the National Commission recommended that the authority accorded to children or their legally authorized representatives to accept risk be strictly limited; any proposal to exceed the threshold of "minimal risk" requires special justification.

Let us now consider in more detail some of the features of the National Commission's recommendations. I shall cover here only the major points that reflect substantial departures from general policies covering the ethical justification of research involving autonomous adults.[3]

As with other classes of research involving human subjects, plans to do research involving children must be reviewed and approved by an Institutional Review Board (IRB). The IRB is required to determine that "where appropriate, studies have been conducted first on animals and adult humans, then on older children, prior to involving infants ..." (Recommendation 2B, [22], p. 2). This recommendation interprets the principle of justice to require that vulnerable persons be afforded special protection from the burdens of research. Adults are perceived as less vulnerable than older children who, in turn, are less vulnerable than infants. Investigators who propose to do research on children without having first done such research on animals, adults, or both will be obliged to persuade the IRB that this is necessary. The strongest justification is that the disorder or function to be studied has no parallel in animals or adults. In such cases, when the research presents any risk of physical or psychological harm or significant discomfort, investigators are expected to initiate their work on older children who are capable of assent before involving infants.

### Assent and Permission

The National Commission abandoned the use of the word "consent" except in situations in which an individual can provide "legally effective consent" on his or her own behalf. As a corollary to this, the term "proxy consent" was discarded.

Recommendation 7 assigns to the IRB the responsibility for determining that adequate provisions are made for: "Soliciting the assent of the children (when capable) and the permission of their parents or guardians ..." ([22], p. 12). The transactions involved in negotiating assent and parental permission are essentially the same as those for informed consent.

According to the National Commission, a child with normal cognitive development becomes capable of meaningful assent at about the age of seven years, although some may be younger and some older. The Department of Health and Human Services (DHHS) did not accept this recommendation regarding the age of assent. Rather, at the time the proposed regulations were published, DHHS solicited public comment on which of three options it would adopt for nontherapeutic procedures, either age 7, age 12, or leaving the age to the discretion of the IRB. The final regulations specify no age of assent. Rather, the IRB is assigned responsibility for "determining whether children are capable of assenting."

As the assent regulation is written, it seems to reflect the presumption that the capability to assent is an all-or-none phenomenon: the child is either capable or incapable of assent. This presumption is incorrect [23, 24] and, I believe, unintended by the regulation writers. However, a literal reading of the regulations seems to create this requirement: If the child is capable of assent – i.e., passes a certain threshold of capability – he or she must be presented with all of the required elements of information.

In my view the regulations are intended to be interpreted to permit the IRB to determine that prospective child-subjects may be capable of understanding some but not all of the elements of informed consent. Thus, for example, it may be appropriate to provide some children with "a description of any reasonably foreseeable risks or discomforts," without providing "an explanation as to whether any compensation ... (is) ... available if injury occurs."

The regulations make no reference to the child's objection except in the definition of assent (Section 46.402): "Mere failure to object should not, absent affirmative agreement, be construed as assent" ([19, p. 9819). However, IRBs should keep in mind that for some protocols in which they have judged the prospective child-subjects incapable of assent, some provisions may still be made to respond to the child's "deliberate objection." The term "deliberate objection" is used to recognize that some children who are incapable of meaningful assent are able to communicate their disapproval or refusal of a proposed procedure. A 4-year-old may protest, "No, I don't want to be stuck with a needle." However, an infant who might in certain circumstances cry or withdraw in response to almost any stimulus is not capable of deliberate objection. In its commentary on Recommendation 7, the National Commission suggested that a child's deliberate objection usually should be regarded as veto to his or her involvement in research. Of course, there are exceptions to this; most importantly, parents and guardians may overrule the young child's objection to interventions and procedures that hold out the prospect of direct benefit to the child and to some research maneuvers designed to evaluate these interventions and procedures.

*Permission*

Unless otherwise specified, the involvement of children as research subjects must be authorized by the permission of the child's parent or guardian. If more than minimal risk is presented by an intervention or procedure that does not hold out the prospect of direct benefit for the individual subject, "permission is to be obtained from ... both parents ... unless one parent is deceased, unknown, incompetent, or not reasonably available, or when only one parent has legal responsibility for the care and custody of the child" (Section 46.408). In other classes of research, "the IRB may find that the permission of one parent is sufficient." "Permission by parents or guardians shall be documented in accordance with and to the extent required by Section 46.117 of Subpart A." Whether or not assent must be documented is left to the discretion of the IRB ([19], pp. 9819–9820).

The purposes of informed consent, parental permission, and assent are entirely different from those of their documentation ([8], Chapter 5). Because the primary purpose of documentation is to protect the interests of the investigator and the institution against those of the subject, it seems generally unnecessary to document assent except in those cases in which the IRB determines that parental or guardian permission is unnecessary. When children are invited to sign forms, as they often and quite properly are, the principal purpose in my view is to enhance their sense of participation in the process.

In its commentary on Recommendation 7, the National Commission indicates that the parental or guardian permission should reflect the collective judgment of the family that an infant or child may participate in research. In research projects for which permission of one parent or guardian is sufficient, e.g., research in which the risks or discomforts are related to an intervention that is designed to be therapeutic, diagnostic, or preventive for that particular child, it may be assumed that the person giving formal permission is reflecting a family consensus.

The requirement for parental or guardian permission assumes that the child is living in a reasonably normal family setting. It further assumes that a normal loving relationship exists between the child and his or her parents. In the event there is probable cause to suspect that no such loving relationship exists, different procedures may be required at the discretion of the IRB.

### *Requirements varying with the degree and nature of risk*

Research that presents to children no more than minimal risk may be conducted with no substantive or procedural protections other than those specified earlier. The following definition of minimal risk may be found in DHHS regulations: "Minimal risk means that the risks of harm anticipated in the proposed research are not greater, considering probability and magnitude, than those ordinarily encountered in daily life or during the performance of routine physical or psychological examinations or tests" (Section 46.102g, [20], p. 8387).

The National Commission provided examples of procedures presenting no more than minimal risk: these are routine immunizations, modest changes in diet or schedule, physical examination, obtaining blood and urine specimens, and developmental assessments. Similarly, many routine tools of behavioral research, such as most questionnaires, observational techniques, noninvasive physiological monitoring, and puzzles may be considered to present no more than minimal risk. Questions about some topics, however, may generate such anxiety or stress as to involve more than minimal risk. Research in which information is gathered that could be harmful if disclosed should not be considered of minimal risk unless adequate provisions are made to preserve confidentiality.

*Minor Increments Above Minimal Risk*

Recommendation 5 deals with research proposals that present to the child minor increments above minimal risk. The risks with which this recommendation is concerned are presented by procedures that do not hold out any expectation of direct health-related benefit for the child.

In order to justify the use of such procedures, the IRB must determine that: "Such intervention or procedure presents experiences to subjects that are reasonably commensurate with those inherent in their actual or expected medical, psychological, or social situations, and is likely to yield generalizable knowledge about the subjects' disorder or condition ..." ([22], p. 8). The requirement that experiences be reasonably commensurate with those inherent in their actual or expected situations requires some clarification. First, it means that the procedures to be followed are those that they or others with the specific disorder or condition under study will ordinarily experience by virtue of their having or being treated for that disorder or condition ([22], p. 9). Thus, it might be appropriate to invite a child with leukemia who has several bone marrow examinations to consider having another for research purposes. It would be much more difficult to justify extending a similar invitation to a normal child. This requirement will make it difficult to develop normal control data for examinations and other procedures that present more than minimal risk.

The requirement for commensurability reflects the National Commission's judgment that children who have had a procedure performed upon them might be more capable than are those who are not so experienced to base their assent on some familiarity with the procedure and its attendant discomforts; thus, their decisions to participate will be more knowledgeable.

The IRB must further determine that: "The anticipated knowledge is of vital importance for understanding or amelioration of the subject's disorder or condition ..." ([22], p. 8). This requirement thus establishes a higher standard than that for research characterized as presenting minimal risk for assessing the importance of the knowledge to be gained. In addition, it strengthens the general requirement to use children as subjects particularly in research that is relevant to their disorder or condition. Thus, it should be very difficult to justify the use of procedures presenting more than minimal risk to develop information irrelevant to disorders or conditions present in the subjects of the research.

These recommendations and the corresponding regulations create a requirement for the IRB to make two difficult judgments for this class of research. What constitutes vital importance? What is the upper limit of "minor" in assessing an increment above minimal risk? I think the National Commission and DHHS each showed wisdom by resisting demands to define the boundaries of these terms. IRBs and investigators are now challenged to explore these concepts and to develop functionally relevant definitions as they consider problems presented by particular protocols. As they share and debate the fruits of their explanations with one another, we can expect a gradual refinement of our understanding of these concepts and how to use them.

Among the procedures that have been approved by Yale's IRB as presenting minor increments above minimal risk are bone marrow aspirations in children with leukemia, single additional spinal taps in adolescents who have already had at least one for a neurologic disorder, and administration of yohimbine in order to gain information about the pathogenesis of a neurologic disorder. The same IRB rejected a proposal to do left heart catheterizations on children at risk for the development of cardiac hemosiderosis.

Proposals to do research presenting more than minor increments above minimal risk must be justified according to even more stringent standards ([8], pp. 249–250). Approval of such proposals may be granted only by the DHHS Secretary after consultation with a panel of experts in pertinent disciplines.

*Interventions Presenting the Prospect of Direct Benefit*

Research protocols that present more than minimal risk of physical or psychological harm or discomfort to children but in which the risk "is presented by an intervention that holds out the prospect of direct benefit for the individual subjects, or by a monitoring procedure required for the well-being of the subjects" may be considered differently. Recommendation 4 requires the IRB to determine that: "Such risk is justified by the anticipated benefit to the subjects;" and "the relation of anticipated benefit to such risk is at least as favorable to the subjects as that presented by available alternative approaches ..." ([22], pp. 5–6).

In this recommendation, the National Commission calls for an analysis of the various components of the research protocol. Procedures that are designed solely to benefit society or the class of children of which the particular child-subject is representative are to be considered as the research component. Judgments about the justification of the risks imposed by such procedures are to be made in accord with other recommendations. For example, if the risk is minimal, the research may be conducted as described in Recommendations 2 and 7 no matter what the risks of the therapeutic components.

The components of the protocol "that hold out the prospect of direct benefit for the individual subjects" are to be considered precisely as they are in the practice of medicine. Risks are justified by anticipated benefits to the individual subjects and, further, by the assent when appropriate of the child and the permission of the parents or guardians.

In Recommendation 7, the National Commission made one further statement relevant to risk presented by potentially therapeutic procedures: "A child's objection ... should be binding unless the intervention holds out a prospect of direct benefit that is important to the health or well-being of the child and is available only in the context of the research" ([22], pp. 12–13).

Section 46.408 (a) of the regulations reflects this concept in a somewhat altered form: "If the IRB determines ... that the intervention or procedure holds out a prospect of direct benefit that is important to the health or well-being of the children and is available only in the context of research, the assent of the children is not a necessary condition ..." ([19], p. 9819).

The general presumption is that parents may make decisions to override the objections of school-age children in such cases. However, in some circumstances the objection of teenagers to decisions on their behalf by parents may prevail. In the practical world of decision-making about who can authorize a therapeutic procedure, whether it be investigational or accepted, it rarely suffices to point to the law and thereby identify the person who has the legal right to make the decision. Many factors must be taken into account in reaching judgments about the capability of various persons to participate in and, in the event of irreconcilable disputes, to prevail in such choices. In general, these judgments become more complicated as the child gets older or as the stakes get higher. For a more complete discussion of these factors, see

Gaylin [5], and Thomasma and Mauer [18]. In short, the necessary considerations have all of the richness and complexity of the same considerations encountered in the course of the practice of medicine. This is because this class of research activities resembles medical practice at least as much as it does the conduct of other classes of research involving human subjects.

The provisions of Recommendation 4 and the corresponding requirements in federal regulations should do much to mitigate the problems associated with the therapeutic orphan phenomenon.

## IV. EPILOGUE

At the beginning of this paper I asked: "Should we do research involving children as subjects?" Then I offered two succinct statements of the meanings of answering "yes" or "no." Now let us reflect further on those two statements.

"If we do, then we shall be doing research on individuals without their informed consent." The implications of such a statement made in the early 1970s were very clear. Use of persons as research subjects without their informed consent was exploitation, a violation of the first principle of the Nuremberg Code, a violation of the requirements of the ethical principle of respect for persons. In short, it was wrong! Now we have what I consider a much more adequate understanding of what it means to show respect for children and other persons having limited autonomy.

"If we do not, then we shall be depriving children as a class of persons of the benefits of research." In the early 1970s, the prevailing perception was that participation in research was a burden. Persons or groups of persons who could not protect themselves from the burdens of research participation through the devices of informed consent were labeled vulnerable and therefore in need of special protections. Now we recognize that some of our past policies and practices designed to protect vulnerable persons from unwanted or unwarranted burdens actually deprived them of important benefits.

The development of federal policy on research involving children reflects these and other refinements in our understanding of how we ought to treat children as individuals and as a class of persons.

*Yale University School of Medicine*
*New Haven, Connecticut*

## NOTES

[1] Portions of this paper are excerpted or adapted from *Ethics and Regulation of Clinical Research* [8], with the permission of the publisher.

[2] For further reading on this matter see Chapters 8 and 9 in the National Commission's Report, *Research Involving Children* [22], *Who Speaks for the Child* [6], *Children's Competence to Consent* [11], and Chapter 10 of *Ethics and Regulation of Clinical Research* [8].

[3] Additional details may be found in Chapter 10 of *Ethics and Regulation of Clinical Research* [8].

## BIBLIOGRAPHY

1. Ackerman, T. F.: 1979, 'Fooling Ourselves with Child Autonomy and Assent in Nontherapeutic Clinical Research', *Clinical Research* **27**, 345–348.
2. Cohen, S. N.: 1977, 'Development of Drug Therapy for Children', *Federation Proceedings* **36**, 2356–2358.
3. Finkel, M. J.: 1978, 'Proposed Regulations for Study of New Drugs in Children', in B. L. Mirkin (ed.), *Clinical Pharmacology and Therapeutics: A Pediatric Perspective*, Year Book Medical Publishers, Chicago, pp. 299–304.
4. Freedman, B.: 1975, 'A Moral Theory of Informed Consent', *Hastings Center Report* **5**:4 (August), 32–39.
5. Gaylin, W.: 1982, 'Competence: No Longer All or None', in W. Gaylin and R. Macklin (eds.) *Who Speaks for the Child*, Plenum Press, New York, pp. 27–54.
6. Gaylin, W. and Macklin, R. (eds.): 1982, *Who Speaks for the Child: The Problems of Proxy Consent*, Plenum Press, New York.
7. Kinney, E. L., Trautmann, J., Gold, J. A., Vessell, E. S. and Zelis, R.: 1981, 'Underrepresentation of Women in New Drug Trials', *Annals of Internal Medicine* **95**, 495–499.
8. Levine, R. J.: 1986, *Ethics and Regulation of Clinical Research*, Second Edition, Urban & Schwarzenberg, Baltimore.
9. Luy, M. L. L.: 1976, 'Package Insert Roulette: The Catch-22 of Prescribing', *Modern Medicine* **44**:5, 23.
10. McCormick, R. A.: 1974, 'Proxy Consent in the Experimentation Situation', *Perspectives in Biology and Medicine* **18**:2, 2–20.
11. Melton, G. B., Koocher, G. P., and Saks, M. J. (eds.): 1983, *Children's Competence to Consent*, Plenum Press, New York.
12. Mirkin, B. L.: 1975, 'Drug Therapy and the Developing Human: Who Cares?' *Clinical Research* **23**, 106–113.
13. Mirkin, B. L.: 1975, 'Impact of Public Policy on The Development of Drugs for Pregnant Women and Children', *Clinical Research* **23**, 233–237.
14. Pence, G. E.: 1980, 'Children's Dissent to Research – A Minor Matter?' *IRB: A Review of Human Subjects Research* **2**:10 (December), 1–4.
15. Ramsey, P.: 1970, *The Patient as Person*, Yale University Press, New Haven.
16. Ramsey, P.: 1980, '"Unconsented Touching" and the Autonomy Absolute', *IRB: A Review of Human Subjects Research* **2**:10 (December), 9–10.

17. Shirkey, H. C.: 1968, 'Therapeutic Orphans', *Journal of Pediatrics* **72**, 119–120.
18. Thomasma, D. C. and Mauer, A. M.: 1982, 'Ethical Implications of Clinical Therapeutic Research on Children', *Social Science and Medicine* **16**, 913–919.
19. U.S. Department of Health and Human Services: 1983, 'Additional Protection for Children Involved as Subjects in Research', *Federal Register* **48**:46 (March 8), 9814–9820.
20. U.S. Department of Health and Human Services: 1981, 'Final Regulations Amending Basic HHS Policy for the Protection of Human Research Subjects', *Federal Register* **46**:16 (January 26), 8366–8388.
21. U.S. Department of Health, Education, and Welfare: 1978, 'Protection of Human Subjects: Research Involving Children', *Federal Register* **43**: 141 (July 21), 31786–31794.
22. U.S. National Commission for the Protection of Human Subjects of Biomedical and Behavioral Research: 1977, *Research Involving Children: Report and Recommendations*, DHEW Publication No. (OS) 77–0005, Washington.
23. Weithorn, L. A.: 1983, 'Children's Capacities to Decide About Participation in Research', *IRB: A Review of Human Subjects Research* **5**:2 (March/April), 1–5.
24. Weithorn, L. A.: 1983, 'Involving Children in Decisions Involving Their Own Welfare', in G. B. Melton, G. P. Koocher, and M. J. Saks (eds.), *Children's Competence to Consent*, Plenum Press, New York, pp. 235–260.

LORETTA M. KOPELMAN

# WHEN IS THE RISK MINIMAL ENOUGH FOR CHILDREN TO BE RESEARCH SUBJECTS?

When should research involving children as subjects be permitted? This difficult and pressing problem is often presented in the form of a dilemma: If we do research involving children as subjects, then we do so using individuals who cannot give informed consent. If we do not, then children as a group are denied many of the benefits of research including therapeutic advances, the possibility of good information about therapies, and the repudiation of dangerous and discredited therapies. Robert Levine shows how the U.S. National Commission for the Protection of Human Subjects of Biomedical and Behavioral Research (The National Commission) tried to find a morally defensible solution to the question of when children may be enrolled as subjects in research. They held that with appropriate review, safeguards, consent from guardians, and assent from the child, children may be enrolled if the study on balance holds out direct and appropriate benefit to them or if the study is not too risky. Levine served as a consultant to the National Commission and shows sympathy (which I share) for the solution proposed.

In this paper, I will focus on the central notion of this policy that children may be involved in many kinds of research if it is not too risky or if on balance it holds out appropriate benefits for them.[1] I will argue that this policy relies on a poorly defined concept of *minimal risk*. Without a consensus on what this means, the general agreement about when children may be research subjects may be illusory; and some empirical evidence exists suggesting that people tailor their notion of minimal risk to fit preconceived ideas about what kinds of studies should be done. After discussing why a standard is needed, I will turn to the question of how the current definition of "minimal risk" is inadequate.

## I. REJECTION OF TWO EXTREME VIEWS

It is necessary to clarify what kind of risk is acceptable for children's studies. Two positions that would allow us to avoid doing so are unacceptable. One such policy permits the same kind of research to be done on children as is done with adults no matter what the risk. This is unreasonable because competent adults may be at liberty to consent to take risks for the sake of others or for their own personal gain when they cannot volunteer others for such risks. Since children do not have the authority or are not generally competent to give consent for risky studies that do not hold out benefit for them, this extreme position is rejected.

Another policy that avoids clarifying what kind of risk is acceptable for children's studies does so because it permits only research that directly benefits the child, as in therapy. Although this position at first seems humane and reasonable, it would, as Shirkey and Levine point out, make children "therapeutic orphans" [11, 12, 14]. It is a flawed policy because it makes it difficult or impossible to conduct controlled testing of therapies for children. Consequently, children would either do without or be given inadequately tested therapies. Moreover, this policy prohibits all non-therapeutic studies, no matter how low the risk. It would even disallow collection and analysis of the growth and development data obtained during children's routine examinations. Altogether, this policy would leave children as a group sorely neglected medically. Suppose, for example, a child at four years of age seems developmentally delayed. We might want to determine if he can stack blocks or ride a tricycle as well as most four-year-olds. Unless we test many four-year-olds, we have no way of knowing the norm. But testing normal four-year-olds to see how well they can stack blocks or ride a tricycle to determine such standards would be a *non-therapeutic* study. All non-therapeutic studies would be disallowed under this policy. Even if the four-year-olds had a wonderful time, we could not record the results if doing so would constitute research. Having a good time, or even uncertain future benefits to a group, does not transform research into therapy. To call something therapeutic, we need to show through testing that it is likely to be useful to treat a certain illness. (Of course, stacking blocks and biking might be shown to be therapeutic, but I suspect that this would be more likely for the typical harassed investigator than for the average four-year-old.)

Rejection of the two extremes, permitting the same sort of research on children as on adults and permitting no research that does not hold out the prospect of direct benefit to the child, leaves a wide middle ground. Thus, it is necessary to offer some uniform standard about risk assessment in children's research in order to have a clear policy. After discussing why a uniform standard is important, I argue that the current, widely adopted standard is poor.

## II. THE IMPORTANCE OF A UNIFORM STANDARD OF RISK

It is important to have a uniform standard to assess the *kind* of risk, whether or not the situation is unusual.[2] For example, children with cystic fibrosis might be questioned about their conception of their illness, treatment, or hospitalization. Such questioning might have an exceptionably low kind of risk even though their situations are unusual. Consider, however, how Saul Krugman, Robert Ward, and others defended their Willowbrook hepatitis studies (where some retarded children being admitted to Willowbrook State School were fed the hepatitis virus). They argued that, while the risk of giving hepatitis to most children would be unacceptable, it was minimal *for these children* because hepatitis was endemic to Willowbrook [9, 18]. This reasoning is problematic. It is unacceptable to place dependent persons in situations where the risk to them is increased and then use that increased risk as reason for redefining what constitutes a minimal risk to them. To use another example, the risks to children living in Beirut and Edinburgh are different; but we would not want to have this automatically influence what sort of research we think would be "not too risky" for them.

If we agree that a uniform standard is needed, then, what should it be and who should apply it? The federal guidelines are designed to be a uniform standard applied by independent, knowledgeable groups charged with estimating the risk and protecting the rights and welfare of the subjects. They and their institutions are accountable for doing this properly and in accordance with the guidelines. These Institutional Review Boards (IRBs) distinguish between different kinds of research in relation to their likely risks and benefits, and protect the rights and welfare of subjects.

An independent classification of studies by likely risks and benefits and an impartial application are obviously important. For example, investigators may be biased; or one might permit therapeutic research but not other sorts of research, or permit it where the alternative is death but not approve it where safer therapies are available. Thus, IRB assessment of risk is an important step in determining if a study is permissible.[3]

Using risk, the federal guidelines distinguish five categories of research to clarify when children may be enrolled as research subjects [16].

(A) Research that is exempt from Department of Health and Human Services (DHHS) requirements, including research related to normal educational practices, observation of public behavior in which the researchers are not involved, educational testing, and collection or study of existing data.

(B) Research not involving greater than minimal risk [46.404].

(C) Research involving greater than a minimal risk but presenting the prospect of direct benefit to the individual subjects [46.405].

(D) Research involving minor increase over minimal risk and holding out no prospect of direct benefit to the individual subjects but likely to yield generalizable information concerning the individual's condition [46.406].

(E) Research not included in the above categories which presents an opportunity to understand, prevent, or alleviate a serious problem affecting the health or welfare of children (this requires the special DHHS approval) [46.407].

Not surprisingly, the last two categories are controversial [13]. DHHS sincerely believed some risky research might be rational and morally justifiable and might not fit the first three less controversial categories. The AIDS epidemic, for example, is causing so many infants to die that DHHS might approve a higher risk research protocol than usual if it might help these children. I do not propose to explore these controversies but to focus on the standard for "minimal risk."

### III. "MINIMAL RISK": THE PIVOTAL CATEGORY

It should be immediately apparent that "minimal risk" is crucial in assigning research to these categories. This is summarized in Table I. IRBs have the responsibility to determine a study's level of risk by using as a guide the definition of "minimal risk." This definition, adopted by the Department of Health and Human Services (DHHS), the Food and Drug Act (FDA), the Public Health Service (PHS), and the National Institutes of Health (NIH), is:

"Minimal risk" means that the risks anticipated in the proposed research are not greater, considering probability and magnitude, than those ordinarily encountered in daily life or during the performance of routine physical or psychological examinations or tests (46.102G, [15], p. 8387).

If a study is found to have no more than a minimal risk, an IRB may ease or waive many demanding requirements. First, it may allow third-party consent for children or those who are mentally impaired. Second, the IRB can alter the lengthy review process by allowing a speedy or expedited review. Third, it can permit some or all of the consent requirements to be waived or modified including information normally required for consent (as in deception studies). This is summarized in Table II.

One would expect that without further clarification there would be

TABLE I
**The final DHHS rules for *Research involving children* [16].**
IRB's RISK/BENEFIT ASSESSMENT of Study:

|  | No more than a minimal risk. | More than minimal risk; but the study holds out the prospect of direct benefit (as in therapy). | A minor increase over minimal risk; but the study holds out no direct benefit to the individual subjects, and is likely to yield important information concerning the individual's condition | Greater than minor increase over minimal risk; but the study is of vital importance and has no direct benefit to subjects, and would be of major benefit to the tested population. |
|---|---|---|---|---|
| Consent is obtained from the legal guardian and the subject assents or does not object | OK | OK | OK | IRB cannot approve: Requires National Ethics Advisory Board Approval |
| Consent is obtained from the legal guardian and the subject refuses | No (if the subject is over seven and if the study does not hold out the prospects of benefit to the subject). | OK | No | Approval, and an opportunity for public participation. |

Consent Status

## TABLE II

Risk assessment where subjects are competent to give consent: Using the federal guidelines the IRBs may decide; but they are legally liable for their judgments, and their institutions may lose federal funding if they fail to discharge their responsibilities appropriately.

If the study has:

| | |
|---|---|
| No more than a minimal risk then: | (1) Some or all elements of informed consent may be waived or modified.<br><br>and/or<br><br>(2) Review may be expedited. |
| More than a minimal risk then it requires: | (1) More safeguards and fuller consent.<br>(2) Full review.<br>(3) More proof of the study's importance.<br>(4) Additional safeguards at the IRB's discretion.<br>(5) Statement of institution's policy on compensation to injured subjects. |
| A higher risk than an IRB can approve then it requires: | (1) National review for approval.<br>(2) Very informed, meaningful and unpressured consent. |

considerable disagreement about some of these categories. For example, what constitutes a "minor increase over minimal risk"? Dr. Levine suggests "inviting" a child who has had many spinal or bone marrow procedures to have extra ones might fit Category-D [11]. Do we agree that this is a "minor increase over minimal risk" and that a good sport should willingly endure it for the sake of science? Both spinal taps and bone marrow procedures are painful. Arguably, it is unfair to subject these children to additional procedures in order to learn about their condition. For while they may already be familiar with these medical procedures, it might seem wrong to make them undergo further suffering when they have already suffered more than most. On the other hand, some might argue that if physicians are going to ask the child to undergo an additional procedure for research purposes, perhaps they will be more careful about giving pain medication and other concerns of comfort for the earlier therapeutic interventions. Moreover, it would be better to have investigators ask permission than to do an extra investigatory procedure "calling" it (or even rationalizing that it is) therapeutic.

## IV. CRITIQUING THE DEFINITION OF "MINIMAL RISK"

"Minimal risk," then, is a most pivotal and important category in the guidelines. By means of it, IRBs are supposed to define "an increase over a minimal risk" and "no more than a minimal risk." Ranking studies in this way is supposed to serve the goals of encouraging important low-risk research while protecting the rights and welfare of subjects. It is designed to balance the social utility of increasing our knowledge by fostering research with protecting subjects by easing consent and other requirements only where the risk is minimal. In what follows, I wish to show that the key concept of "minimal risk" is poorly defined in the guidelines.

The definition of minimal risk is stated disjunctively; this means that only one part needs to be fulfilled for minimal risk to be established. However, both parts are problematic.

Let us consider the *first* part: "Minimal risk means that the risks anticipated in the proposed research are not greater, considering probability and magnitude of harm, than those ordinarily encountered in daily life ..." One insight this captures seems to be: There might be a risk of catching a cold from the researcher; but that could happen anywhere, so such risks should not count against the study or be a part of the estimation of possible risk. What is important is that there is no addition to life's risk as a result of participating in the study.

This first part, however, is unsatisfactory. It may be interpreted in one of three ways, all of which have difficulties. It may mean minimal risk is defined as the probability and magnitude of harm determined by estimating: (A) all the risks ordinary people encounter; or (B) the risks all people ordinarily encounter; or (C) the minimal risks all ordinary people ordinarily encounter.

### Alternative (A)

Included among the activities of ordinary people are riding in cars, playing football, hang gliding, parachuting, and flying. But the risks inherent in such activities cannot be assumed to be minimal, for they are not; nor can they be thought to be an appropriate guide for determining the acceptable or minimal nature of the risk of certain studies. Therefore, the first interpretation will not do because it would allow studies with significant risk (e.g., like the risk of parachuting) to be said (incorrectly) to have a minimal risk.

### Alternative (B)

The second interpretation is that this first disjunct means the risks all people ordinarily encounter. Many people do not go hang gliding, parachute or play football. These are not routine activities for all of us, so according to this interpretation we should not focus on them. This interpretation assumes that we know the kinds of risks we all ordinarily encounter and their probability and magnitude. Neither is obvious. Most of us drive cars, walk across busy streets, fly in airplanes, and live in a world filled with nuclear weapons. Are these the everyday risks the definition refers to? How do we determine what risks are encountered routinely by all of us and estimate the probability and magnitude of these risks? Even if we could, does this have to be done before we can determine that a study has no more than a minimal risk? It seems easier to determine that a study of how many blocks four-year-olds can stack is a minimal risk study than it is to determine the nature and magnitude of risks all persons normally encounter.

### Alternative (C)

The third interpretation, that the meaning of minimal risk should be understood as the minimal risks ordinary people ordinarily encounter, is obviously not illuminating. It is included because the definition is sometimes defended along these lines. To argue that the guidelines are by definition only referring to minimal everyday risks converts the definition's first part into a useless tautology. Therefore, whichever interpretation of the first part is adopted, it is inadequate to clarify minimal risk for the purposes of the guidelines.

The *second* part of the definition, "Minimal risk means that the risks anticipated in the proposed research are not greater, considering the probability and magnitude than those ordinarily encountered ... during the performance of routine physical or psychological examinations or tests," is also problematic. Routine examinations are a source of anxiety for many; it might be unacceptable to regard a study as having no more than a minimal risk that would similarly provoke such anxiety.

Moreover, routine physical and psychological examinations often disclose sensitive psychosocial information. This definition does have the qualifying terms "ordinarily encountered" and "routine," thereby ruling out physically risky procedures such as x-rays, bronchoscopy or cardiac catheterizations. But beyond non-routine physical procedures, it leaves us (and IRB members) to ponder what is "ordinarily encountered" or "routine" in the office visits of persons to psychologists, psychiatrists, pediatricians, internists, or family practitioners, because almost every kind of human need and problem is frequently encountered. This is why confidentiality has been consistently regarded as an important value for the health professions. Thus, while certain physically risky procedures are ruled out, it does not rule out what is frequently as great a worry in determining risk; namely, the invasion of privacy, breach of confidentiality, labeling, and stigmatization. This is a major deficiency, since it is in these areas that policy needed to be clarified and since elsewhere in the guidelines they are signaled as important considerations about risk.

I believe most IRBs do consider psychosocial risks of labeling, confidentiality, stigmatization, overtesting, and privacy in their estimation of risk. Thus, they apparently do better than to be guided by this standard of "minimal risk." Working together as informed people of good will, IRB members, I believe, generally recognize that they have a moral responsibility to protect subjects and that the decision about risk is a moral one.

But there is reason to believe the lack of a good standard is creating difficulties. A survey by Janofsky and Starfield [6] of pediatrics department chairs and pediatric research center directors found considerable differences on estimates about whether the procedures such as venipuncture, arterial puncture, and gastric and intestinal intubation were regarded as risky keep[4] in. Most, for example, regarded arterial puncture to have "greater than a minimal risk"; but between 8% and 24%, depending on the child's age group, thought it had less than a minimal risk. An editorial in the same issue of the *Journal of Pediatrics* found such variation "cause for concern" and concluded that IRBs need better standards about risk assessment for research [10].

To summarize, the position of DHHS shared by Levine, myself, and many others is that with adequate review and appropriate consent, assent, and safeguards, children may be research subjects if the study holds out a suitable prospect of direct benefit to them or if the study is not too risky. But this consensus is illusory if we mean very different things by "not too risky," as the Janofsky-Starfield study suggests.

I have argued the definition of "minimal risk," a pivotal concept in the federal guidelines, is poorly defined. But if the guidelines have such a disastrous central concept, why do they work as well as they do? Perhaps it is because the IRBs do better than the definition of "minimal risk." In addition, as Beecher [1] pointed out and data supports [17], most research is not very risky, and most researchers are decent people who put the welfare of subjects first. A few do not, and that is why peer review, consent, and guidelines are important. Still, we could use better standards. As the pivotal federal standard to make risk judgments consistent, "minimal risk" is vague and problematic, perhaps intentionally. Other standards defined in terms of minimal risk, such as "a minor increase over minimal risk" and "greater than a minimal risk," are no less flawed.

*East Carolina University School of Medicine*
*Greenville, North Carolina*

## NOTES

[1] Portions of this paper are excerpted or adapted from "Estimating Risk in Human Research" [8].

[2] It was apparent from the beginning this would be a pivotal notion. The National Commission argued long and hard about what "minimal risk" should mean [13].

[3] The federal guidelines propose safeguards for children [16] in addition to the general recommendations [15] about attending to the worthiness of the study, to its scientific soundness and significance, to preceding the study by appropriate animal studies, and to minimizing the risks and providing for confidentiality and fairness in selecting subjects. In addition to gaining adequate consent from parents or guardians, the guidelines seek to gain the assent of the child where appropriate. Therapeutic studies involving more than minimal risk are justified by their probable direct benefit to the child. But in addition, Institutional Review Boards (IRBs) may exempt studies from consent and approve studies that present a minor increase over minimal risk to subjects.

[4] This was using a definition almost identical to the current definition of "minimal risk." The differences between them are trivial.

## BIBLIOGRAPHY

1. Beecher, H. K.: 1966, 'Ethics and Clinical Research,' *New England Journal of Medicine* **274**, 1354–1360.
2. Brock, D. W.: 1989, 'Children's Competence For Health Care Decision Making', in this volume, pp. 181–212.
3. Holder, A. R.: 1989, 'Children and Adolescents: Their Right to Decide About Their Own Health Care', in this volume, pp. 161–172.
4. Holmes, R. L.: 1989, 'Children and Health Care Decision Making: A Reply to Angela Holder', in this volume, pp. 173–179.
5. Holmes, R. L.: 1989, 'Consent and Decisional Authority in Children's Health Care Decision Making: A Reply to Dan Brock', in this volume, pp. 213–219.
6. Janofsky, B. A., Starfield, B.: 1981, 'Assessment of Risk in Research on Children', *Journal of Pediatrics* **98**, 842–841.
7. Kopelman, L.: 1978, 'Ethical Controversies in Medical Research: The Case of XYY Screening', *Perspectives in Biology and Medicine* **21**, 196–204.
8. Kopelman, L.: 1981, 'Estimating Risk in Human Research', *Clinical Research* **29**, 1–8.
9. Krugman, S. and Giles, J. P.: 1973, 'Viral Hepatitis Type B (MS-2 Strain): Further Observations on Natural History and Prevention', *New England Journal of Medicine* **288**, 755–260.
10. Lascari, A. D.: 1981, (Editorial) 'Risks of Research in Children', *The Journal of Pediatrics* **98**, 759–760.
11. Levine, R. J.: 1989, 'Children As Research Subjects', in this volume, pp. 73–87.
12. Levine, R. J.: 1986, *Ethics and Regulation of Clinical Research*, Second Edition, Urban and Schwarzenberg, Baltimore.
13. McCartney, J. J.: 1978, 'Research on Children: National Commission Says Yes, If ...', *Hastings Center Report*, **8**, 26–31.
14. Shirkey, H. C.: 1968, 'Therapeutic Orphans', *Journal of Pediatrics* **72**, 119–120.
15. U.S. Department of Health and Human Services: 1981, 'Final Regulations Amending Basic HHS Policy for the Protection of Human Research Subjects', *Federal Register* **46** (January 26), 8366–8392.
16. U.S. Department of Health and Human Services: March 8, 1983, 'Additional Protections for Children Involved as Subjects in Research', *Federal Register* **48** (46), 9814–9820.
17. U.S. Department of Health and Human Services: 1982, 'Protection of Human Subjects; Compensation for Research Injuries; Requests for Comments on Report of the President's Commission', *Federal Register* **47** (November 23), 52880–52930.
18. Ward, R., Krugman, S., Giles, J. P., Jacobs, A. M., and Bodansky, O.: 1958, 'Infectious Hepatitis: Studies of Its Natural History and Prevention', *New England Journal of Medicine* **258**, 407–416.

# SECTION II

# CHILDREN, ILLNESS, AND DEATH

# INTRODUCTION

Children, no less than adults, need emotional support and comfort in the face of serious illness or impending death. Professionals and families may, however, find it difficult to accept the fact of a child's critical or terminal illness or be reluctant to discuss it with the child. Some claim that withholding bad news will protect children from psychological harm; others argue that harm is more likely to result from keeping such information from them when they want it. Failing to communicate with children about what they need or want is likely to increase their feelings of isolation, fear, confusion, and mistrust. Even when information is withheld from them, older children and adolescents are likely to learn about their condition by overhearing others' conversations, observing changes in their care, or recognizing changes in their bodies. Although they recognize the difficulties, our authors conclude it is important to develop ways to communicate effectively with children and adolescents about their illnesses, to help them articulate their own conceptions of what illness means and to help them cope with events of serious illness in a way that affirms their personal worth. One technique for accomplishing these goals has focused on the use of children's literature. Literary depictions of death as a natural and peaceful event have been used in hospital wards to help children come to terms with terminal illness.

In 'Death and Children's Literature: *Charlotte's Web* and the Dying Child', Rosalind Ladd asks why dying children are attracted to this classic by E. B. White. She answers that they may find not only an absorbing and well-written story, but also "clear values about life and death and an attitude toward dying that is of classical and universal appeal" (p. 108). Arguing that similar philosophical assumptions and views may be found in Plato's discussion of the death of Socrates, she compares Wilbur the pig, who is not prepared for his death, with both Charlotte the spider and Socrates, who are. Both authors, she holds, believe that "what is natural is good and that a dying which is natural and in accordance with nature is a good dying and to be accepted but that a death which is unnatural is a bad death and to be protested" (p. 109). Killing Wilbur the pig for food is an unnatural death and, therefore, to be resisted. In contrast, Charlotte the spider's death, after her rich, full life and after she has created her egg sac, is a natural death. R. Ladd compares the characteristics of Socrates' clearly exemplary dying to the "model of a naturalistic death in *Charlotte's Web*." Socrates prepares himself intellectually and emotionally for his death, and he has the opportunity to retain some control over his life until the end. Ladd argues, given that a person has come to the point of death, a good dying may be characterized by openness, the loving support of family and friends, the hope of a kind of immortality, and, in short, a peace and harmony with nature. In addition, the natural deaths of both Socrates and Charlotte have purpose and are without pain; both are portrayed as a gradual weakening and a peaceful fading away.

Loretta M. Kopelman comments that while the analysis of the good death given by Rosalind Ladd is important, it is problematic because she offers no analysis of natural. None seems obvious because, Kopelman writes, whether understood as what is found in the totality of nature, what is part of the nature of a thing, or what is expected in nature, good deaths do not necessarily seem natural. Kopelman compliments Ladd on reminding us of the importance of a death where individuals are treated respectfully, with control over their lives, and the opportunity to continue to develop their potential to the end. The peaceful, pain-free death where the dying person is in control and surrounded by loved ones is one, however, that is becoming increasingly rare in our society. R. Ladd, says Kopelman, reminds us of its importance for adults as well as for children. Her comparison between E. B. White and Plato, and between Charlotte the spider and Socrates, is illuminating and instructive. Kopelman, however, suggests a difference between E. B. White and Plato concerning how to understand the problem of why innocent children suffer pain and early death. She argues that however similar they may be in their account of the good death of Socrates and of Charlotte the spider, they offer different solutions to the problem of evil. Plato suggests that pain and suffering simply are some of the conditions of finitude. E. B. White celebrates the wonders of nature and suggests the view that pain and suffering always have a purpose. Kopelman suggests this may be why his work is appealing and comforting, especially to the dying.

Literature may be used to escape from or to face reality. Like adults, children may seek to face or to deny their problems. Thus, a recurring problem is distinguishing between what we think children understand or can cope with and children's actual abilities to do so. These decisions are problematic even when made for competent adults. For children, we often turn to developmental psychology to help us with the first problem: What can they understand? In 'Children's Conceptions of Illness and Death', Gareth Matthews presents the standard account in developmental psychology of children's understanding of death and then examines the implications of this account for issues of disclosure and decisionmaking in the health care of seriously ill children. According to this account, children pass through three stages in their understanding of death: young children are said to view death as a sleep-like state from which one can awaken, children in their early elementary years view death as a sleep-like state which is irreversible, and children around age nine or ten come to understand death as a total cessation of physical and mental function. Upon first analysis, Matthews suggests, this account would seem to support a paternalistic stance with regard to disclosing information about death to seriously ill children or involving them in decisions about health care, since it would seem that children under age ten are incapable of understanding the significance of this information and hence incapable of participating in decisions about their own treatment. Upon closer examination, however, Matthews argues that although one may not succeed in communicating the full import of a terminal prognosis to a seriously ill child, the message one can communicate based on that child's understanding of death may be less threatening and less emotionally disturbing to the child than the "mature" understanding of death. Hence, it may not be appropriate to withhold this disclosure on the grounds that the child cannot cope with this information emotionally.

In the final section of his paper, Matthews questions whether the standard developmental account correctly captures the understanding of death in children with life-threatening or terminal illness. Matthews cites Myra Bluebond-Langner's study of leukemic children; she describes five stages in which terminally ill children came to understand their situation, including a final stage in which these children understood that they were dying. Matthews emphasizes Bluebond-Langner's finding that chronological age was irrelevant to the progression of the children she studied through the five stages of awareness; rather, the children's understanding was dependent on the extent of their experience with their disease and its treatment, and with the progress of other children with the same disease. Matthews concludes, in view of this finding, that the standard developmental account is largely irrelevant to the issues of disclosure and decisionmaking for terminally ill children. This does not settle the issue of what terminally ill children should be told or how they should be involved in decisions about their care, but it does remove one objection to giving terminally ill children a more active role in their care.

In his commentary on Matthews, John Moskop further examines the implications of the Bluebond-Langner study for determining the competence of children and for informing children about their illness and care. He suggests that Matthews' concern in exposing the limitations of the Standard Developmental Account might be to prevent its misapplication to the situation of critically ill children. For this group of children, Bluebond-Langner's findings suggest that the ability to understand their situation depends less on age than on experience with their disease and its treatment. Moskop notes that, in addition to understanding, competence for decisionmaking depends on the ability to reason and on the possession of a set of values or goals. He discusses the extent to which seriously ill children possess these other two requirements for competence. Finally, Moskop takes issue with Bluebond-Langner's suggestion that terminally ill children benefit from a practice of mutual pretense which avoids all discussion of their illness. Against this view, he cites studies which have reported good results from a policy of communicating openly with terminally ill children.

ROSALIND EKMAN LADD

# DEATH AND CHILDREN'S LITERATURE: *CHARLOTTE'S WEB* AND THE DYING CHILD

An observation by an anthropologist studying the interactions of children in a hospital leukemia ward suggested the topic of this paper: the treatment of death in children's literature, with specific reference to children with life-threatening illness. In *The Private Worlds of Dying Children*, Myra Bluebond-Langner notes:

> The most popular book among these children [in a hospital ward with terminal leukemia] was *Charlotte's Web*. When Mary and Jeffrey reached stage 5 [the final stage of awareness of illness], it was the only book they would read. Several children at stage 5 asked for chapters of it to be read to them when they were dying. But as one parent stated, "They never chose the happy chapters." They always chose the chapter in which Charlotte dies. After any child died, the book had a resurgence of popularity among the others ([1], p. 186).

Recent psychological studies have revised our ideas of how children come to understand the concept of death, and their responses to it. We now know that young children have ideas and worries about death, even before they are old enough to verbalize them, and that fatally ill children have awareness even if they have not been told their diagnosis and never talk about it.[1] Indeed, Bluebond-Langner cites the choice of *Charlotte's Web* as one indication of their non-verbal awareness.

The enormous power of literature to affect people – to shape their attitudes and values – has been universally recognized, even to the extent that Plato would have "unworthy" kinds of poetry and stories banned from his ideal Republic. Children's literature, in particular, serves the function of transmitting a society's values from one generation to another. The recent rash of children's books dealing with "real life" issues such as divorce, sexuality, drug abuse, and death, both fiction and non-fiction, seems to recommend particular attitudes and makes one wonder if contemporary children's authors accept the view that the purpose of literature is primarily didactic.

However, a number of older, much-beloved children's classics, which are not "about death," nevertheless talk about death. *Little Women*, for example, includes an episode of an infant's death from scarlet fever and the long invalidism and eventual death of one of the central characters. These events are treated with an appropriate sensitivity and described in terms children can understand, but with an openness and matter-of-fact acceptance characteristic of the world before vaccinations and antibiotics, where children had to be prepared to face the realities of infant mortality and the consequences of untreatable childhood diseases.

What a child learns about death and dying from books is more important than it may seem. Many families are still reluctant to discuss these things with children, so what they learn from books may be their only form of death education. Moreover, adults are said to revert to their childhood conceptions when crisis strikes.

Thus, it is important to examine the attitudes and values about death and dying embodied in the books children read. This is the task this paper will undertake. Focusing on *Charlotte's Web*, I want to ask these questions: What are the values being expressed in this book? Why do children with life-threatening illness find it so appealing? And finally, are the values expressed "good" values, i.e., ones we would want our children to adopt?

My thesis is that dying children are attracted to and comforted by *Charlotte's Web* for good reason, that it embodies, within a context of an absorbing story and the skillfully wrought language of literature, clear values about life and death, and an attitude toward dying which is of classic and universal appeal.[2] More particularly, we may find in *Charlotte's Web* many of the same philosophical assumptions and conclusions that we find in the classic source of these ideas, Plato's *Apology, Crito, and Phaedo*.

*Charlotte's Web* is not, in any obvious way, "about" death. Yet, the theme of death and dying is introduced on the very first page of the book and, more importantly, introduced in a context of raising questions about values. "Where's Papa going with that ax?" ([10], p. 1), Fern asks, and then launches into a full-scale eight-year-old's protest against killing the runt of the pig litter on the farm. "Control myself?" yelled Fern, "This is a matter of life and death ..." ([10], p. 2).

Interestingly enough, the killing of the runt of the litter because "The pig would probably die anyway", and because "A weakling makes trouble" ([10], pp. 1, 3), parallels the ethical problem we recognize in contemporary pediatric medicine as "letting defective newborns die," and Fern offers a strong philosophical argument against it: "But it's unfair ... The pig couldn't help being born small, could it? ... This is the most terrible case of injustice I ever heard of" ([10], p. 3).

The saving of the runt pig, later to be named Wilbur, sets the story in motion. Perhaps to most children, it is just an exciting beginning to a story;

but to a child attuned consciously or unconsciously to matters of life and death, i.e., a child with life-threatening illness, it introduces a particularly significant theme.

As the story of Wilbur the pig and his friend Charlotte the spider unfolds, the main value statement that the book as a whole makes is this: What is natural is good and a dying which is natural and in accordance with nature is a good dying and to be accepted, but a death which is unnatural is a bad death and to be protested.

The theme that what is natural is good is established explicitly and implicitly in the setting of the book. The farm, where most of the action of the story takes place, is described as a place with order and regularity, where the needs of the various animals are met and each thing is cared for: the cat is given a fish-head to eat, the hay is regularly pitched down to the cows and horses and sheep, and the saddles and harnesses are polished ([10], p. 13–14). "Every day was a happy day, and every night was peaceful." In his naturalistic setting, even the barn smells, which in a less natural setting we might find offensive, are described as wonderful and sweet ([10], p. 11).

In contrast to what is natural and good, there are two examples of bad deaths in the book. The first, already described, is the proposed killing of the runt pig. This death is successfully averted by Fern's indignant protest and her promise to care for the little pig.

The second example of an unnatural and bad death makes up the focus of the main plot: the farmer's intention to kill Wilbur when he is a full-grown, fattened pig. The unnaturalness of this death is underlined by the words by which it is described: they are going to "Turn you into smoked bacon and ham" the old sheep tells Wilbur. "Almost all young pigs get *murdered* by the farmer ... There's a regular *conspiracy* around here ... Everybody is in the *plot*" ([10], p. 49, italics added). The word 'butchering,' which is certainly appropriate here, calls forth sinister and violent connotations, following the words 'murder' and 'conspiracy'.

Wilbur's response to the news about his fate is instructive. He immediately protests and articulates what must be in any person's mind and heart when he becomes aware of a fatal diagnosis: "I don't want to die! Save me, somebody! Save me!" ([10], p. 50). Charlotte succeeds in quieting his initial outburst, but days later, as he is trying to go to sleep, he reacts again, this time in a quiet and thoughtful way: "The thought of death came to him and he began to tremble with fear. 'Charlotte,' he said, softly ... 'I don't want to die'" ([10], p. 62). At the same time he thinks about death, he thinks about how much he loves life.

Charlotte, the understanding friend and counselor, responds to Wilbur's outcries both times. Although she tries to comfort him, she does not try to stop him from talking about death. She does not deny the facts, try to hide the truth, cover it over with optimism, or ignore it in hopes it will go away. Nor does she criticize his emotional response. The message to the sensitive child-reader is: It's OK to talk about dying, it's OK to be angry, it's OK to have fears. In a society in which, until very recently, it has been taboo to talk about death and dying, especially for a child, this is a powerful message. It also coincides with most recent psychologists' views that truth-telling and open communication best fulfill the dying child's needs.

The plot moves on as Charlotte resolves to help Wilbur escape his threatened death. Implicit in her response – which, as we will see, is quite different from her later response to her own impending death – is this line of reasoning: Wilbur's death at the hand of the farmer is not natural; therefore, it is not necessary. Therefore, it can be averted. And again: Wilbur's death is not natural; only what is natural is good; therefore, his death is not good; therefore, it is justified to try to avert it.

By contrast, Charlotte's own death – for spiders biologically can live through just one season – is described as a part of the natural process of living, growing, and dying, and is never protested, feared, or averted, though the sadness of it is not denied or played down. Because it is a part of, or the end of, the natural life cycle, it is not a surprise or a shock, but is anticipated, planned for, and accepted.

Unless one makes a simple identification of all life as good and all death as bad, which this book does not lead us to do, it is meaningful to distinguish a good death from a bad death and a good dying from a bad dying. Insofar as Charlotte's death is natural and inevitable, it presents us with the model of a good death.

The reader is prepared for the death of Charlotte at the end of the summer in a number of ways. There is foreshadowing early in the book, when the song sparrow is described as one who "knows how brief and lovely life is" ([10], p. 43). Later, there is a clear allusion to the fact that the spider knows what the end of summer will bring. The narrator says, "Summer was half gone. She knew she didn't have much time" ([10], p. 67). And the coming of the end of summer is introduced by the crickets' "sad, monotonous" song. "'Summer is over and gone,' they sang ... 'Summer is dying, dying'" ([10], p. 113).

Charlotte does not announce her imminent death to Wilbur, but leads him gradually to the realization of what is happening. She talks of how tired she is, how little energy she has; he sees her as swollen and listless; she naps frequently – all symptoms of physical decline that an ill child will recognize.

Yet, the overall tone remains one of peacefulness and even a kind of contentedness. "Her heart was not beating as strongly as usual and she felt weary and old, but she was sure at last that she had saved Wilbur's life, and she felt peaceful and contented" ([10], p. 153). And again later, she says she feels "A little tired, perhaps. But I feel peaceful" ([10], p. 163).

The theme of the peacefulness with which one can approach one's dying is important for the young reader, and seems to arise from the naturalness of the whole birth, life, death cycle. Charlotte sums it up in a rather Stoical way: "After all, what's a life, anyway? We're born, we live a little while, we die" ([10], p. 164). Her use of "we" reminds us that life and the fact of death are universal and apply to us all.

Charlotte's actual dying is also described in a very quiet and peaceful way. Wilbur has a chance to say good-by – with a wink, "in the only way he could" ([10], p. 171). She uses all her strength to wave one of her front legs at him. And then a rather stark paragraph: "She never moved again." The Fair Grounds are cleared out and deserted, and then the chapter ends with: "No one was with her when she died" ([10], p. 171).

Charlotte's contentedness in the face of death is attributed to the fact that she has accomplished her mission in life: she has saved Wilbur's life and she has laid her eggs and spun the silken egg sac that will protect her children. There is no unfinished business in Charlotte's life, a comforting psychological consideration.

Charlotte has also managed to exert a measure of control over her own dying. She has chosen to go to the Fair with Wilbur, even though she knows that means she will not ever be able to return to her comfortable home in the barn, and she has chosen to continue working for Wilbur's safety, even though she is tired and weak. Although she cannot control whether or not she will die, she can control where and how she spends her last days, and that is important to her sense of well-being, as it must be to all people in that ultimate situation.

Moreover, the possibility of immortality is suggested clearly not only by the 514 little spiders that will hatch in the spring, but in the thoughts and memories that Wilbur always will have of her. On the very last page of the book, the reader is assured that "Wilbur never forgot Charlotte" ([10], p. 184).

Thus, surrounded by friendship and love up to the end, accepting her death as part of the natural life cycle, choosing for herself the circumstances of her death, having accomplished her life's work, and assured of some kind of immortality, Charlotte is able to die a good death.

Other themes important to the psychological and philosophical aspects of dying a good death emerge in the novel. The theme of the importance of friendship and love, for example, is developed in a way that reflects the mind-body distinction, the physical vs. the spiritual. Wilbur, when he first arrives at the farm, spends two days miserable and crying, not even eating, which is pretty unusual behavior for a pig. Finally, he realizes that he doesn't want food, he wants love, he wants a friend ([10], p. 27). Wilbur, not the proverbial pig of Mill's philosophy, is willing to give up the satisfaction of immediate, physical pleasures for the higher pleasure of friendship. The theme of sacrifice for the sake of friendship and loyalty runs through all of Charlotte's efforts to save Wilbur. It is echoed again at the end of the book in Wilbur's willingness to bribe Templeton, promising the rat first choice of everything in his feeding trough every day in the future if Templeton will help him save Charlotte's egg sac ([10], p. 168).

Templeton is the "bad guy" of the novel. He is described as having "no morals, no conscience, no scruples, no consideration, no milk of rodent kindness, no compunctions, no higher feeling, no friendliness, no anything" ([10], p, 46). He is the pure hedonist, and although he plays an important role in saving both Wilbur and Charlotte's babies, he has to be bribed to do it. A foil to the others, he does not see the higher value of friendship vs. the lower pleasures. His philosophy of life is summed up when he invokes the philosophical distinction between quality and quantity of life. Chided for eating too much and endangering his health, he replies, "Who wants to live forever? ... I get untold satisfaction from the pleasures of the feast" ([10], p. 175). But for the others, friendship is the highest value: Wilbur "realized that friendship is one of the most satisfying things in the world" ([10], p. 115), and Charlotte says, "You have been my friend. That in itself is a tremendous thing .... By helping you, perhaps I was trying to lift up my life a trifle" ([10], p. 164).

Related to the theme of friendship is a less obvious theme, the caring relationship between mother-figure and child. Wilbur is always in the role of the child, the uncertain, somewhat blundering, but good-willed child in need of both physical and emotional support. First Fern, the little girl, nurtures him like a baby. She feeds him from a bottle, and pushes him in her doll carriage. He is described as her infant or her baby. Later, it is Charlotte the spider who assumes the mother role for Wilbur, as well as being his friend. She calms his fears, advises him to get plenty of sleep, and to chew his food thoroughly

([10], p. 64). "Never hurry and never worry," she says. She tells him stories and even sings him a lullaby ([10], p. 64).

Thus, the sick child who may identify with Wilbur, the "child" with life-threatening problems, may find comfort in the thought that friends will stick with you through your troubles and that there are adults, especially mothers, who will guide you and take care of you.

Another aspect of the physical-spiritual distinction appears in the suggestion of a mystical, mysterious dimension of existence in addition to the often-emphasized beauty of the physical world. When Charlotte manages to weave the words "Some Pig" into her web, and the handyman sees it shining with dew in the early morning, he responds by falling to his knees and uttering a short prayer; when he brings the farmer to see it, they both begin to tremble, and the farmer describes it to his wife as a sign, a miracle, and then drives over to tell the minister about it ([10], pp. 78–80). Yet, the miracle aspect of Charlotte's web remains only a suggestion, and the narrative continues in a basically naturalistic mode.

An interesting perspective on the nature of miracles is offered by the old country doctor Fern's mother consults about the children. Talking about the words spun in the spider's web, he says, "When the words appeared, everyone said they were a miracle. But nobody pointed out that the web itself is a miracle" ([10], p. 109). Both the physical and the spiritual are found in the world of nature, it is suggested.

There are other, lesser themes that help explain the interest of dying children in *Charlotte's Web*. There are continuing reminders that, as nature changes with the seasons, children are growing up, caught between the safety of childhood and the striving for independence. Early in the book, the little pig pushes through a loose board in the fence and gets out into the outside yard. "How does it feel to be free?" the goose asks. "I like it," says Wilbur. "That is, I *guess* I like it!" Actually, the narrator continues, "Wilbur felt queer to be outside his fence, with nothing between him and the big world" ([10], p. 17). At the Fair, Fern and her brother are allowed to go off by themselves, "into the wonderful midway where there would be no parents to guard them and guide them" ([10], p. 131). Leaving the security of parents and venturing out alone is a challenge every child must face, but one that presents itself with a special poignancy to a dying child.

As if to counter the sense of insecurity children often feel, the author of *Charlotte's Web* several times expresses his faith in the abilities of children, compliments them on their capabilities. Against mothers' worries that their children will fall off the barn swing, the author says, "Children almost always hang onto things tighter than their parents think they will" ([10], p. 69). And the doctor, reassuring Fern's mother, says: "Children pay better attention than grownups. If Fern says that the animals in Zuckerman's barn talk, I'm quite ready to believe her" ([10], p. 110).

There is, of course, an irony in all of this, an absurdity – in the existential sense – in saying that a dying child can find comfort in reading the description of a good death which is good because it is natural, a part of nature's cycle of being born, living for a while, and then dying. The irony is that the death of a child is *not*, in our day, in our society, seen as natural in this way. Infant mortality is not a familiar fact of life. Reliance on vaccines, miracle drugs, sophisticated surgery, and other technology leaves us with an impatience and profound sense of unfairness about the diseases of childhood which are not yet avoidable or curable.

I turn now to the classic philosophical literature. The prime exemplar of a good death and a good dying is that of Socrates, as described in *The Apology*, *Crito*, and *Phaedo* [3]. Although there are other kinds of good deaths in history and literature – e.g., the death of a hero in war, the sacrifice of a mother for her children, the death of a martyr for a cause – Socrates' death is an exemplar for two reasons in particular. First, it is the kind of dying which every person might strive to achieve for himself, while only those thrust by fortune into a heroic situation can die a hero's death; and secondly, it is explained and justified by the philosopher's own metaphysical and ethical beliefs and arguments about the values of human life and death. Socrates' belief is that the whole of a philosopher's life is, or should be, a preparation for death, and we see that he dies according to the same principles by which he lived. Socrates shows us how to live a good life and he shows us how to die a good death.

In what follows, I shall focus on those characteristics of Socrates' dying which contribute to its status as exemplar, compare it to the model of a naturalistic death in *Charlotte's Web*, and make some comments about how all of this may apply to the situation of a dying child.

How, then, does Socrates die?

One of the most significant things about Socrates' dying is that he was well prepared for it intellectually and emotionally, and he approached it with full knowledge. The philosopher's life-long preparation for death is based on three philosophical beliefs: that mind and body are distinct, that the life of the mind is infinitely superior to the life of the body, and that the life of the mind continues after the life of the body ends. The philosopher thus de-values the pleasures of the physical life which are lost at death and approaches death, which he sees as the liberating of the soul from the "tomb" of the body, without fear. Whether or not these beliefs hold up under criticism is not the issue here; the point is that there is both a logical and a psychological connection between holding these particular beliefs and the attitude and approach one adopts toward death and dying.

Beyond his general philosophical beliefs, Socrates is also well-prepared for death in practical terms. Unlike many modern medical situations where truth-telling is an issue, Socrates knows exactly when, why, and how he is to die. He is able to approach his final days with full knowledge. *The Phaedo* describes his asking and getting information; he is told the exact course of effects that the hemlock will have on his body. "Just drink it and walk around until your legs feel heavy, and then lie down and it will act of itself."

One result of Socrates' awareness is that he is able to talk about his death openly with his friends, and he is able to plan for it. He says his good-byes and he settles his debts. "Crito, we owe a cock to Asclepius," he says, and we are told these were his last words. There is no "unfinished business" hanging over Socrates.

Related to the fact that Socrates knows well what is in store for him is the fact that, right up until the end, he is able to make choices, exercise his autonomy, and keep some control. His first great choice, of course, is to stay and take his punishment instead of trying to escape from prison, as his friends urge. He also could be said to "choose" the attitude with which he faces his death. As the Stoics later made clear, although we cannot control the events of our life and death, we can control our attitudes and will. Socrates chooses to go willingly to his death, not trying to extend even his last hour.

Many of these same things characterize the death of Charlotte in *Charlotte's Web*. She is aware of what is happening to her, she plans for it, lays her eggs, talks with her friends about it and says her good-byes. Much of the peacefulness that surrounds her death may be attributed to her foreknowledge and planning. Although the life-span of a spider cannot be changed, Charlotte is still able to exercise choice over some aspects of her dying. Her first great choice is to go with Wilbur to the Fair, even though that means she will die there instead of at home. Then she chooses to use her last days spinning words into her web to help Wilbur.

Families and physicians are finding ways in which even young children who are dying can continue to make some choices for themselves. The possibility of dying at home instead of in the hospital is a significant example and in itself opens up greater opportunities for exercising choice: what to wear, when to eat or sleep, what music to listen to. Small choices, perhaps, but choices nonetheless. Adolescents who are encouraged to make wills or express their wishes about funeral arrangements or, more significantly, those who are given a deciding vote about when to stop therapy are exercising autonomy in ways that are important both ethically and psychologically.

One remarkable aspect about the descriptions of the death of both Socrates and Charlotte which must be considered if we are to take either of them as an exemplar of a good death is the complete absence of pain. The effect of the hemlock is that the body gradually becomes cold, numb, and without feeling. The natural death that Charlotte experiences is a gradual weakening and feeling tired, but it is a peaceful feeling, she says, and there is no hint of real discomfort.

Whether or not Socrates' death was actually as peaceful as it is described in *The Phaedo* is challenged in a rather convincing article by William B. Ober, M.D. The actual course of hemlock poisoning, as described in both classical sources and in contemporary manuals on toxicology, involves nausea and vomiting, the gasping respiration of asphyxia, and sometimes convulsions. Ober concludes: "It is almost certain that Plato ... chose to edit the facts and present a literary version which was in harmony with Socrates' philosophical ideas" ([7], p. 269).

If Ober is right that *The Phaedo* presents a white-washed version of the death of Socrates, all the stronger is the argument that the dying of Socrates is – and was meant to be – the exemplar of a good dying. The classic view is not that pain and suffering are ennobling, as is perhaps suggested in the story of the dying of Christ, but that it is in keeping with the ideal of the rational man to be conscious and rational and in control while dying.

Surely it is unrealistic to assume that the illnesses which prove fatal to children will not cause pain, either in themselves or in the treatments administered. Yet, many deaths of young children from leukemia, for example, are described as relatively peaceful in their final days, a gradual weakening and slipping away into coma from sepsis or from pneumonia. With sophisticated pain medications, such as carefully controlled intravenous infusion or self-administered doses of morphine, it is often possible for the child to maintain a comfortable enough level of consciousness in which he can interact with other people and make his wishes known. The importance of appropriate pain medication should be seen not just as a way of avoiding bad pain, but as a way of contributing to the dignity of the dying child.

The last characteristic of the good death which I want to talk about is shared by both Socrates and Charlotte, but presents problems for the case of the dying child. Perhaps the most notable thing about Socrates' death, especially in the idealized version, is that it is a meaningful death: he dies for a cause he believes in, he chooses not to escape because he thinks it is the right thing to do, and he is remembered for the principles by which he lived and died. The death of Charlotte, too, is a meaningful one in the same sense: she has devoted her life to helping her friends and she will be remembered for that, and she lays eggs for the life of the next generation as she is coming to the end of her own life.

By contrast we ask, What is the meaning of the death of a child? Parents who ask why and children who may see death as punishment are trying to find an answer. Some theologians try to explain it in terms of God's will and some philosophers see it as evidence that there is no benevolent God. One of the most recent, popular responses to this question is the reiteration of an old response: that there is no explanation, that it is beyond human understanding [4]. Many parents succeed in giving their child's death meaning by allowing him to participate in research which may benefit other children, by contributing to an organ bank, or by establishing scholarships or charitable programs in his memory.

What the child cannot do, as both Socrates and Charlotte can, is look back over accomplishments of a full lifetime with satisfaction and accept the inevitability of death more graciously because of it. The philosopher Thomas Nagel says that the evil we see in death is the loss of good which continued living would have given us [6]. The death of a child, then, it would follow, must always be more evil than the death of an adult; his loss of future possibilities of happiness is greater simply because of his younger age.

What this paper has been discussing, however, is not the evilness of death but the characteristics of a good dying. Given that a person has come to the point of his death, what factors can make his dying a good dying? The values expressed in *Charlotte's Web* can give a child an understanding and appreciation of dying which may contribute to the good quality of his own dying. That a good dying is in harmony with nature, is anticipated without fear, is peaceful, and is approached with the support of loving friends and family and with hope of some kind of immortality is an ideal also expressed in the death of Socrates. If *Charlotte's Web* is a tale for children, Socrates' death in *The Phaedo* is a tale for adults – and they share very similar values and perceptions.

## RESPONSE TO PROFESSOR KOPELMAN

Loretta Kopelman raises an important question when she asks: What makes a good death natural? This theme is central to my reading of *Charlotte's Web*, to the comparison with Socrates' view of death in *The Phaedo*, and to the view that this is an attitude helpful to the dying child. I shall try here to clarify the sense of 'natural' which is intended.

The claim that I attribute to *Charlotte's Web* is the claim that what is natural is good and that all natural deaths are good. This does not mean, however, that all good deaths are natural. Some 'non-natural' deaths may be called good, for either of two reasons. First, they may happen to satisfy in general the criteria of natural death. For example, if an artistic or musical genius has completed a life's work at an early age and is beginning to lose his productivity, if he then dies of tuberculosis without much suffering, surrounded by friends, without the sense of a burning desire to finish one more picture or concerto, etc., we could say his death is a good death, even though he is young.

The second reason for not making the claim that all good deaths are "natural" is perhaps more important because that would suggest that there can be only one kind of good death. We need, however, to keep open the possibility of pluralistic views of the good death. Socrates' Stoic acceptance of the inevitable is the direct opposite of Dylan Thomas' powerful plea *not* to go gently into that good night, and we should be able to acknowledge both as valid. Despite Kübler-Ross's empirical findings that many people arrive at acceptance as the last stage of dying, that is not necessarily the only good way of dying or a good way for all people. The concept of the good death is a value concept and a pluralistic view of values means that there are different legitimate ways to live one's life and die one's death.

Thus, there is an important difference between saying that *Charlotte's Web* suggests that natural death is good, which explains why reading this book may be helpful for the dying child, and saying that all good death must be natural in Charlotte's or Socrates' sense. A dying adolescent, especially, may find reading Dylan Thomas more akin to his own values and psychological state than reading about Charlotte or Socrates.

Kopelman is right in saying that all kinds of things are found in nature, including pain and suffering, and thus the sense of 'natural' in the claim that what is natural is good cannot mean simply 'what is found in nature.' But there is what I will call an Aristotelian sense of 'natural' which is richer and captures the sense that fits the *Charlotte's Web* context. Aristotle says that all things in nature – plants, animals, and humans – live, change, and grow according to their own nature. There is a plan for each kind of thing and this plan is fulfilled when it reaches its goal or entelechy, which is at the stage of being able to reproduce itself. Charlotte, the grey spider, fits this pattern exactly; her cycle of life is completed with the laying of her eggs and that is when, like all spiders of her kind, she must die. Socrates, it is true, died of a dose of hemlock, not a gradual fading away of his physical powers, like Charlotte; but given the nature of his life as a martyr-like figure, his accomplishments, his age, etc., it does not strike us as unnatural in the Aristotelian sense that he dies.

*Charlotte's Web* is not saying, as Kopelman suggests, that pain, suffering, and evil are illusions. They are real, but not part of nature's intended plan. Some deaths are bad, and the book gives us two clear examples of bad deaths (though averted): the farmer's proposed killing of the little runt pig at the beginning of the book, and the later proposed slaughter of the fattened Wilbur. Given the necessity of death, some deaths are bad and some are good, or at least better than others. The moral action in the book consists of averting those that are bad and accepting those that are good.

*Wheaton College*
*Norton, Massachusetts*

## NOTES

[1] The psychological literature on the dying child is well summarized in [8] and [9].
[2] For other discussions of this theme in *Charlotte's Web* see [2] and [5].

## BIBLIOGRAPHY

1. Bluebond-Langner, M.: 1980, *The Private Worlds of Dying Children*, Princeton University Press, Princeton.
2. Delisle, R. G. and Woods, A. S.: 1976, 'Death and Dying in Children's Literature: An Analysis of Three Selected Works', *Language Arts* **53** (6), 683–687.
3. Grube, G. M. A.: 1975, *The Trial and Death of Socrates*, Hackett Publishing Co., Indianapolis.
4. Kushner, H. S.: 1963, *When Bad Things Happen to Good People*, Schocken Books, New York.
5. Lukens, R. J.: 1982, *Critical Handbook of Children's Literature*, 2nd ed., Scott Foresman Co., Glenview, IL.
6. Nagel, T.: 1979, 'Death', *Mortal Questions*, Cambridge University Press,

Cambridge.
7. Ober, W. B.: 1979, 'Did Socrates Die of Hemlock Poisoning?', *Boswell's Clap and Other Essays*, Illinois University Press, Carbondale, IL, pp. 262–270.
8. Schowalter, J. B.., Patterson, Tallmer, M. *et al.* (eds.): 1983, *The Child and Death*, Columbia University Press, New York.
9. Wass, H. and Corr, C. A.: 1984, *Childhood and Death*, Hemisphere Publishing, Washington.
10. White, E. B.: 1952, *Charlotte's Web*, Harper and Row, Evanston, IL.

LORETTA M. KOPELMAN

## CHARLOTTE THE SPIDER, SOCRATES, AND THE PROBLEM OF EVIL

Why do children, especially those facing death, find consolation in E.B. White's novel *Charlotte's Web?* [12]. Anthropologist Bluebond-Langner confirmed a long-held belief among pediatric health professionals that dying children find this novel comforting. It is, she notes, their most popular book. "They always chose the chapter in which Charlotte dies, After any child died, the book had a resurgence of popularity among the others" ([2], p. 186).

They may find it important because this charming story teaches some timeless lessons. It shows, for example, how to distinguish between what one can and cannot control, recommending the stoic message that one accept what one cannot change. It celebrates the miracle of life and "the glory of everything" ([12], p. 183), even spider-webs and manure. Death and pain, it suggests, are not random events, and evil itself, with the proper perspectives, can be understood as a necessary part of reality. To say something is appealing, of course, is no argument that it is cogent. This view of evil and the world, however comforting, has certain problems.

As *Charlotte's Web* opens, a little girl, Fern, saves the runt of the litter, Wilbur the pig. Instead of being killed, he is sold to a neighbor who plans to slaughter Wilbur for Christmas dinner. Wilbur's idyllic barnyard life is destroyed when he learns of this. Charlotte the spider decides to trick the humans into thinking Wilbur is special by writing messages about him into her web. Thus, she rescues Wilbur from this untimely death in order to help her friend and to ennoble her life. Charlotte lives knowing that life is short, and reflecting on what is important. Helping her friends and preparing her egg sac, for example, are as central in her life at its prime as when she dies. The story ends with Wilbur living unafraid and coping with the loss of his beloved Charlotte.

Rosalind Ladd [8] suggests that the reason children are attracted to *Charlotte's Web* is that Charlotte's death is portrayed as a natural and good death, similar to Plato's account of the death of Socrates [9]. Ladd characterizes a good death as pain-free, peaceful, and coming surrounded by loved ones; it follows a virtuous and full life where the dying person is treated respectfully and, to the extent possible, exercises control over his or her life.

There are certain limitations to Ladd's instructive comparison. *First*, this is not the only vision of a good death. Some might prefer dying heroically, with "their boots on," or without knowing death was near. And children dying of cancer might find such deaths more appealing than those facing them. Still, all other things being equal, we generally prefer to avoid pain, anxiety, abandonment, broken commitments, and unfulfilled tasks. The death Ladd describes as good, then, also indicates what most of us would regard as a good life: one that is rich, long, and fulfilled.

*Second*, this group of children probably knows more intensely than the rest of us that only some of us are lucky enough to die without pain, with a feeling of accomplishment, at a great age and with friends. Thus, learning of this is unlikely to be the appeal of *Charlotte's Web*. Ladd's analysis, moreover, principally instructs the living about how to provide care for the dying: Be truthful, do not abandon them, and relieve their pain. But the question we sought to answer is why *Charlotte's Web* may offer comfort to dying children themselves. I believe that to answer this question it may be more useful to focus on what these works say about how we ought to live.

*The Phaedo* and *Charlotte's Web* encourage us to live virtuously and to see what is valuable with the eyes of someone who has only a short time to live. Their inspiring messages are not that some of us are lucky in how we die; rather that dying well reveals how we ought to live. Charlotte and Socrates had good deaths, they teach, not only because their deaths happened to be portrayed as pain-free and with friends, but because they lived virtuously. Perhaps to stress this point, Plato took the liberty of inaccurately describing the death from hemlock as pain-free.

In their accounts of how one ought to live, both *The Phaedo* and *Charlotte's Web* teach first to distinguish between events one can and cannot influence. Charlotte and Socrates show us how to accept with grace what cannot be changed. Socrates, seeing no honorable way to avoid his death sentence and flee his unjust condemnation by the Athenians, is described as seeming "quite happy ... both in his manner and in what he said; he met his death so fearlessly and nobly" ([9], 58e). While the other barnyard animals openly discuss grief, loss, fear, and death, Charlotte offers useful strategies for coping with disruptive events. Where no control is possible, however, she calmly resigns herself, for example, to her weakening and death. This candor may, in itself, appeal to dying children because adults find it so painful to discuss these topics with them. Furthermore, they see how Charlotte, who is wise and virtuous, prepares her life for death.

*Third*, the characterization of a good death as natural or a natural death as good does not seem intuitively plausible and Ladd offers no analysis of *natural* to help us make it so. She writes "what is natural is good and dying which is natural and in accordance with nature is a good dying and to be accepted but death which is unnatural is a bad death and to be protested" ([8], p. 109). Ladd says that killing Wilbur for food would be an unnatural death, and therefore should be resisted. In contrast, Charlotte's death after her rich, full life is a natural death and, therefore, should be accepted. Ladd compares Socrates' clearly exemplary dying to the "model of a naturalistic death in *Charlotte's Web*" ([8], p. 114).

Let us consider different meanings of natural to see if the case can be made that a natural death is good or a good death is natural. First, 'natural' can mean *found in nature*. Unjust, early, and painful deaths, however, occur in nature. The Book of Job has one of the most eloquent statements that in nature, suffering, pain, and grief fall on the innocent as well as the evil, the mighty as well as the weak:

But ask the beasts, and they will teach you; the birds of the air, and they will tell you; or the plants of the earth, and they will teach you; and the fish of the sea will declare to you (12:7–8).

Second, 'natural' has sometimes been taken to mean *what is in a thing's nature*. Charlotte says, "I just naturally build a web and trap flies and other insects" ([12], p. 39). Wilbur, taking the insects' point of view, sees their death as bad, untimely, and to be avoided, saying that eating insects is miserable, cruel, and bloodthirsty. Charlotte replies that she must live by her wits or starve:

"I have to think things out, catch what I can, take what comes. And it just happens, my friend, that what comes is flies and insects and bugs. And furthermore," said Charlotte, shaking one of her legs, "do you realize that if I didn't catch bugs and eat them, bugs would increase and multiply and get so numerous that they'd destroy the earth, wipe out everything?"

"Really?" said Wilbur, "I wouldn't want that to happen, Perhaps your web is a good thing after all."

The goose had been listening to this conversation and chuckling to herself. "There are a lot of things Wilbur doesn't know about life," she thought. "He's really a very innocent little pig, he doesn't even know what's going to happen to him around Christmastime; he has no idea that Mr. Zuckerman and Lurvy are plotting to kill him" ([12], p. 40).

White points out that spiders' eating of the superabundance of insects is natural and good, not only from the spider's view but from a wider perspective of things. What appears to be evil to the innocent may really be a necessary part of the order of things.

Ladd says butchering Wilbur for food would be bad, unnatural, and should be properly resisted. While we object to killing our friend Wilbur, no one in the story objects that other pigs are going to be slaughtered. White suggests that it is natural for spiders to eat insects as well as for humans to eat flesh. To say, as Ladd does, that what we ought to do or resist is based on what fulfills a nature, offers a problematic explanation. It runs the risk of circularity or of explaining the obscure (what things are good or bad) by the more obscure (what is in something's nature). For example, if it is in the nature of spiders to trap and eat young, juicy insects, why is it not in the nature of humans to eat flesh as well? Ladd suggests that we may need to consider what is in a thing's nature to answer when a death is good or natural. But how can we know this? If we know this from what creatures *do*, then it *is* in the nature of humans to eat flesh. If it is from what they *ought to do*, then we should admit this is a normative judgment, not a description about something's nature. Thus, it seems hard to understand how a good death is natural or a natural death is good in terms of what is in a thing's nature.

Third, natural can mean *what is to be expected*. Accordingly, a natural death could mean what is predictable either in terms of the life-expectancy of some individual's species, or in terms of their particular biological organization. We say, for example, that it is natural for humans to out-live spiders because of their biological organization. We know this from studying the different species. This view, however, is problematic. For example, the genetic composition of some individuals is such that even without external assaults, their life-expectancies are shorter than those of the typical member of their species. Hence on this view their subsequent early death is both *natural* (expected given their biological organization) and *unnatural* (different from the life-expectancy of their species.) Moreover, some early childhood deaths are due to genetic factors, and although expected, it would be problematic to say that they were natural if that meant that we should not fight the death, seek a cure, or find it a tragedy. Genetic and biochemical compositions of individuals seem increasingly to play a role in the understanding of disease. Thus, the description of life-expectancy simply in terms of what one could expect for members of a species is of limited value. Furthermore, it is not clear what assaults on the integrity of an organism should be viewed as external to it. Some of us are more susceptible to illnesses than others. The problem with this view is that sadly we sometimes expect shortened life, pain, and suffering for those unfortunate to have certain make-ups or who get certain diseases. Moreover, we can predict that a certain number of persons, including children, will contract fatal illnesses.

Thus, a good death is not natural where nature is understood as the totality of what is, or what is in the nature of things, or what is to be expected. Indeed, it would be suspicious to reach a normative conclusion about what deaths are good or bad solely from premises describing what we find in nature.

A fourth way to use natural is in contrast to the *supernatural*. Plato sometimes differentiates between natural and supernatural realms, presenting the natural world as miserable and inferior. This helps him to account for why there is suffering and pain among the innocent of the world.

Death and suffering are conditions of finite existence, argues Plato in *The Phaedo*. Socrates says that he does not fear death since it will either be a dreamless sleeping through eternity or a transformation to a higher, better reality. Sleep will be nothing to fear compared to this life. And if there is a life beyond death, as Socrates argues in *The Phaedo*, then it will be better.

So long as we keep to the body and our soul is contaminated ... the body provides us with innumerable distractions in the pursuit of our necessary sustenance, and any diseases which attack us hinder our quest for reality. Besides, the body fills us with loves and desires and fears and all sorts of fancies and a great deal of nonsense, with the result that we literally never get an opportunity to think at all about anything ([9], 66b).

In this dialogue, Plato presents the natural world as necessarily deficient, finite, incomplete, and unstable. At best, it is a poor copy of what is good, just, and beautiful. Pain, death, and suffering exist because that is how a limited existence must be. Death is at worst a release from it, and at best a transformation to a better form of existence.

Therefore, Plato does not, as Ladd suggests, conclude what is natural is good. Nature just *is*; when compared to what is really good, just, and fair, it falls far short. Moreover, Plato offers a very different account of why there is suffering and early death from that suggested by White.

Plato, in *The Phaedo*, says that physical reality is "contaminated." E. B. White, in *Charlotte's Web*, presents it as wonderful. He glorifies life, and the importance of the most common things in nature, including "the nearness of rats, the sameness of sheep, the love of spiders, the smell of manure, and the glory of everything" ([12], p. 183). Through his portrayal of the overall goodness and the purposefulness of the farm community, White suggests the view that the universe is good and that evil either has a purpose or is an illusion. Butchering Wilbur is murder from one point of view; but, as the other farm animals point out, it is a part of the order of animals eating animals to survive. Charlotte's eating young insects is bad from their perspective (I suppose) but, White says, good for her and for our planet. Even the thoroughly evil character of the story, Templeton the rat, turns out to be a good fellow. He is introduced as having "no morals, no conscience, no scruples, no consideration, no milk of rodent kindness, no anything" ([12], p. 46). Yet this supposedly immoral, evil character is not only the reader's comic relief but repeatedly salvages Wilbur's and Charlotte's plans. In his own sleazy, rat-like ways, he does good deeds. Without his usual bribe, for example, he bites Wilbur's tail so he will awaken and get his award. Templeton's cynicism is refreshing amid so much sweetness. White shows us that in a fuller view Templeton is not evil, but good and necessary to the narrative.

White, then, does not seem to allow that nature is ever really bad. While it may be a more comforting view than Plato's, ought we to believe it? In the next section, I suggest that they present different solutions to the problem of evil. A paradigm of this problem is the suffering and death of children.

## THE PROBLEM OF EVIL

The problem of evil arises for those who want to believe the universe was designed by an all-knowing, perfectly good and powerful Creator. If the Creator is good, omniscient, and powerful, why do the innocent suffer pain and early death? The dilemma arises as follows: If God is all-knowing and powerful, God must be able to prevent evil. If God is perfectly good, God must want to prevent it. But the existence of evil suggests the dilemma that either God is not all-knowing and powerful or God is not perfectly good. The problem may be solved in various ways: by denying that God is all-powerful, as Plato does, or by holding that pain and suffering are either illusory or necessary, as White suggests.

Different kinds of evil should be distinguished. There is moral evil or the bad use of free will. Children sometimes suffer because, for example, they are the victims of child abuse. Thus, some pain and suffering are the result of moral evil and could be eliminated if people were better. But this does not account for other kinds of evil, the pain and suffering of an unpreventable disease, for example.

Plato and White seem to picture suffering and evil in the world differently. For Plato, the non-moral evils of pain, untimely death, and loss cannot be eliminated from the world because they are part of what is. There is no way to make the chaotic world entirely fit real notions of goodness and justice. No one is to blame; this is a necessary feature of finitude. But the picture suggested in E. B. White's *Charlotte's Web* is that the world is not only as good as it can be (a view with which Plato might agree) but that when viewed from the right perspective, evil has a purpose or disappears as illusory; life is a triumph, miracle, and glory.

Mr. Zuckerman took fine care of Wilbur all the rest of his days, and the pig was often visited by friends and admirers, for nobody ever forgot the year of his triumph and the miracle of the web. Life in the barn was very good – night and day, winter and summer, spring and fall, dull days and bright days. It was the best place to be, thought Wilbur, this warm delicious cellar, with the garrulous geese, the changing seasons, the heat of the sun, the passage of swallows, the nearness of rats, the sameness of sheep, the love of spiders, the smell of manure, and the glory of everything ([12], p. 183).

In short, Plato and White seem to offer different solutions to the problem of evil or how to understand the evils of pain and suffering among the innocent. White's solution (it always has a purpose) is perhaps more comforting than Plato's (it just is how things are). This in turn suggests an important difference about how to understand the problem of why the innocent suffer pain and early death. For those who spend any time on a pediatric service, this is a potentially soul-crushing matter. Plato suggests nature is "contaminated," and pain and suffering simply are part of the conditions of finitude. The child who gets sick and suffers is "dealt a bad hand" by nature. White suggests another view, however, one which may explain why children facing death find *Charlotte's Web* comforting. In this story, fears are expressed but we are reassured: everything is as good as it can be, evil and suffering are explained as necessary, death is not painful, and those who do not have to die (Wilbur) are saved. One is not abandoned since one is "never without friends." The dying person is central, good, and wise. She is always loved and remembered as a good friend and a person of accomplishment. There is rebirth and continuity: "each spring there were new baby spiders."

Wilbur never forgot Charlotte. Although he loved her children and grandchildren dearly, none of the new spiders ever quite took her place in his heart. She was in a class by herself. It is not often that someone comes along who is a true friend and a good writer. Charlotte was both ([12], p. 184).

## CHILDREN'S SUFFERING AND DEATH

It is beyond the scope of this paper to evaluate the various solutions to the problem of evil, especially when considered in relation to suffering and dying children. For some, it is hard to believe children's death and suffering *cannot* have a purpose. For others, it is hard to believe it *can*. That is, it is hard for them to accept that such evil fulfills a legitimate purpose as someone's punishment, or as the occasion for the rest of us to show heroic virtues, or as an object lesson, or as otherwise necessary in a perfectly good and just world. It would, for example, seem to be a cruel God who taught us lessons by having little children suffer. Moreover, if we look at who thrives and who perishes in the world, it does not seem reasonable to conclude that the wicked perish and the good thrive in accordance with any justice of rewards and punishments. The good sometimes suffer and die young while the wicked seem to thrive and lead long and healthy lives.[2]

A common response is that we do not know enough to judge why there is such evil in the world. This may be a statement of faith. Instead of an explanation, this response claims that none is possible for us. For some, however, this view that the suffering of children is always purposeful requires too great a leap of faith. Bluebond-Langner says that everyone on the pediatric service, herself included, at some point raged at the children's pain and early death [2].

Still, it is no doubt a comforting view that from the right perspective all suffering always has a purpose or evil is an illusion. If it is true, we can always conclude that someone's suffering an early death has a cosmic purpose. It is a comfort perhaps because it fits our usual defense mechanisms. It gives us the feeling of the existence of goodness, power, and control in what seems otherwise a complicated, chaotic, or brutal world. Even if we do not understand something, we can believe that only if our perspective were great enough, we would see its necessary purpose. We have a ready answer to "why did this happen to us?" or "what does this suffering mean?"

However comforting, there are certain dangers to the view that evil always has a purpose or is illusory. People, in their enthusiasm for it, may deny the reality or intensity of pain, failing to give the sufferers aid. For most of our history, the death of children was a common event. Families were not estranged by such tragedy, and children were routinely exposed to the death of those close to them. A product of today's medical advances is that death, illness, and pain are closely associated with old age. If we do not acknowledge the reality of pain, sickness, and death among children, however, we do not respond appropriately to it. First, we may isolate in silence the very people who need to talk about the tragedy, the children and their families [2–7].

Second, in our denial we may forget to finance children's programs properly. Barbara Starfield has pointed out that in the United States, health care programs for the elderly are funded far more generously than for children equally in need, even though children will have benefit from health care decades longer [11]. Third, we may not make the special effort to help these children by giving them special opportunities to make decisions for themselves or to develop their potential as best they can under very trying and desperate conditions [3, 5–6].

Fourth, if we do not acknowledge that children suffer, then we may be less humane in our treatment of them. Too little attention, some charge, is paid to giving pain medication to children [1,3,4,7]. Also, if aggressive treatment will only result in pain and merely postponing death, it may be rational to treat palliatively. Some writers and policymakers have minimized this sort of overwhelming physical suffering to the degree that it is not permitted to count in making treatment decisions. For example, the recent "Baby Doe" regulations require that infants must be treated to the maximal extent unless they are dying, irreversibly comatose, or unless such treatment would be virtually futile in terms of the survival of the infant and inhumane [9]. But some children can live for a time with maximal treatment though they would be in uninterrupted pain. In writing this policy they gave insufficient attention to the suffering some treatments cause [7]. Thus, it is important to acknowledge the reality of pain and suffering among children.

We began by asking why children facing death find *Charlotte's Web* comforting. I have suggested that in addition to being a charming story it teaches some important lessons about how one ought to live. Ladd offered an interesting comparison of this novel and Plato's *The Phaedo* to explain why *Charlotte's Web* is appealing to dying children. She argues, both works describe a good death and portray a natural death as good or a good death as natural. The analysis of the good death given by Ladd and her portrayal of a good death as natural, however, are problematic. First, while a death that is comfortable, expected, timely, and with friends is generally desirable, it is not the only vision of a good death. Second, the inspiring message of *Charlotte's Web* and *The Phaedo*, I believe, is that they teach us that a death is good when part of a good life. It is not just that some of us are lucky in how we die (at an old age, without pain, without unfulfilled promises, and with friends). These works teach about how one ought to live; they both suggest that dying well is a part of living well. They do this, in part, in teaching us the importance of distinguishing between what is and is not within our control. Third, good deaths, while desirable, are not natural nor natural deaths good on any understanding of natural that I can think to consider. Finally, I suggested that there is a great difference between how the problem of evil, epitomized by the suffering and death of children, is addressed by Plato and White. These different views may have important consequences for how we treat others.

*East Carolina University School of Medicine*
*Greenville, North Carolina*

## NOTES

[1] I discuss this in more detail elsewhere [6].
[2] Other views about the nature of good and evil are easily found in children's literature. One of the most popular is that evil is a real force and not just an illusion or condition of nature. As such, it must be constantly battled. Comic strips and television are filled with stories where superheroes and the forces of good constantly battle the super villains and the forces of evil. The forces are portrayed as nearly equal.

## BIBLIOGRAPHY

1. Anand, K. J. S. and Hickey, P. R.: 1987, 'Pain and its Effects in the Human Neonate and Fetus', *The New England Journal of Medicine* **317** (21), 1321–1329.
2. Bluebond-Langner, M.: 1978, *The Private Worlds of Dying Children*, Princeton University Press, Princeton, NJ.
3. Fletcher, A.: 1987, 'Pain in the Neonate', *The New England Journal of Medicine* **317** (21), 1321–1329.
4. Kopelman, L. M., Irons, T. G., and Kopelman, A. E.: 1988, 'Neonatologists Judge the "Baby Doe" Regulations', *The New England Journal of Medicine* **318** (11), 677–683.
5. Kopelman, L.: 1978, 'On the Right to Information and Freedom of Choice for the Dying: Is It for Minors?' in O. J. Saheer (ed.), *The Child and Death*, C. V. Mosby, St. Louis, pp. 238–247.
6. Kopelman, L.: 1985, 'Paternalism and Autonomy in the Care of the Catastrophically and Chronically Ill Child', in N. Hobbs and J. Perrin (eds.) *Issues in the Care of Children With Chronic Illness: A Sourcebook of Problems, Services, and Policies*, Jossey-Bass Publishing Company, San Francisco, pp. 61–86.
7. Kopelman, L.: 1988, 'The Punishment Concept of Disease', in C. Pierce and D. VanDeVeer (eds.), *AIDS: Ethics and Public Policy*, Wadsworth Publishing Company, Belmont, CA, pp. 49–55.
8. Ladd, R.: 1989, 'Death and Children's Literature', in this volume, pp. 107–120.

9. Plato: 1961, *Phaedo*, H. Tredennick (trans.), in E. Hamilton and H. Cairns (eds.), *Plato*, Princeton University Press, Princeton, NJ.
10. U.S. Department of Health and Human Services: 1985, Final Rule, 'Child Abuse and Neglect Prevention and Treatment Program', *Federal Register* **50** (April 15), 14878–14892.
11. Starfield, B.: 1989, 'Child Health and Public Policy', in this volume, pp. 7–21.
12. White, E. B.: 1952, *Charlotte's Web*, Harper and Row, New York.

GARETH B. MATTHEWS

# CHILDREN'S CONCEPTIONS OF ILLNESS AND DEATH

In her recent book, *Conceptual Change in Childhood* [2], Susan Carey reports that "there is a robust clinical literature on the child's understanding of death" ([2], p. 60). She says that in reviewing the literature published in English during the last 80 years, "A remarkably consistent picture emerges from this research. All authors agree on three periods in the child's emerging understanding of death" ([2], p. 60).

It would, indeed, be remarkable to find complete agreement on anything of importance in developmental psychology. It is especially noteworthy to find agreement on the difficult question of how we come to understand the ultimate threat to our own existence. Carey reports,

In the first period, characteristic of children age 5 and under, the notion of death is assimilated to the notions of sleep and departure. The emotional import of death comes from the child's view of it as a sorrowful separation and/or as the ultimate act of aggression. ... In this period death is seen neither as final nor as inevitable. Just as one wakes from sleep or returns from a trip, so one can return from death. Although children associate death with closed eyes and immobility, as in sleep, they do not grasp the totality of the cessation of function. Nor do they understand the causes of death. Even though they might mention illness or accidents, it is clear that they envision no mechanisms by which illness or accidents cause death ([2], p. 60).

Carey quotes typical exchanges or protocols of conversations with children to illustrate this first stage. Here is one taken from M. H. Nagy ([3], p. 10):

T. P. (age 4:10) A dead person is just as if he were asleep. Sleeps in the ground, too.
   G. Sleeps the same as you do at night, or otherwise?
   T. P. Well – closes his eyes. Sleeps like people at night. Sleeps like that, just like that.
   G. How do you know whether someone is asleep or dead?
   T. P. I know if they go to bed at night and don't open their eyes. If somebody goes to bed and doesn't get up, he's dead or ill.
   G. Will he ever wake up?
   T. P. Never. A dead person only knows if somebody goes out to the grave or something. He feels that somebody is there, or is talking.
   G. Are you certain? You're not mistaken?
   T. P. I don't think so. At funerals you're not allowed to sing, just talk, because otherwise the dead person couldn't sleep peacefully. A dead person feels it if you put

something on his grave.
> G. What is it he feels then?
> T. P. He feels that flowers are put on his grave ... .
> G. What do you think, wouldn't he like to come away from there?
> T. P. He would like to come out, but the coffin is nailed down.
> G. If he weren't in the coffin, could he come back?
> T. P. He couldn't root up all that sand.

Perhaps I should stop a moment to register some nervousness about whether this and other protocols really provide solid evidence for the Stage-One conception of death. This particular protocol is headed by Carey, "Death is reversible, assimilated to sleep and departure." Morever, as I have just reported, the way Carey describes the first stage, which would be the normal stage for a four-year-old like T. P., it includes the idea that "death is seen neither as final nor as inevitable." Carey adds: "Just as one wakes from sleep or returns from a trip, so one can return from death." And yet when this child, T. P., is asked of a dead person, "Will he ever wake up?" the reply is, "Never." Admittedly, when the child is encouraged to expand on the dead person's plight, we get responses like "The coffin is nailed down" and "He couldn't root up all that sand," comments that hardly make the task of reversing death seem totally impossible. But the way Stage One is described, one would have expected the child to talk about reversing death, awaking from that deep sleep, rather than saying that the sleeper will never awake. And anyway, what counts for us as a thoroughly surmountable obstacle, such as removing nails or rooting up sand, may well seem to a child quite insurmountable.

I think it is very important to raise skeptical questions like this concerning the evidence for an account of cognitive development. But obviously in this case, since the literature on the subject is so vast, my qualms will not count for much unless they can be shown to have application to a substantial number of experimental reports. I shall not attempt such an effort here. I mention these qualms just to indicate that perhaps the evidence may be open to an interpretation different from the one Carey tells us all authors agree on.

Here is Carey's account of Stage Two:

> The second stage (early elementary years) in the child's understanding of death is transitional and is characterized differently in the different studies. All authors agree that children now understand the finality of death, and that they understand the sense in which a dead person no longer exists. However, children still see death as caused by an external agent. In Nagy 1948 [i.e., the article from which the protocol above is taken] death was personified in this stage as "Father Death" who caught people. Safier ... and Koocher ... report little personification, but agree that their subjects thought of

death as external to the dying person ... The child does not yet conceptualize death in terms of what happens within the body as a result of these external events ([2], pp. 61, 63).

That brings us to Stage Three. "In the final stage," Carey reports,

death is seen as an inevitable biological process. Such a view of death first becomes evident around age 9 or 10. ... To Koocher's question about the causes of death, one sage 12-year-old answered, "When the heart stops, blood stops circulating, you stop breathing, and that's it. ... Well, there's lots of ways it can get started, but that's what really happens" ([2], p. 64).

"Cultural differences," Carey admits,

can be seen in the details of the children's beliefs. Only the Hungarian children personified death in the second stage: these children also blended the biological and the religious in the third stage. That is, many knew that the body suffers biological death, but considered that the soul lives on. Surprisingly, this religious notion is little evident in the American protocols. Underneath the cultural variation, however, the same developmental sequence over the years 4 to 10 emerged in each study: from seeing death in terms of human activities and emotions to seeing it as the cessation of bodily processes ([2], p. 64).

Not all readers of this literature draw the same conclusion as Carey. For instance, Mark W. Speece and Sandor B. Brent, in a very useful survey, first isolate three components of the concept of death and then maintain that according to virtually all studies, most children understand all three components by age 7 [6]. One of their three components is irreversibility – the notion that one cannot go from being dead to being alive again. A second is nonfunctionality – the idea that in death we cease to function altogether, even to hear and to think. The third is universality – the idea that everything that is alive eventually dies, even I. All three of these components, Speece and Brent maintain, are revealed by the literature to be mastered by most children by age seven.

No doubt the conclusion of the Speece-Brent survey can be reconciled with the Carey three-stage account, even though the impact on a reader of the Speece-Brent survey is likely to be rather different from the impact of reading Carey. It might be worthwhile to examine the difference in these and other presentations of the literature on children's emerging understanding of death and to reflect on the difference it makes in our attitudes toward children. I shall not, however, attempt such a thing here. Instead I shall take Carey's presentation as what I shall call "the Standard Developmental

Account" of the way the concept of death is acquired in children.

It is possible this standard account is flawed by sampling errors. For example, Carey comments that "surprisingly," as she puts it, the "religious notion" that, though "the body suffers biological death ... the soul lives on" is "little evident in the American protocols" ([2], p. 64). I, myself, suspect that whether it is evident in conversations or protocols is something that will vary significantly with the population being polled. In particular, children in staunchly Catholic communities or in heavily fundamentalist Protestant communities might be expected to talk about surviving death, perhaps even about the bodily resurrection of the dead.

There is, however, a more interesting question than the one about how many American ten-year-olds, or how many adults for that matter, really talk outside church about, and seem to believe in, "the resurrection of the dead and the life of the world to come" (to use the words of the Nicene Creed, repeated by countless Christians of all ages every Sunday). This more interesting question arises from the fact that the Standard Developmental Account seems to have as an implication that anyone, adult or child, who believes in the immortality of the soul, let alone the resurrection of the body, is cognitively retarded. Can that be right?

I suspect that most staunch Christian believers today must learn to deal with two belief systems at once, a secular system and a religious one. Cognitive maturity requires not only mastering each one separately but also working out some way of understanding their interconnection.

In addition to the staunch believer, there is also the steadfastly hopeful person to consider. Socrates, after being sentenced to death by a court of his peers, had, according to Plato's *Apology*, these memorable words to say:

Death is one of two things. Either it is annihilation, and the dead have no consciousness of anything, or, as we are told, it is really a change – a migration of the soul from this place to another. Now if there is no consciousness but only dreamless sleep, death must be a marvelous gain ... because the whole of time, if you look at it in this way, can be regarded as no more than one single night. If, on the other hand, death is a removal from here to some other place, and if what we are told is true, that all the dead are there, what greater blessing could there be than this? ...

Now it is time that we were going, I to die and you to live, but which of us has the happier prospect is unknown to anyone but God ([5], 40c5–42a5, in part).

Could there be an intelligent agnostic today able to echo Socrates' words with both understanding and sincerity? Some people might say, "No." I'm inclined to say, "Yes." If I am right, there is not only a problem in relating the

Standard Developmental Account to the case of an intelligent Christian of very orthodox persuasion but even a problem in doing so for a modern Socrates.

This is important because if the Standard Developmental Account describes a development in cognitive maturation that applies fully and without qualification only to the thoroughly secular and religiously skeptical segment of society, then the story is not one of cognitive development simpliciter. Rather, it is at best an account of cognitive development within a certain subculture of our society, even if that subculture is a dominant one. The issue, though, is not really one of numbers. It is one about whether the changes the developmental psychologist charts are more ingrained in human nature or more culture-relative in such a way that the whole process is best looked upon as a kind of socialization.

Suppose, however, Carey's account, or what I am calling "the Standard Developmental Account," is correct; and the Stage-Three concept of death the development culminates in is the right concept for a cognitively mature person to have. What are the implications of all this for the care of injured or ill children, and especially for the care of children with a life-threatening injury or illness?

One set of ethical issues important in the treatment of almost all patients concerns *disclosure*. How much should doctors tell patients about diagnoses and prognoses? The Standard Developmental Account suggests that for Stage-One or Stage-Two patients who are suffering from a life-threatening accident or illness, disclosure is a non-issue. Since these patients have only a proto-concept of death and, therefore, only a defective conception of the threat posed by life-threatening accident or illness, they simply aren't in cognitive position to have the seriousness of their situation disclosed to them. Perhaps the medical team needs to try to deal with these children's fears in the way one might try to deal with an adult's phobias; but these patients are not cognitively competent to have their situation disclosed to them.

A related issue is *decisionmaking*. What part, if any, should a seriously injured or ill child play in decisions concerning treatment? Even though consent to treatment is not a legal requirement for child patients, perhaps some sort of involvement in the decision process is *morally* required if the child is to be respected as a person in her or his own right. But what sort of involvement? And with children of what ages?

The relevance of the Standard Developmental Account is, it seems, equally obvious here. To play any sort of meaningful role in making decisions concerning one's own treatment, one has to understand something of the seriousness of one's injury or illness and appreciate to an important extent what success and failure in treatment might amount to. A child with a limited or defective concept of death is simply not equipped to understand the seriousness of life-threatening illness or injury and, therefore, cannot play any rational role in choosing the course of treatment.

The Standard Developmental Account thus seems to underwrite a completely paternalistic approach to children under ten years old with respect to (1) disclosure of diagnosis and prognosis and also with respect to (2) consent to treatment. Whatever one wants to say about these issues as regards mentally deficient or impaired adults, one is encouraged to think that, although the management of child patients ought to minimize patient distress, it need not, indeed, it cannot really, respect patient autonomy; the cognitive competence required for autonomy is simply missing in young children.

Someone might object that even mentally competent adults are often in positions relevantly comparable to that assigned to young children by the Standard Developmental Account. Suppose, for example, I am diagnosed as having terminal cancer, with less than a year to live. Suppose I am told that chemotherapy will extend my life by a few weeks or months, and I consent to the treatment. I may later regret my decision. I may say I would never have agreed to that treatment if I had known how awful it would be. Can anyone, I may ask myself, who has not experienced such suffering know what it will be like to be in position to make a rational decision about whether the benefit of a few extra months of life, let alone only a very few weeks, will outweigh the cost of that pain? And if the answer in this and similar cases is 'No', then aren't adults often cognitively disadvantaged in treatment decisions just as much as any child might be?

There is merit to this objection. Many of our decisions are made in at least partial ignorance of the quality of possible outcomes. And some of our decisions are made in serious ignorance of what the possible outcomes would be like. Still, a conceptual deficit seems much more crippling to genuine autonomy than a mere lack of experience. If I don't have a proper concept of death, there is no way I can take my death into proper consideration. I am incapable of success in even thinking about it. There is no way you can warn me concerning it or, for that matter, assure me that I have a good chance of postponing it for many years. Nor can I make, or help make, any rational decisions concerning it. Or so it seems.

Perhaps, however, we are moving too fast here. I mentioned two sorts of issues that might be affected by accepting the Standard Developmental Account: disclosure issues and decisionmaking or consent-to-treatment issues. Further reflection might lead us to conclude that the Standard Developmental Account has a rather different relevance to one set of issues than to the other.

Before going any further, it might be useful to characterize the three concepts that make up the Standard Developmental Account with some help from the Speece-Brent analysis. In particular, let's use *irreversibility* as the key component of the Stage-Two concept of death and nonfunctionality as the key component of the Stage-Three concept. We might then come up with these three definitions to capture the three death concepts:

Stage One: x is dead$_1$ =$_{DF}$ x is in a very deep sleep-like state from which it is possible to awake.

Stage Two: x is dead$_2$ =$_{DF}$ x is in a very deep sleep-like state from which it is not possible to awake.

Stage Three: x is dead$_3$ =$_{DF}$ x is in a state in which all functions, including thinking and feeling, stop, never to begin again.

At best, these definitions are an oversimplification. For one thing, there is no mention here of universality – the idea that everyone, including me, will die. Then, although the first two definitions are very clearly related, it isn't sufficiently clear how the third definition can be closely enough related to the others to be a more mature concept of the same thing; namely, death.

Through these three definitions are not entirely satisfactory as a way of capturing the Standard Developmental Account, perhaps they are good enough to use as a first approximation. I don't think that the further refinements called for would undermine the discussion to follow.

Armed with this additional apparatus, let's return to the issue of disclosure. There could be at least two quite different sorts of worry about disclosing the diagnosis of a terminal illness to a young child. One sort of worry might be the worry about success. If a child is told in so many words, "Probably you will be dead soon," and all the child can understand by this message is, "Probably you will soon go into a very deep sleep-like state from which it is possible to awake," then we must say, given the self-imposed non-religious parameters of our discussion, there simply cannot be full disclosure. A child who has no concept of a person's going into a state in which *all* functions, including thinking and feeling, stop, never to begin again, cannot understand that is what will soon happen to her or him.

A second sort of worry one might have about disclosing the prognosis to a terminally ill or injured child is a worry about whether the child will be able to deal with the psychological and existential threat such a prognosis poses. Can the child understand the information and manage to go on living with enough confidence and hope to make the remainder of life worthwhile? Whether any of us can deal with such a prognosis is for us all an ultimate question, whether we are four, or ninety-four, or somewhere in between. It might be thought especially hard for a child to deal with such a threat. Even if it is not especially hard for a child (or at least no harder for a child than for an adult), it might be thought especially cruel to aim at full knowledge in the child of this tragic fact when, perforce, the child understands relatively little of the rest of the world anyway.

A little reflection should reveal, however, that contrary to what I suggested the first time around, the Standard Developmental Account should make us less reluctant, rather than more reluctant, other things being equal, to say things to a child that, for an adult, would constitute the disclosure of a terminal illness. This is so because, according to the Standard Developmental Account, what words like "Probably you will soon be dead" can mean to a Stage-One or Stage-Two child will be something less threatening than what they would mean to a typical adult.

So the implications of the Standard Developmental Account for the issue of disclosure of terminal illness are twofold: first, one will simply not succeed in getting across the mature message; but, second, the message one *can* succeed in getting across is, in any case, less threatening than the mature message the adult would receive in such circumstances.

Why, then, are we so reluctant to utter to a child words to the effect, "Soon you will die" – so reluctant, in fact, that we will grasp at any straw to justify nondisclosure? No doubt, the reluctance we feel in the face of a child is related to the general reluctance we feel to disclose such a prognosis to anyone of any age. With the child we have the excuse for nondisclosure, "Oh, she wouldn't understand anyway."

Yet there are additional factors at work in the case of the child. For one thing, almost all of us feel instinctively a sense of great injustice when a young child is doomed to imminent death. We don't have to be particularly religious to feel this sense of injustice. Even the most secular among us naturally think the lottery of life shouldn't be so grossly and obviously unfair to the very young.

Then, there goes with the knowledge that a child is terminally ill a sense that innocence in general has been violated. It is important for us as adults that there be innocence in the world, somewhere. Where could it be found, if not in children? A child who is mangled by an accident or threatened with death by illness is then naturally seen by us as a horrible violation of what little innocence there is to be found in the world.

Having these attitudes we grasp at straws. "The child will not understand anyway," we may say; "developmental psychologists tell us that children of this age simply can't understand." But the justification is defective. If the Standard Developmental Account is right, disclosure is not as threatening to the Stage-One, or even to the Stage-Two, child as it would be to us.

To say this is not, of course, to give a full justification for complete candor. It is just to eliminate one among several possible arguments for keeping quiet or dissembling. Thus, some people say that the emotional resources of young children are just too limited to handle even the Stage-One version of a terminal prognosis. I, myself, doubt that this is true as a general thing. But I shall not discuss that issue here. My point is just that the Standard Developmental Account, itself, does not provide a satisfactory basis for a general policy of nondisclosure of impending death to young children.

Decisionmaking seems to present a strikingly different problem. One who thinks of what is called "death" as a mere sleep-like state from which one may awake can be expected to make quite different treatment decisions than one who views death as a state in which all functions, including thinking and feeling, stop, never to begin again. I am, therefore, inclined to stand by my earlier assessment of the bearing of the Standard Developmental Account on decisionmaking. If that account is correct, then there simply is no way for a Stage-One or a Stage-Two child to be fully engaged in treatment decisions. The requisite conceptual machinery is missing.

The discussion so far rests on an unexamined assumption that I want now to question. It is the assumption that the Standard Developmental Account, if it is a correct account of standard cognitive development, will have a direct application to a child suffering from a life-threatening accident or illness.

It is worth reminding ourselves that the only experience with death many children between the ages of four and ten have is an experience with the death of a pet. Even if a child in this age range has experienced a death in the family, it will much more likely be the death of a grandparent than that of a parent or sibling.

Of course, there are children in this age range who lose a parent, sibling, or best friend. And there are children in this age range who, themselves, suffer life-threatening accident or terminal illness. But these latter cases especially, being exceptional, will not play a significant role in the outline of the Standard Developmental Account. We need to ask specifically whether there is good reason to think that the Standard Developmental Account applies to them.

I turn now to terminally ill children in this age range and what we know specifically about their understanding of death. A group of such children has been studied with wonderful sensitivity by Myra Bluebond-Langner in her pioneering work, *The Private Worlds of Dying Children* [1].

The children Bluebond-Langner studied were victims of acute lymphocytic leukemia. At the time of her study, 1971–2, the prognosis for such patients was almost hopeless. Fifty patients were included in her study, thirty-two as "informants," eighteen as "primary informants." Their ages ranged from three to nine. Of the eighteen primary informants, six survived at the end of the nine-month study. By the time Bluebond-Langner finished up her book some five years later, none was still living ([1], p. x).

These leukemic children were marked off from other children and from "healthy" society at large in countless ways. Bluebond-Langner describes some of the ways:

Although, like other children, [dying children] are sensitive, intelligent, kind, willful, and young, they will not "become," they have no future. Participation in the institutions that mark one as children in this society are not for them. They do not go to school, they go to the hospital. They do not wear sexually, or even socially, appropriate clothing, but nondescript hospital gowns that are either immodestly short or unbecomingly large, and they go about without underwear. Boys are not discouraged from masturbating, and girls are not discouraged from punching. Little is withheld materially, and reprimands usually come only when tempers are short.

Dying children are more like the elderly than their own healthy peers – without futures, worried, often passive, unhappy, burdened with responsibilities for others and their feelings. They even resemble old people – either bald or disheveled, emaciated or bloated, incapacitated, generally sickly, and most of all losing, failing with time. Their worth can only be measured by what they do now, unlike other children, who have time to prove themselves ([1], p. 213).

It should come as a surprise to no one that the Standard Developmental Account of how children come to acquire a mature concept of death does not fit the leukemic children Myra Bluebond-Langner studied. I do not mean to suggest that the children understood their condition right away, let alone accepted it. Typically, terminally ill adults don't understand or accept their condition right away either. But by the measure of the Standard Developmental Account, these children were precocious indeed.

Bluebond-Langner describes five stages in which terminally ill children come to understand their situation:

Each stage was marked by the acquisition of significant disease-related information.

The children first learned that "it" (not all the children knew the name of the disease) was a serious illness. [Stage 1] At this time they also accumulated information about the names of the drugs and their side effects. By the time children reached stage 2, they knew which drugs were used when, how, and with what consequences. The third stage was marked by an understanding of the special procedures needed to administer the drugs and additional treatments that might be required as a result of the drugs' side effects. The children knew which symptoms indicated which procedures, and the relationship between a particular symptom and procedure. But they saw each procedure, each treatment, as a unique event. Not until they reached stage 4 were they able to put treatments, procedures, and symptoms into a larger perspective. By then, the children had an idea of the overall disease process – that the disease was a series of relapses and remissions, that one could get sick over and over again in the same way, and that the medicines did not always last as long as they were supposed to, if at all. But it was not until the fifth stage that the children learned the cycle ended in death. They realized that there was a finite number of drugs and that when these drugs were no longer effective, death became imminent ([1], p. 167).

Of course, any child who reached the fifth stage in this process had a conception of death as the irreversible cessation of all biological functions. And any child who reached that stage knew that death would come to her or him, not sometime in the unreal future, but soon. Thus, every leukemic child in Bluebond-Langner's study who reached Stage 5 in the process she describes would have a conception of death that includes all the elements in the last-stage concept of the Standard Developmental Account.

It is a striking feature of Bluebond-Langner's analysis that the five stages she identifies in the children's emerging understanding of their illness are classic Piagetian stages *except* for one crucial aspect: chronological age is irrelevant. Thus, the order of the stages is irreversible. Moreover, no stage may be skipped, though one child might linger in a given stage for a year while another child passes through the same stage in a week. The understanding a child achieves at a given stage is a prerequisite for the understanding to be achieved at the next stage. Progress in the development is directed by the child's own need to make sense of the world as she or he experiences it. Cognitive dissonance plays a role here, as in Piaget, for the discrepancy between what parents and medical staff *say* and what actually happens, including the way parents and staff actually behave, is sometimes what prods the child to move on to the next stage.

As I have said, the feature of Bluebond-Langner's stages that makes them non-Piagetian is the utter irrelevance of chronological age. This is also the feature that puts her account at odds with the Standard Developmental Account.

Bluebond-Langner explains the irrelevance of age in terms of differences among the children's experiences:

> The place of experience in the socialization process helps illuminate why a child could remain at a given stage without passing to the next for what seemed an unusual length of time. Tom, for example, remained at stage 4 for a year, whereas Jeffrey remained at stage 4 for only a week. Since passage to stage 5 depended on the news of another child's death, and none had died after Tom reached stage 4, he could not pass to stage 5. When Jennifer died, the first child to die that year, all the children in stage 4, regardless of how long they had been there, passed to stage 5.
>
> The role of experience in developing awareness also explained why age and intellectual ability were not related to the speed or completeness with which the children passed through the stages. Some three- and four-year-olds of average intelligence knew more about their prognosis than some very intelligent nine-year-olds, who were still in their first remission, had had fewer clinic visits, and hence less experience ([1], p. 169).

Concomitantly with their passage through the above five stages in understanding their illness, these children, according to Bluebond-Langner, passed through a succession of five self-concepts. In Stage 1 they thought of themselves as simply seriously ill. In Stage 2 they thought of themselves as "seriously ill and will get better." In Stage 3 they thought of themselves as "always ill and will get better." In Stage 4 they thought of themselves as "always ill and will never get better." In Stage 5 they thought of themselves as "dying" ([1], p. 169).

I want to reflect a little now on the bearing of Myra Bluebond-Langner's study on the Standard Developmental Account of how children acquire a mature concept of death, and especially on what Bluebond-Langner's study should tell us about the relevance of the Standard Developmental Account to the treatment of terminally ill children.

Even if the Standard Developmental Account were an entirely satisfactory presentation of standard development (I have suggested that there are problems with it), it clearly could not be a satisfactory basis for understanding the conception of death that terminally ill children in extensive treatment programs have. Their life experience and their efforts to make sense of their treatment experience make them unnervingly precocious in matters of illness and death. Their precociousness can encompass a detailed understanding of how their illness may soon result in the deterioration of organic functions in their own bodies and eventually the full cessation of all that functioning.

The Bluebond-Langner study suggests an even more profound conclusion. It suggests that if we placed children on a continuum from *No direct experience with death at all* on one extreme to *Terminally ill and in a treatment program with other terminally ill children* on the other extreme, the experience of the vast majority of children would place them somewhere between the two extremes; but most of them would be closer to the innocence extreme than to the extreme of terminal illness.

If the Standard Developmental Account were satisfactory as an account of anything, it would be as an account of the typical child's developing comprehension of death. But the typical child does not have a life-threatening illness or injury and is not in an extensive life-or-death treatment program. The Standard Developmental Account is, therefore, largely irrelevant to the ethical issues of disclosure and decisionmaking in the most excruciating cases; namely, cases of the most serious injuries and illness. In fact, if my picture of the continuum is appropriate, the Standard Developmental Account becomes increasingly inappropriate just in proportion as the child's medical situation becomes more desperate and the issues of disclosure and participation in decisionmaking become more daunting to parents and medical staff.

So what is the conclusion? What should a terminally ill or injured child be told? What part, if any, should a four-year-old or a nine-year-old play in treatment decisions? My paper does not equip us to answer those questions – far from it. All I have done is to rule something out. I have ruled out as inappropriately based the simplest answer; namely, "Don't tell children anything or involve them in any way in treatment decisions, because at their age, they can't understand what death is."

Among the options left is the possibility of open discussion and full involvement in treatment decisions. In response to an earlier version of this paper, John Moskop called my attention to a fascinating report on the effort by Oklahoma Children's Memorial Hospital to treat terminally ill patients between the ages of 6 and 20 openly and with real respect for their autonomy [4]. The results were remarkable. Even the youngest patients in this group were able, it seems, to participate fully in the decision as to whether they should be given "phase II drugs." Those who elected to receive only "supportive care" remained, with one exception, strikingly healthy of mind and free from severe depression.

Though I have not actually supported this option, I have, I hope, removed an automatic objection to it. Since that objection has been widely thought to have the full weight of developmental psychology behind it, removing it is, I trust, a matter of non-trivial significance.

Here is a final comment about listening to children. For several years now, I have been trying to awaken parents and teachers to the possibilities of doing real philosophy with young children. Despite what our best developmental theories suggest, many children do naturally reflect on the very deepest questions of space, time, and existence; and they are, many of them, quite able to develop lines of reasoning that can excite the interest and that deserve the respect of professional philosophers. Yet it is very hard for even well-meaning adults to listen without condescension to what children have to say or to pursue discussions with them when the adults are not able to control those discussions fully or settle authoritatively the issues they raise.

No doubt, some children who face the real prospect of imminent death also have, for that very reason, important things to tell us and to discuss with us if we are only strong enough to listen and to share. But having a good discussion with such a child requires both an openness to the child and an openness to thinking about death that we adults find extremely difficult to manage. A terminally injured or ill child is the ultimate threat to our paternalistic pretensions. If we can learn to deal honestly with that threat and to deal respectfully as well as lovingly with such a child, we will have taken a major step in the development of our own maturity.

*University of Massachusetts at Amherst*
*Amherst, Massachusetts*

## BIBLIOGRAPHY

1. Bluebond-Langner, M.: 1978, *The Private Worlds of Dying Children*, Princeton University Press, Princeton, New Jersey.
2. Carey, S.: 1985, *Conceptual Change in Childhood*, MIT Press, Cambridge, Massachusetts.
3. Nagy, M. H.: 1948, 'The Child's Theories Concerning Death', *Journal of Genetic Psychology* **73**, 3–27.
4. Nitschke, R., Humphrey, G. B., Sexaner, C. L., Catron, B., Wunder, S., and Jay, S.: 1982, 'Therapeutic Choices Made by Patients with End-stage Cancer', *The Journal of Pediatrics* **101**, 471–476.
5. Plato: 1961, *Apology*, H. Tredennick (trans.), in E. Hamilton and H. Cairns (eds.) *The Collected Dialogues of Plato*, Princeton University Press, Princeton, New Jersey.
6. Speece, M. W. and Brent, S. B.: 1984, 'Children's Understanding of Death: A Review of Three Components of a Death Concept', *Child Development* **55**, 1671–1686.

JOHN C. MOSKOP

# TERMINALLY ILL CHILDREN AND TREATMENT CHOICES: A REPLY TO GARETH MATTHEWS

Gareth Matthews' paper [5] is a penetrating discussion of two closely related topics. In the first part of the paper, he offers a description and analysis of what he has called the "Standard Developmental Account" of children's conceptions of death. In the second part of the paper he focuses on an in-depth study of terminally ill children and compares its findings with the Standard Developmental Account. My commentary will explore the implications Matthews draws from these studies for the questions of whether and how to inform seriously ill children about their disease and help them participate in decisions about its treatment.

After describing the standard account of the development of children's understanding of death, Matthews considers the bearing of this account on questions of disclosing information to seriously ill children and involving them in decisions about their health care. He concludes that the Standard Developmental Account does not justify a policy of nondisclosure, but that it does support the claim that children cannot participate fully in decisions about their care (because they lack the requisite conceptual apparatus). Matthews then goes on to question a previously unexamined assumption, namely, the assumption that the Standard Developmental Account does indeed apply to seriously or terminally ill children. His answer, based on the Bluebond-Langner study, is that it does not.

If this is the case, however, why does Matthews spend so much time with the Standard Developmental Account in the first place? Why doesn't he move directly to the special situation of gravely ill children? Two reasons suggest themselves.

First, Matthews may be concerned that developmental psychologists view their results as establishing a universal, closely age-related developmental process. If this were the case, it would be important to show how Bluebond-Langner's study confounds their view. It is not clear, however, why these psychologists would hold a view that goes so far beyond what their empirical studies can establish. In fact, one of the reports Matthews cites, that of Speece and Brent, is careful to point out that it reviews the literature regarding *healthy* children's understanding of death ([8], p. 1671).

Even if research psychologists do recognize that standard or typical developmental processes cannot be imputed either to specific individuals or to non-standard groups, Matthews may still wish to point out the limitations of the standard account in order to prevent *others* from applying it uncritically. Health professionals, for example, may inappropriately use this account to guide their communication with seriously ill patients. *If* it is the case that many health professionals are misled by a superficial understanding of standard developmental processes (Matthews offers no evidence for this), it would, of course, be important to point out the limitations of the Standard Developmental Account and the importance of determining how much each individual child understands about his or her illness.

Bluebond-Langner's conclusion that terminally ill children do possess a mature understanding of death, however, does not by itself establish their competence to make health care decisions. As Matthews is careful to point out, his paper does not enable us to say what a terminally ill child should be told or what role that child should have in treatment decisions. Rather, he reaches the more modest conclusion that a decision not to inform the child cannot be justified on the grounds that the child simply cannot understand what death is. The ability to understand and communicate is, in fact, only one of three requirements for decisionmaking capacity identified by the President's Commission for the Study of Ethical Problems in Medicine [7]. The other two requirements, according to the Commission, are an ability to reason and deliberate about one's choices, and possession of a set of values and goals ([7], p. 57). In order to conclude that seriously ill children should play a significant role in decision making about their care, then, we would at least need to show that they satisfy these other two requirements for decisionmaking capacity in addition to that of understanding. In his review of the developmental psychology literature on this subject, Brock concludes that in the 11- to 13-year-old period children generally acquire various reasoning skills and the sense that they have control over what happens to them ([2], p. 186–187). Like Matthews' Standard Developmental Account of understanding, however, Brock's conclusions are based on studies of the general population of children and may not apply to the special circumstances of terminally ill children. If this group of children is precocious in their ability to understand, perhaps they are also precocious in the development of reasoning skills.

What of the third requirement for decisionmaking competence, namely, possession of a set of values and goals? Children, even young children, clearly express values and goals, but there are at least two problems with simply deferring to those values in decisionmaking. The first problem is that children's values and goals tend to change in significant and predictable ways as they grow to adulthood. We may, therefore, wish to protect children from the risks inherent in choices based on an immature set of values. As Robert Holmes points out in this volume ([4], p. 216), however, adults also often make risky choices based on what might be considered immature values, yet we generally respect those choices. Moreover, terminally ill children will never have the chance to develop a more mature set of values. We may, therefore, best promote their happiness and well-being for the brief remainder of their lives by honoring the values and desires they have now, at least when doing so is not unduly burdensome or dangerous to the children or others.

The second problem with relying on children's values in decisionmaking is that parents have substantial interests and goals for their children and a strong tradition of authority to act for their children in protecting their welfare. Thus, even if children were judged competent on the basis of the three requirements we have been considering, it is not clear how their interests in self-determination should be weighed against parental interests in rearing and protecting their children. In the case of terminally ill children, however, it would seem that parental and child interests may tend to converge in a shared concern about the short-term goal of enhancing the time remaining and easing the suffering of the child.

Thus far I have considered some reasons for and against finding children, especially terminally ill children, competent to assume a significant role in decisionmaking about their health care. Even if we conclude that such children are not capable of making these decisions, health care professionals must still face the question of informing, namely, should children be informed about their disease, treatment, and prognosis? If so, how?

A striking conclusion of the studies carried out by Bluebond-Langner and others is that terminally ill children come to know that they are dying whether or not they have been told ([1], p. 198). This knowledge appears to emerge out of patients' growing understanding of the nature and course of their disease, conversations with other child-patients, overheard remarks of parents and professionals, and especially knowledge of the death of another child with the same disease.

If a child knows she will die it might seem self-evident that parents and caregivers should discuss this with her. Such discussions would seemingly forge a closer bond with the child, strengthen trust, overcome the need for pretense on all sides, and allow the child and family to prepare for their inevitable separation.

I believe such discussions *are* beneficial, although I no longer think that their value is self-evident. In fact, Bluebond-Langner suggests that open communication about death with the child may do more harm than good. This is so, Bluebond-Langner argues, because acknowledging death openly deprives all those involved of their established social roles. Children can no longer be viewed as in the process of growing up, parents can no longer be viewed as rearers or protectors of their children, and staff can no longer be viewed as healers. Bluebond-Langner concludes that it was in order to preserve these roles and the relationships they entailed that the overwhelming majority of children, parents, and staff in her study persisted in avoiding all discussion of the disease and its prognosis as the child's death approached ([1], pp. 229–233).

Bluebond-Langner may very well be right about why her subjects shrank from open communication, but I am not convinced that this situation is itself unavoidable. For one thing, Bluebond-Langner points out in her study that the policy of pretense and avoiding open communication with children was initiated by staff members of the hospital and clinic at their first encounter with the children, and thereafter continually reinforced by the staff and parents ([1], pp. 200–201). If the health care providers had had a policy of communicating openly with children from the outset, the value of maintaining a mutual pretense may not have seemed nearly so great for all concerned later on. Rather, roles and relationships might have had a chance to evolve gradually in order to accommodate the changing circumstances. After all, at least some of the psychological and physical needs of a terminally ill child are different from those of a child with a reversible illness, but those needs cannot be met, or at least cannot be met as effectively, if no one is willing to acknowledge the difference between the two. Bluebond-Langner suggests that children sometimes chose to maintain silence in order to spare their parents the necessity of facing the child's death ([1], p. 229). Something seems to me to be amiss here if the dying child is expected or required to be stronger and better able to sacrifice for the needs of others than either parents or caregivers. Perhaps maintaining a pretense is the only way to insure continued parental support for these children in some cases, but I believe that most parents and professionals can approach the emotionally wrenching reality of death openly, with support and strength for the dying child rather than withdrawal or abandonment. In fact, a group of physician-researchers in Oklahoma has reported good results with a policy of communicating openly and sharing treatment decisions with children dying of cancer [6]. Spinetta's review of research on communicating with seriously ill children supports this result, concluding that "the evidence is beginning to mount that the children, siblings, and parents would be best served by being encouraged to bring into the open their anxieties about the illness and its possible consequences" ([9], p. 48).

Because I believe that open communication about dying is generally beneficial, I would take issue with the often-heard claim that we should let the patients tell us what they want to know and when. In a setting where communication about one's illness is discouraged, like that described by Bluebond-Langner, children rarely asked staff and family for information. Thus, waiting for the sick child to ask authority figures for information may place too great a burden on the child, reinforce the existing atmosphere of silence, and serve as an excuse for not attending to the patient's need to discuss his condition and exercise some control over his care. Professionals, not patients, should, I believe, take the initiative in establishing open communication. If communication is well established and child-patients sense others' willingness to share information with them, then I would agree that their questions will be important indicators of what they want to know and when.

In his final comment, Gareth Matthews invites us to undertake the difficult task of listening to what children who face death tell us. It may also be a difficult task to involve these children in decisions about their health care, but if, as I suspect, this task fosters greater openness, honesty, respect for children's views, and responsible control by children over their own lives, it will be well worth the effort.

*East Carolina University School of Medicine*
*Greenville, North Carolina*

## BIBLIOGRAPHY

1. Bluebond-Langner, M.: 1978, *The Private Worlds of Dying Children*, Princeton University Press, Princeton, New Jersey.
2. Brock, D.: 1989, 'Children's Competence for Health Care Decisionmaking', in this volume, pp. 181–212.
3. Holder, A.: 1989, 'Children and Adolescents: Their Right to Decide about Their Own Health Care', in this volume, pp. 161–172.
4. Holmes, R.: 1989, 'Consent and Decisional Authority in Children's Health Care Decisionmaking: A Reply to Dan Brock', in this volume, pp. 213–219.
5. Matthews, G.: 1988, 'Children's Conceptions of Illness and Death', in this volume, pp. 133–146.
6. Nitschke, R. *et al.*: 1982, 'Therapeutic Choices Made by Patients with End-stage

Cancer', *Journal of Pediatrics* **101**, 471–476.
7. President's Commission for the Study of Ethical Problems in Medicine and Biomedical and Behavioral Research: 1982, *Making Health Care Decisions*, Washington, D.C.
8. Speece, M. W. and Brent, S. B.: 1984, 'Children's Understanding of Death: A Review of Three Components of a Death Concept', *Child Development* **55**, 1671–1686.
9. Spinetta, J. J.: 1978, 'Communication Patterns in Families Dealing with Life-threatening Illness', in O. J. Z. Sahler (ed.), *The Child and Death*, C. V. Mosby Company, Saint Louis, pp. 43–51.

# SECTION III

# CHILDREN'S AND PARENTS' ROLES IN MEDICAL DECISIONMAKING

# INTRODUCTION

Which children are capable of understanding certain information and making judgments for themselves? When are they old enough or in a condition to be emancipated from parental authority? Although children achieve full civil rights at the age of majority, it is, of course, a legal fiction that persons suddenly become fully mature on a particular birthday. Perhaps to remedy this, legislation and litigation have increasingly granted to older children rights to make decisions for themselves independently of their parents. Federal policies encourage even young children's assent or participation in consent for treatment and for research on the grounds that they are generally better off if they are informed about their illness and dealt with truthfully.

Granting rights to children is frequently defended on utilitarian grounds, when it has been shown to be in their best interest. Children might go without treatment for venereal disease, for example, rather than face disapproving parents. It is also a way to encourage the development of responsibility and autonomy in children. Several difficult questions emerge, however: How do we adequately protect and properly teach responsibility and moral integrity to children who face serious or chronic illnesses? When and how do we help older children deal with issues of consent or assent to therapy or to participation in research? How do we deal with the mature or emancipated minor? And what treatment, counseling and services ought to be available to minors?

If parents are typically the primary advocates for their child, what are the limits of their authority to act on behalf of the child? According to the legal *parens patriae* doctrine the state may intervene to protect the child's best interests, thus limiting the authority of the parents. Paternalistic legislation protects children from abuse, neglect and incompetent caretakers. Still other laws try to give children equal opportunities by requiring provision of care essential for their development such as immunizations, eye tests, screening and other physical examinations. All of these grant rights to children quite independently of what their parents may wish and are regarded as important ways to protect children. They do, however, in some basic sense, limit the authority of parents to determine what they will provide for their children.

The policy of limiting parental authority raises ethical and social issues regarding what criteria should be used to do so. One standard way to decide legal and moral questions about health care for children is to determine what is in their best interests. The appeal of the best interest standard is that it directs us toward an important group of considerations about what will benefit the person for whom others make choices. It is, therefore, a valuable standard in protecting and caring for children. Disagreements have arisen, however, about what serves children's best interests and about how much authority the state and the physician should have in relation to the parents. In addition, some have argued that an exclusive focus on one individual's best interests can neglect others unfairly. Thus, it may be difficult to strike a just balance between these competing interests.

When faced with difficult treatment decisions for children, physicians often reflect upon what they would do for their own child were he or she in a similar situation. This too may be a useful moral test of the merits of their own judgments (are they universalizable?), or it may be an excuse to escape open discussion and impose their views unjustifiably. Similarly, parents often ask physicians what they ought to do, or what the physicians would do as a parent faced with a similar choice. This too may be a useful moral test of the merits of their judgment, or an attempt to evade their responsibilities to their child. This third section is devoted to an examination of some of these issues.

Angela Holder's 'Children and Adolescents: Their Right to Decide about Their Own Health Care' reviews current legal standards governing the authority of minors to make health care decisions for themselves. Holder points out that although the common law once gave parents (actually, the father) total control over treatment decisions for children under twenty-one, that is still the rule only for young children in non-emergency situations. Courts and legislatures have granted significant decisionmaking authority to adolescents by means of minor treatment statutes and the concepts of "emancipated minor" and "mature minor." Holder notes that the extent of that authority, however, may depend upon the kind of treatment in question; for example, adolescents have the legal right to refuse elective treatment such as plastic surgery, but not necessary or life-saving treatment.

Holder goes on to examine several specific areas of health care in which special questions about minors' consent to treatment may arise. She argues that teen-agers have a right to consent to psychiatric care, but not to treatment or counseling from non-medical practitioners whose training may be questionable. Minors' rights to challenge commitment to mental hospitals by their parents are more limited, however. Though some state laws give minor patients a right to a hearing if they object to their admission to a mental hospital, the Supreme Court has ruled that a state's failure to provide such a mechanism does not violate a minor's constitutional rights. Minors also have a right to obtain contraceptives and abortions without parental consent, though in the case of abortion, a girl may be required to convince a judge that she is mature enough to make an informed decision. In sum, Holder cites a trend toward increased authority in medical decisionmaking for minors, and she also notes a number of areas in which minors' rights to obtain medical care remain more limited than those of adults.

In his commentary on Holder, Robert Holmes maintains that questions about the treatment of children generally, and not just in health care, lack a clear and coherent framework. Parents may face a conflict in raising children between providing for their well-being and nurturing their developing autonomy. States may face a similar conflict between acting directly to protect or promote children's well-being and promoting the independence and strength of familial relationships. Holmes uses examples taken from Holder's review of the law to illustrate the difficulty of balancing these conflicting interests on the part of both parents and the state. He asks why, if parents can commit their children involuntarily to mental hospitals because of concern over their behavior, they should be barred from regulating their children's access to contraceptives. Holmes argues that parents should be given greater authority to control the sexual practices of their children, in view of the fact that teen-agers rarely reflect on the profound significance of becoming a parent.

Dan W. Brock, in 'Children's Competence for Health Care Decisionmaking,' examines the general legal and medical policy presumptions about minors' incompetence to make health care decisions. He argues that they ought to be revised to reject the presumption that older children are incompetent to make decisions concerning their medical care. In both the legal and medical arenas, he argues, there is considerable confusion about the conceptual issues concerning competency determinations for children and for adults. Brock argues that competency decisions are always task-specific, but that competency itself may be understood as a threshold concept. Competency decisions are task-specific since they "involve matching the capabilities of a particular person at a particular time and under particular conditions with the demands of a particular decisionmaking task" (p. 183). In considering what capacities are needed for competency, he finds it useful to distinguish three sorts of capabilities: "1. capacities for communication and understanding of information; 2. capacities for reasoning and deliberation; [and] 3. capacity to have and apply a set of values or conception of the good." (p. 184). Brock summarizes results from child developmental studies and theories that are relevant to general assessments of these capacities in children. This evidence, he argues, "supports the conclusion that children by age 14 or 15 usually have developed the various capacities necessary for competence and health care decisionmaking to a level roughly comparable to that attained by most adults" (p. 188).

For adults, where the presumption is that they are competent, the crucial question is, How do we rebut the presumption of competence? For children, because of the law's presumption of incompetence, it is, How do we rebut the presumption of incompetence? Brock argues that "for both adults and children, there are two important values at stake: respecting their interest in self-determination while also protecting and promoting their health and well-being" (p. 190). He argues that the determination that a child is competent does not mean the child should have final decisionmaking authority, since parents also have a legitimate interest in making decisions for their minor children. Though there may be no way to balance the central values, Brock finds some decisions inappropriate, such as giving in to every choice made by the child or being unduly paternalistic. Rather than adhering to a fixed age cutoff, we should look at the nature of each child's reasoning. As a general guideline, however, we may be best served by allowing children at around fourteen years of age to participate in medical treatment decisions if they wish to do so.

Robert Holmes agrees with Brock that the general presumption of incompetence on the part of minors to decide medical care is untenable. He criticizes Brock's view that only children over about fourteen years of age may be competent to engage in medical decisionmaking. Holmes suggests that even younger children may have "at least as much competence in this area as most adults." He proposes an analogy: an unskilled chess player can follow the moves of one who is a master, although unable to work them out for himself. Both adults and children, he claims, are analogous to the unskilled chess player in their ability to make medical decisions; though they cannot formulate treatment strategies for themselves, they can follow and assent to the strategies proposed by their physicians. Holmes tentatively suggests that in many cases children by the time they reach the age of ten could follow such discussions. He qualifies this conclusion, however, saying that more research is needed and that we would want to carefully consider the situation and circumstances as well. Like Brock, he does not think competence to share in decisionmaking or to give assent means children should necessarily have the final decisionmaking authority on the grounds that parents also have legitimate and significant interests in providing for the welfare of their children.

William Ruddick takes up the question of parents' responsibility for their children's health care in 'Questions Parents Should Resist'. Ruddick argues that we should *resist* asking the following three questions in choosing health care for children: "(1) 'What would I do if it were my child?' asked by pediatricians of themselves; (2) 'What would you do if it were your child?' asked of pediatricians by parents of ill children; [and] (3) 'What is in the child's best interest?' asked by both physicians and parents" (p. 221).

To be justified in using the first question to determine how to care for a child, a pediatrician must assume one of three things: the parents are incompetent to make these decisions, the parents share his or her own values, or if not, his or her own values should be decisive. Ruddick questions each of these assumptions. Parents should be reluctant to ask the second question, according to Ruddick, because they will rarely know how physicians feel about their own children and because physicians' judgments may be distorted by professional optimism about the results of treatment. The third question, commonly used by judges and other officials, tends to view children as separable from their families. It is difficult, Ruddick argues, for parents to view their children in this way, and this question may distort and over-simplify the situation.

Instead of the above questions, Ruddick proposes that parents ask "How will treatment affect our lives?" This question emphasizes rather than suppresses the fact that the lives of parent and child are intertwined. Since their welfare is indivisible, Ruddick concludes, asking this question would benefit both children and parents.

In his commentary, H. Tristram Engelhardt, Jr., endorses Ruddick's effort to move beyond the best interest of children as the sole criteria in choosing their health care. Engelhardt argues that best interest considerations must be balanced by concern for the costs and burdens incurred by families and society. Engelhardt finds unambigious support for this position in the distinction between ordinary and extraordinary care found in traditional Catholic moral theology. According to this doctrine, the obligation to provide medical treatment for a child could be defeated if it required families to assume grave burdens or great costs. In contrast to this approach, Engelhardt points out, the current federal Baby Doe/Child Abuse Regulations severely restrict parents' ability to participate in health care decisionmaking for severely ill newborns. Because they limit parental authority and require families to assume great costs in the care of a child, these regulations, Engelhardt argues, represent an assault on the integrity of the family unit.

ANGELA R. HOLDER

# CHILDREN AND ADOLESCENTS: THEIR RIGHT TO DECIDE ABOUT THEIR OWN HEALTH CARE

What limitations should be placed on the right of a specific child or adolescent to make decisions about his or her own health care? This is a continuing issue in the everyday practice of pediatrics [14, 17, 28, 30]. My friends who practice amubulatory pediatrics tell me that the number of adolescents who present themselves for treatment is increasing, they are becoming younger, and the situations in which they find themselves are more complicated.

At common law, a child, which meant anyone under 21, not 18, was a chattel of his or her parent – actually, of his or her father. A father had the right to sue a physician who treated his son or daughter without his permission, even if the treatment had been perfectly appropriate, because such an intervention contravened the father's right to control the child.

That is still the rule with a young child in a non-acute situation. If a three-year-old is visiting his grandmother who brings him to a plastic surgeon to remove a birthmark the parents had decided to leave alone, the plastic surgeon is at substantial risk of suit from the parents if he operates without parental knowledge [6, 45].

In an emergency, any child, no matter how young, may be treated without parental consent [27, 41]. An "emergency" is any condition which requires prompt treatment, and does not mean solely a condition which may cause death or disability. If a four-year-old is brought to the Emergency Room by a 12-year-old baby sitter because he has cut his foot, if immediate attempts to locate a parent are unavailing, it is perfectly legal to sew up the child's foot. In fact, I suspect that it is more likely that a suit would arise from letting a child lie around in an emergency room in pain because a surgeon could not find a parent and was too afraid of a suit to take care of the child, than it is that a parent would think of seeing a lawyer because a compassionate physician took care of a hurt or frightened child.

The emergency exception to parental consent would also cover the situation where a child or young adolescent presents for treatment of a relatively minor illness such as a sore throat or earache. Although my pediatrician friends tell me that ten-year-olds in fact do not come to the doctor by themselves, if one did, as long as the therapy presented few risks and the child understood the problem, if a parent could not be located it would be perfectly all right to provide treatment.

To proceed to the more usual situation, courts in a great many contexts are permitting adolescents much more decisionmaking autonomy than they had in earlier times. Society has changed within the past twenty-five years and teenagers are much more independent in all areas of their lives. Courts have not been unaware of this.

## I. THE EMANCIPATED MINOR

For at least two hundred years courts in the Anglo-American system have recognized the concept of an "emancipated minor." This means one who is living on his or her own (in the earlier days of the law, it was on *his* own), is self-supporting, and is not subject to parental control. The old concepts of emancipation included minors in the military (of which there were numerous when the age of majority was 21) or married minors. The concept has evolved nowadays to include college students, even when parents are completely responsible for paying the bills, or unmarried minor mothers [3] – and in some states, a pregnant minor is considered completely emancipated. A runaway is also considered an emancipated minor, no matter how young he or she may be.

## II. MINOR TREATMENT STATUTES

Because physicians were concerned about the legality of treating teenagers, many states have enacted what are known as minor treatment statutes. These provide a specific age, usually 16 but in some states it may be as young as 14, at which a minor may be considered completely independent for health care purposes and treatment may be given as if he or she were an adult. All states now have statutes giving a physician the right to treat any minor for VD without parental knowledge. Almost all states permit treatment of any minor for drug or alcohol problems without parental knowledge. Most of the VD, drug, and alcohol statutes in fact forbid informing parents without the child's permission, since teenagers would not seek treatment for these problems without knowing that their parents could not find out about it. The North Carolina statute [32] provides that a physician may treat a minor of any age for VD, drug or alcohol abuse, pregnancy, or emotional disturbance without parental consent, although the statute explicitly excludes abortion, sterilization, or commitment to a mental hospital from its terms. I point out, however, that if a parent receives a bill, he or she will find out what the problem was.

Thus, even if the statute does not specifically forbid sending a bill to parents (as Connecticut's does) they should not get one if the confidentiality of the patient is to be preserved.

### III. THE MATURE MINOR

Even in the absence of statutes, there have been no cases reported within the past 30 years in which a parent successfully sued a physician for non-negligent care of an adolescent without the parent's knowledge [8, 26, 44]. Courts have in effect bowed to modern reality and decisions apply what has come to be known as the "mature minor" rule. In effect, the legal principle now applied is that if a young person (of 14 or 15 or over) understands the nature of proposed treatment and its risks and can give the same degree of informed consent as an adult patient, and the treatment does not involve very serious risks, the young person may validly consent to receiving it. In determining whether allowing a particular minor to make a particular decision is reasonable, therefore, the age and maturity of the patient must be considered, but the nature of the illness and the risks of the therapy are also very relevant [17, 39, 43]. A particular fourteen-year-old might be perfectly capable of giving informed consent to treatment of acne; no oncologist would consider treating the same 14-year-old for leukemia without his family's knowledge.

The issue arises in a real-world sense only in the context of ambulatory care. Parents are not financially responsible for non-emergency care to which they do not consent [2, 25, 37]. Since a teenager who wants expensive elective treatment (such as plastic surgery on her nose) is the sole person responsible for the bill if her parents do not agree to the operation [10], most hospitals will not admit a minor for non-emergency treatment unless the parent (who is almost always the only one with health insurance) signs a financial responsibility form.

### IV. REFUSAL OF TREATMENT

Thus, as we can see, in terms of general consent to medical care, there is a decided trend in the law toward greater autonomy for increasingly younger adolescents. Pre-adolescents, however, do not have such a right.

Where there is capacity and a right to consent to treatment, one must presume that there is a corresponding capacity and right to refuse. Competent adults certainly have a universally recognized right to refuse treatment they don't want, even if it is considered likely to save their lives. While there are few, if any, cases on a minor's right to refuse treatment outside the context of abortions, I would predict that any court would rule that a minor has the right to refuse any treatment to which he has the right to consent. Thus if a 15-year-old could present himself to an out-patient clinic and be treated for acne or a sore throat without parental permission, he could refuse to accept therapy for those same conditions if his mother brought him to the physician. In fact, the teenager's right to refuse completely elective treatment is probably greater than his or her right to consent. If, for example, a surgeon refuses to accept a teenager for plastic surgery without parental consent, which would undoubtedly be the case, even if the young person's mother brought him or her to the surgeon for purely cosmetic surgery, the adolescent would undoubtedly have the right to refuse.

On the other hand, where treatment is necessary, even if not life-saving, I doubt seriously that a minor has the right to refuse it. If a teenager has appendicitis and his parent consents to the surgery, I do not think that a court would consider it battery to take the minor to the operating room against his or her will. In situations where the minor's life is at stake, I have no problem with the idea that people have no right to refuse life-saving treatment until they are adults. Many 16-year-olds with bone cancer would probably feel that they would rather be dead than lose a leg; assuming parental consent (or a court order if the parent backs up the patient) I do not think any court would agree that the teenaged patient has the right to make that decision.

I might add that in the context of cancer chemotherapy at Yale we have an impressive number of teenagers who want to stop treatment before the physicians think they have had a complete course. These tend to be cases (such as chemotherapy following surgery for Wilm's tumor) in which the prognosis with therapy is excellent. Either the patients don't feel all that terrible from the disease but the chemotherapy makes them sick and they don't like it, or they are embarrassed that their hair has fallen out. Pediatric oncologists learn how to deal with these situations so well that rarely do we have persistent noncooperation from a patient, but in the few cases in which we have had, I have had no hesitation in advising that the patients should be told that they are going to have chemotherapy whether they like it or not. In a different situation, where the likelihood of success from the chemotherapy is really quite limited and the patient is almost certainly going to die even if treatment is continued, of course, concern for the patient, regardless of age, means that his or her wishes should be respected.

To turn to specific situations in which the issue of a minor's consent or refusal of treatment is most likely to arise, I will discuss the minor in the mental health system, the minor and contraception, and the minor and abortion.

## V. THE MINOR IN THE MENTAL HEALTH SYSTEM

Although parent-child problems can surface at any age, adolescence is usually the time in which conflict between parent and child may end in the psychiatrist's office and then even in the courts. Collisions between teenager and parent may run the gamut from not doing homework on time to involvement with cults, drugs, and serious alcohol problems. Serious mental illness such as schizophrenia and depression may begin to manifest during adolescence.

### A. *The Minor in the Outpatient Psychiatric System*

Since minor treatment statutes and the mature minor rule apply to all forms of care provided by physicians, I assume that the same rules apply to an adolescent who goes to see a psychiatrist as apply to a visit to a dermatologist. If the psychiatrist wishes to treat the teenager, it is difficult to see how a parent could object to standard therapies. Since private psychotherapy is, however, extremely expensive, the issue is most likely to be raised when the teenager goes to a mental health center or community counseling service. Where the community facility is one in which a qualified psychiatrist is on staff and the other mental health professionals such as psychologists, nurse clinicians, or psychiatric social workers provide therapy as a team effort with physicians, I think there is no question that the minor has the right to seek treatment. There are, however, many so-called "mental health facilities" where no physicians are involved or available. In particular, many drug treatment centers provide therapists who are drug addicts themselves and no medical backup is available. This may be a viable method of providing drug treatment or other specialized therapies, but I do not think the legal right of a minor to seek medical care necessarily applies to such facilities. This is because of the potential harm that could result from such programs that fall outside the usual standard of practice.

The intent of legislatures and courts who broadened the adolescent's right to seek medical care was to get young people to physicians, not to well-meaning but non-medical counselors whose training may be questionable. To say that a minor's right to seek counseling is dependent on the type of license the counselor holds may seem an elitist view, overly weighted in favor of "mainline medicine," but the process of medical training and the training of clinical psychologists, nurse clinicians, and psychiatric social workers who work with psychiatrists does at least provide determinable and enforceable standards of care within a patient population whose ability to differentiate among the qualifications of caregivers may be quite unsophisticated. There are no cases of which I am aware on the point, but in summary, it is my opinion that an adolescent has the same right to seek treatment from a psychiatrist as he or she does from any other physician, and the care may be given by qualified non-physicians such as psychologists who work as part of "a treatment team" with the psychiatrist. North Carolina, incidentally, has the only state Attorney General's opinion I have ever seen which, in interpreting the minor consent statute, indicates explicitly that social workers and psychologists working with physicians may treat minors who may consent for themselves [31].

I think a teenager clearly does not have the right to seek counseling from a place where no one has any psychological credentials, such as a drug-addiction center run by addicts, and except in North Carolina, the law is not now clear about the teenager's right, without parental consent, to seek counseling from qualified psychologists, social workers, and nurses who operate in a practice not shared with a physician.

The issue of refusal of outpatient mental health treatment which the parent wishes the adolescent to have, but the adolescent does not want, is a legally fascinating one but for very practical reasons is not an issue in the real world. If a parent takes a young person to a psychiatrist and the patient does not want to cooperate, all he or she has to do is to refuse to talk to the psychiatrist. It is physically possible to hold someone down and administer medical care if one must do so; it is not possible to engage in psychotherapy with someone who refuses to talk. (Presumably those adolescents who are sufficiently ill to require electro-convulsive shock therapy or medications and will not cooperate require hospitalization.)

### B. Commitment of the Minor Patient

Adult mental patients have certain legal protections against unjustified or malicious commitments to mental hospitals that minors do not necessarily have. Adults may voluntarily go to a mental hospital and, if they wish to leave, may do so unless the hospital concludes that they are so ill that they may be involuntarily committed. In that case, they have a right to a judicial hearing. Commitment as an involuntary patient, if one is an adult, may only follow a judicial hearing at which one is, among other things, entitled to counsel. The standard of commitment of an adult is that he or she is dangerous to self or others or is gravely mentally disabled, meaning that the person is too mentally ill to provide himself or herself with the basic necessities of life [33].

Minors, however, fall into an entirely different category. Many states allow "voluntary" commitment of minors by their parents, which means that although the young person cannot leave the hospital, he is considered a "voluntary" patient and has no right to a hearing at which a judge can rule on the merits of the commitment. If all families were healthy and happy, that might be a reasonable solution to the problem, but in any number of instances, minors have been "dumped" into mental hospitals when the parent was probably the one who needed treatment [23].

Nowadays it seems that frightened parents, probably in good faith in most instances, are having teenagers in record numbers admitted to for-profit drug and alcohol hospitals and treatment centers without very clear indications that in-patient treatment is required. Moreover, if recent media reports are correct, there are some centers where the quality of care and the training of therapists is open to very serious question [1].[1]

Beginning in the 1960s, after a fair number of young people were quite literally committed to mental hospitals by parents who were upset with their anti-war protests, long hair, and general hippiness, most states did enact statutes providing for judicial intervention in the commitment process when minors were patients. In other states, courts ruled that these minor patients had the right to some constitutional protections. In North Carolina, for example, it was held in a 1975 case that a 15-year-old boy who had been committed by his mother had a right to a hearing within 72 hours of admission, although the young person could still be signed in by a parent in an emergency [22].

Many of the states' new statutes provided that a child under thirteen or fourteen could be admitted by parents without judicial intervention. Most, however, specified that minors over that age might continue to be admitted by their parents, but once in the hospital, the patients had a right to a prompt hearing with the assistance of counsel if they wished to object to their admission.

On June 20, 1979, however, the Supreme Court ruled [34, 42] that if a state legislature did not choose to enact procedural rights for minor mental patients, a minor's federal constitutional rights were not violated if his parents admitted him to a mental hospital over his objection.

Thus the rights of children and adolescents committed to mental hospitals are, at this time, exclusively derived from state law and no federal constitutional protections are given them.

## VI. THE MINOR AND CONTRACEPTION

Beginning in 1965, the United States Supreme Court struck down state statutes making the prescription or use of contraceptives a criminal offense [15]. In 1972 the Court held that the constitutional right of privacy, which precluded state intervention in this matter, applied to unmarried adults as well as married ones [13]. In 1977 the Court considered a New York statute that made it a criminal offense to sell non-prescription contraceptives to minors and held that minors do have a right of privacy that extends to contraception [9].

In that case, Mr. Justice Stevens wrote one of my favorite passages about the use of the legal system to enforce what someone thinks is "proper" conduct by young people:

Although the state may properly perform a teaching function, it seems to me that an attempt to persuade by inflicting harm on the listener is an unacceptable means of conveying a message that is otherwise legitimate ... It is as though a State decided to dramatize its disapproval of motorcycles by forbidding the use of safety helmets. One need not posit a constitutional right to ride a motorcycle to characterize such a restriction as irrational and perverse ([9], p. 715).

Title X of the Public Health Service Act of 1970 has a provision stipulating that federally funded family planning agencies are required to provide services without regard to age or marital status [38].

On January 26, 1983, however, the Department of Health and Human Services published regulations requiring federally-funded family planning facilities to notify a minor's parents within 10 days of her visit to the facility that she has been given contraceptives. This was known as the "squeal rule" and had provoked more than 120,000 comments when published as proposals in 1982. The regulations were scheduled to take effect in February, 1983, but New York State filed one suit and Planned Parenthood filed another in the District of Columbia. In both cases the courts found the rules unconstitutional, so they were never implemented.

Thus, it is finally clear that a teenager who requests contraception has a constitutional right to have it, if she is seeking it at a federally funded facility.

A private physician has the right to refuse to provide it, but if the girl is a regular patient, he or she may have an obligation to refer her to Planned Parenthood or a public clinic.

One thing we should remember if we conclude that parents should have a voice in a girl's decision to obtain contraception – if they veto that choice and she has a baby (except in Wisconsin which enacted a "grandparent responsibility" statute a few months ago), her parents are not responsible, financially or otherwise, for their grandchild. In many states in which a girl is emancipated by operation of law when she becomes a mother, her parents are perfectly free legally to evict her and her baby from their home. Thus if a teenage girl's parents are allowed to determine her access to contraception, it has always seemed to me that the law should require them to assume some responsibility for the result if they refuse.

## VII. THE MINOR AND ABORTION

The Supreme Court, of course, in 1973 struck down state laws making abortion a criminal offense [12, 40]. The court held that statutory restrictions on first trimester abortions were unconstitutional except that, like any other surgical procedure, it was legitimate to require that they be performed by a licensed physician. As soon as those cases were decided, many states enacted statutes requiring parental consent before a minor could obtain an abortion. In a series of cases the Supreme Court held that a statute that required parental involvement without an alternative was an unconstitutional violation of the girl's right of privacy [4, 5, 7, 11, 16, 35, 36].[2]

What the Supreme Court seems to have done is to hold that state statutes will be constitutionally acceptable if they provide that a girl has access to a judge. The judge's role is limited to deciding if she is mature enough to make an informed decision about the abortion, if she does not choose to involve her parents. State statutes vary widely. In all of them, however, if an adolescent can pay for an abortion, a not inconsiderable barrier to her access to one, she may not be denied one simply by reason of parental veto.

A situation my pediatrician friends tell me is much more frequent than many people would believe is the one in which a mother brings her pregnant teenager to a gynecologist because she thinks the girl should have an abortion, but the girl wants to have and keep her baby. Only a few of these cases have been litigated, probably because it does not reasonably occur to most physicians to perform abortions on people who do not want them, so the issue never gets to a court, but all courts that have considered the matter have concluded that the same right to privacy that allows a woman of any age to conclude that she wants an abortion will protect a decision to give birth [19, 24].

In dealing with a sexually active minor, of course, physicians and other health care professionals should urge the girl to talk to her parents. If she refuses, however, there is solid constitutional support for providing her with contraception and, if there is no state statute to the contrary, in allowing her to choose to have an abortion.

Times have changed. Adolescents are, throughout various aspects of their lives, exercising many more freedoms than middle-aged people ever dreamed of having as teenagers. The difference between being an adolescent in the 80s and having been one in the 50s is not only that the world is a different (and perhaps more threatening) place nowadays, it is that not only one's antediluvian parents but one's antediluvian health care providers seem to have forgotten what it was like to be teenager. It isn't that we don't remember, it is that we remember a different world. The role of young people during the Viet Nam War, among other factors, forever changed institutional and public perceptions of their maturity. They now vote at 18. The legal system has basically bowed to reality in recognizing the rights of young people in the 1980s.

*Yale University School of Medicine*
*New Haven, Connecticut*

## NOTES

[1] For example, there was a 20-minute segment about parental admissions of adolescents to these facilities on ABC-TV's program "20/20" [1].
[2] An excellent discussion of the law on this point, although it was published before the 1983 Supreme Court decision, is [7].

## BIBLIOGRAPHY

1. ABC-TV's program '20/20', August 21, 1986.
2. *Accent Service Company v. Ebsen*, 306 NW 2d 575 (Neb 1981).
3. *Bach v. Long Island Jewish Hospital*, 267 NYS 2d 289 (NY 1966).
4. *Bellotti v. Baird*, 424 US 952 (1976).
5. *Bellotti v. Baird*, II, 443 US 622 (1979).
6. *Bonner v. Moran*, 126 F 2d 121 (CA DC 1941).
7. Buchanan, E.: 1982, 'The Constitution and the Anomaly of the Pregnant

Teenager', *Arizona Law Review* **24**, 553–610.
8. *Carter v. Cangello*, 164 Cal Rptr 361 (Cal 1980).
9. *Carey v. Population Services International*, 431 US 678 (1977).
10. *Cidis v. White*, 336 NYS 2d 362 (NY 1972).
11. *City of Akron v. Akron Center for Reproductive Health*, 462 US 416 (1983).
12. *Doe v. Bolton*, 410 US 170 (1973).
13. *Eisenstadt v. Baird*, 405 US 438 (1972).
14. Gaylin, W. and Macklin, R. (eds.): 1982, *Who Speaks for the Child? The Problem of Proxy Consent*, Plenum Press, New York.
15. *Griswold v. Connecticut*, 381 US 479 (1965).
16. *H. L. v. Matheson*, 450 US 398 (1981).
17. Hofmann, A. D. and Pilpel, H. F.: 1973, 'The Legal Rights of Minors', *Pediatric Clinics of North America* **20**, 1001–1008.
18. Holder, A. R.: 1985, *Legal Issues in Pediatrics and Adolescent Medicine*, Chapter 5, 2nd Edition, Yale University Press, New Haven.
19. *In the Matter of Mary P.*, 444 NYS 2d 545 (NY 1981).
20. *In re Anonymous*, 248 NYS 2d 608 (NY 1964).
21. *In re G.*, 104 Cal Rptr 585 (Cal 1972).
22. *In re Long*, 214 SE 2d 626 (NC 1975).
23. *In re Sippy*, 97 A 2d 455 (DC Mun Ct App, 1953).
24. *In re Smith*, 295 A 2d 238 (Md 1972).
25. *Ison v. Florida Sanitarium and Benevolent Association*, 309 So 2d 200 (Fla 1974).
26. *Lacey v. Laird*, 139 NE 2d 25 (Ohio 1956).
27. *Luka v. Lowrie*, 136 NW 1106 (Mich 1912).
28. Melton, G. B., Koocher, G. P., and Saks, M. J. (eds.): 1983, *Children's Competence to Consent*, Plenum Press, New York.
29. Morrissey, J. M., Hofmann, A. D., and Thrope, J.C.: 1986, *Consent and Confidentiality in the Health Care of Children and Adolescents: A Legal Guide*, The Free Press, New York.
30. Munson, C. F.: 1981, 'Toward a Standard of Informed Consent by the Adolescent in Medical Treatment Decisions', *Dickenson Law Review* **8**, 431–454.
31. North Carolina General Statutes, § 90–21.1, 47 N.C.A.G. 83 (1977).
32. North Carolina General Statutes, § 90–21.5.
33. *O'Connor v. Donaldson*, 422 US 563 (1975).
34. *Parham v. J. R.*, 442 US 584 (1979).
35. *Planned Parenthood Association of Missouri v. Danforth*, 428 US 52 (1976).
36. *Planned Parenthood of Kansas City, Missouri v. Ashcroft*, 462 US 476 (1983).
37. *Poudre Valley Hospital District v. Heckart*, 491 P 2d 984 (Colo 1971).
38. Public Health Service Act, PL–91–572, 84 Stat. 1504, 42 *USC* Section 300–300a (8), 1970.
39. Rait, G. E., Jr.: 1975, 'The Minor's Right to Consent to Medical Treatment', *Southern California Law Review* **48**, 1417–1456.
40. *Roe v. Wade*, 410 US 113 (1973).
41. *Sullivan v. Montgomery*, 279 NYS 575 (NY 1935).
42. Tiano, L. V.: 1980, *'Parham v. J. R.:* '"Voluntary" Commitment of Minors to

Mental Institutions', *American Journal of Law and Medicine* **6**, 125–149.
43. Wadlington, W. J.: 1978, 'Minors and Health Care: The Age of Consent', *Osgood Hall Law Journal* **11**, 115–125.
44. *Younts v. St. Francis Hospital and School of Nursing*, 469 P 2d 330 (Kans 1970).
45. *Zaman v. Schultz*, 19 Pa D & C 309 (1933).

ROBERT L. HOLMES

# CHILDREN AND HEALTH CARE DECISIONMAKING: A REPLY TO ANGELA HOLDER

The question of whether children should be given a role in decision-making concerning their own medical care is part of a broader question of how children should be treated in general. And for answering that we lack a clear and coherent framework. Falling as they do somewhere between adult humans and animals in terms of their capacities, but normally having from the moment of conception the potential to become adults, children constitute a category of living things unlike any other from the standpoint of the moral and conceptual problems their treatment raises. Improperly cared for, they tend to grow up incapable of relating normally and happily to other adults; overly cared for and kept dependent, they tend to grow up incapable of relating normally and happily to other adults. Few can agree what constitutes just the right kind and quality of care.

Historically, the problem lies with traditional neglect of the moral issues concerning the treatment of children. It has been said that the concept of childhood did not even exist prior to the 17th century. Children were long regarded as mere property of their parents, to be disposed of at birth if they were unwanted (often at the pleasure of the father, but sometimes of the state as in ancient Sparta), or abandoned to whatever fate might befall them. Psychiatrist Henry F. Smith writes that when statistics began to be kept in France in the 18th century, it was determined that between 1773 and 1790 close to 6,000 children were abandoned each year in Paris alone, not all of them by any means children of the poor [2]. Frequently they were sold into prostitution or sold to beggars who would often cripple them so as to make them more effective in eliciting alms from passersby.

Many of the abuses of children over the centuries are no longer moral problems today. Those abuses can be seen to be wrong. While there may be questions about how precisely to eliminate them, there is agreement that it should be done. More interesting is the area in which it is unclear precisely how children should be treated, and in which sensitive and concerned people may reasonably disagree.

One such area concerns the extent to which children should be treated as autonomous agents, competent to make basic decisions concerning their life and well-being. This is an area in which children come into conflict with their parents. Many of the issues faced by the courts, as Angela Holder makes clear, concern just this problem. When these issues do get into the courts, there are three principal parties involved: the child, the parents, and the government (state or federal).

For parents there is a genuine dilemma. On the one hand, they are expected to be responsible for the well-being, and in varying degrees, for the very conduct of their children. There is a morally grounded paternalistic concern for the physical and mental well-being of children as well as for their moral well-being. On the other hand, as children grow older a new and conflicting interest comes into existence, set against the first one. It is that of the child to be autonomous in its decisionmaking. The child, for its part, particularly during the teen years, when it gets one foot in the door of adulthood, usually wants as much freedom as it can get. And the common measure of its success in getting it is the extent to which it can chart a different course from that wished for it by its parents. Thus responsible parents find themselves caught in the middle, between guiding and supervising their children, on the one hand, and letting them enjoy more freedom of choice, on the other.

Any social and legal system that values the family as ours does tacitly acknowledges that in undertaking parenthood people have the right to have their values and conception of the good, as well as their religious and moral views, predominate in the raising of offspring. Otherwise we should have the state involved in all of these areas, as Plato thought it ideally should be, from an early age. This right is not an imprescriptible one, of course, and there are clear limitations to the ways parents may treat their children. But the latitude is great. And families that are viewed as happiest are most often those in which there are close bonds and shared values.

So there are two moral issues here: the first concerning the degree of autonomy parents should grant their children, the second concerning the degree of justifiable intrusion by the state into the relationship between parents and children. The first is largely a question of autonomy versus paternalism in the care of children. The second involves both that issue and the issue of the state's paternalistic interest in the well-being of children within its borders. It also involves constitutional issues, such as privacy, which apply to adults as well as children. Both highlight aspects of the question of how children should be treated, and both are central to the specific problems regarding children's health-care decisions discussed by Holder, namely, mental health, contraception, and abortion.

Holder characterizes the law as it presently exists in these areas. But I am unsure whether she thinks it is good law and reflects the proper balance between autonomy and paternalism or among the interests of parents, children, and the state. I shall pose three questions arising out of what I take to be her understanding of the law, and a fourth which concerns parental responsibility in circumstances in which the parents veto the use of contraceptives by their child.

The first problem concerns the consistency of the law in two of the areas which have been discussed. On the one hand, we are told, many states allow so-called "voluntary" commitment of children by their parents, "which means that although the young person cannot leave the hospital, he is considered a 'voluntary' patient and has no right to a hearing at which a judge can rule on the merits of the commitment" ([1], p. 167). Apparently many of these commitments have arisen from parental concern over behavior like that characteristic of many anti-war protesters in the '60s – long hair and general "hippiness." Today it is more often a concern with drug and alcohol use. And we are told that it has frequently happened that "minors have been 'dumped' into mental hospitals when the parent was probably the one who needed treatment" ([1], p. 167). Children who are committed by their parents thus have no constitutional protection, and are entirely under highly variable state laws.

On the other hand, minors, legally speaking, have a right to privacy which extends to contraception, so that "it is finally clear that a teenager who requests contraception has a constitutional right to have it if she is seeking it at a federally funded facility" ([1], pp. 168–169). As the point usually of seeking contraceptives is to allow sexual activity without the attendant risks, and as sexual activity among teenagers is one of the areas parents traditionally have been sensitive about, often for reasons deeply-rooted in religious and moral convictions, this means that the state is prepared to intrude into the parent-child relationship in these circumstances in favor of the child. If, as has been recently stated in the controversy over prescribing and dispensing contraceptives in health clinics in New York City schools, sex is a health-care issue, this means further that the state is prepared to wrest from parents the final decision in this area of health care when parents and children differ about what is in the children's best interest.

Parents can commit their children involuntarily to mental hospitals, but they cannot have the final legal say in whether they have access to contraceptives. Is this consistent from a moral point of view? If a minor is to have the right to determine the nature and extent of his sexual activity, then why not the same with many of the sorts of conduct which apparently lead to the commitment of minors to mental hospitals? Or, on the other side, if the law allows parents involuntarily to confine their children in circumstances of the sort Holder describes, but effectively prohibits them from attempting to regulate their children's sexual activity in the areas indicated, does this make sense?

Second, I would like to question one aspect of the "mature minor" principle. As characterized by Holder, it says "that if a young person (of 14 or 15 or older) understands the nature of proposed treatment and its risks and can give the same degree of informed consent as an adult patient, and the treatment does not involve very serious risks, the young person may validly consent to receiving it" ([1], p. 163). While I cannot be certain from her characterization, I take it this means that a minor can seek out and receive such treatment without the parents' knowledge. If this is the case, and if such an implication is to be regarded as following from the mature minor principle, then the principle needs further justification. It is one thing to have the competence to consent to medical treatment; it is another to have the right to determine whether you will have that treatment; and yet another, if one has the second right, to have the right to choose the treatment without the knowledge of one's parents. And, we might add, if one has a legal right in all of these areas, it is another thing still to have the moral right. On the face of it this marks an abridgement of the role of parents in raising children by depriving them of knowledge of medical treatment given their children even when they do not have a legal right to prohibit the treatment.

Third, with regard to dealing with sexually active minors, Holder says that "... physicians and other health care professionals should urge the girl to talk to her parents. If she refuses, however, there is solid constitutional support for providing her with contraception and, if there is no state statute to the contrary, in allowing her to choose to have an abortion" ([1], p. 170). She also says that courts that have considered the matter have concluded that "... the same right to privacy that allows a woman of any age to conclude that she wants an abortion will protect a decision to give birth" ([1], p. 170).

Here Holder does express a clear personal opinion. She says that if a teenage girl's parents are allowed to determine her access to contraception "it has always seemed to me that the law should require them to assume some responsibility for the result if they refuse" ([1], p. 169). As things stand, if a minor becomes pregnant and has a baby (except in Wisconsin), "her parents are not responsible, financially or otherwise, for their grandchild. In many states in which a girl is emancipated by operation of law when she becomes a mother, her parents are perfectly free legally to evict her and her baby from their home" ([1], p. 169).

This is a difficult issue and I'm not sure where, on balance, I come down on it. But let me suggest another side to it.

Becoming a parent is one of the most momentous events that can occur in a person's life. There is virtually no other situation in the ordinary course of things when one comes to have as close to absolute control for many years over the life of another human being. Moreover, in virtually no other situation does one come to acquire the same set of moral responsibilities for the emotional, psychological, and physical well-being of another person. For these reasons, the decision to become a parent should be attended by profound reflection on its significance. This should also be the case for sexual activity which may result in parenthood. This, however, rarely happens, and certainly not with teenage pregnancies. So if parents do not want their 12- or 13-year-old daughter engaging in sexual activity, and if in addition they have strong opposition to abortion, it seems to me to require a strong justificatory argument to defeat the presumption that the state should not interfere in their attempt to raise their child according to their best moral and religious judgments.

If the state does intrude into the family situation, saying that the child has a constitutionally protected right to contraception, or to abortion or to give birth if she chooses, such intrusions seem to make sense only on the assumption that the child has the maturity to make these decisions and to enter into motherhood if she chooses. Otherwise she should be under the guidance of her parents. But if she has that degree of maturity, and if she does become pregnant, perhaps the assumption should be that she has voluntarily chosen this as an autonomous moral agent. Then, if the parents choose not to support yet another child whom they did not want and did not bring into the world and which was conceived through activities which they forbade, it is unclear why they should be required by law to be responsible for that new life. Loving and caring parents, I should hope, would do this; or at least would not evict their child and her baby from the home. But if they do not, I do not see the justification for their being compelled by law to do so.

While one may view it that if the parents refuse to consent to contraception for their daughter, her subsequent pregnancy is a result of their action, another way if viewing it is that it is the result of the *daughter's* action, including her refusal to abide by parental counsel. The very reasoning by which the child is tacitly deemed responsible enough to be sexually active, to have access to contraception, and to have an abortion or to give birth as she chooses, would seem to argue that she is responsible and mature enough to look after the child herself, or at any rate, not to expect the state to compel her parents to support it. (I would in any event suggest that if the parents of a pregnant girl who vetoed her use of contraceptives are to be held financially responsible for the resultant offspring, then the parents of the boy by whom she has become pregnant should share in that responsibility).

On the other hand, in the sort of case in which the state has not conferred the right to have access to contraception on the child, along with the right to give birth if she chooses, then I agree with Holder that parents should bear some responsibility for the results if they refuse to allow their daughter to have access to contraceptives. For then the very reasoning by which it is argued that she is insufficiently responsible to be sexually active, which will typically be the parents' reason for denying her such access, will argue that she is insufficiently responsible to look after her child on her own. Just as parents are quite properly often held accountable for mischief or harm caused by their children, so they may be deemed to share responsibility in this instance of a child's violation of their own prohibitions regarding sexual activity.

This is not to say that parents may not often make bad decisions in these situations, or give bad advice. Or that they may often give no guidance at all, which often is the case and part of a larger problem of parental neglect. That they may do this goes without saying. But that they *may* (and often enough do) do this does not mean that they need to do it, nor does it mean that they always do this. To give the state the right to intervene between parents and child under these circumstances is to give it the authority to intervene both where the interests of the child, and perhaps those of the parent-child relationship as well, are served, as well as in those cases where it is not.

Laws intended to protect the child who is neglected or abused will "protect" also the defiant, irresponsible child who wants only to thwart the very constraints responsible parenting is expected to impose. Does the state (meaning state or federal government) have the wisdom to produce outcomes of these dilemmas which are better on balance, morally, all things considered, than those which currently result? Perhaps, but that, too, is far from obvious.

Perhaps the state would do better to put its effort into providing an educational system that would enable children to appreciate what responsible parenting is, in the hopes thereby of helping them to avoid situations which put both them and their parents in these difficult situations. For the children of these same boys and girls are the very ones who will be at the center of identical problems in a few years (no more than 11 or 12 years, in some cases), as they seek help for unwanted pregnancies. And it will be *their* mothers (and fathers, in some cases) – the children whose problems are now at the center of attention – who will have failed them because of irresponsible parenting. This is the longer-range problem, and one that I can foresee being adequately dealt with only by what is arguably within the proper domain of the state's activity, education.

*The University of Rochester*
*Rochester, New York*

## BIBLIOGRAPHY

1. Holder, A.: 1989, 'Children and Adolescents: Their Right to Decide About Their Own Health Care', in this volume, pp. 161–172.
2. Smith, H. F.: 1984, 'Notes on the History of Childhood', *Harvard Magazine* **86** (6), 64A–64H.

DAN W. BROCK

# CHILDREN'S COMPETENCE FOR
# HEALTH CARE DECISIONMAKING

What role should children play in decisionmaking about their health care?[1] The doctrine of informed consent requires that medical treatment only be given to adults with their competent, informed and voluntary consent. The law, as well as common medical practice, presumes that adults are competent to decide about their medical treatment, though this presumption can be rebutted in particular instances. On the other hand, the law presumes that minors, who in most states are persons below the age of 18 years, are *not* competent to decide about their medical care; though with some exceptions to be noted below. For minors, the law generally holds that others, usually parents or guardians, are to decide for them about their medical treatment. The general presumption then in legal policy is that adults are entitled to decide about their medical care while children are not. However, medical practice, as we shall see, for good reason has often involved minors in decisionmaking about their care to a greater extent than the law seems to require.

One purpose of this paper is to address whether the general policy presumption of minors' incompetence for health care decisionmaking is sound, or whether it ought to be revised. Questions about the competence of both adult and minor patients arise not just at the level of general legal policy, however, but also in concrete circumstances concerning a particular patient's decision for or against a specific treatment. The other broad purpose of this paper is to clarify the nature of judgments about the competence of a minor to make a particular treatment decision. Clarity on the nature and basis of competence judgments is important both for assessing whether current legal policy concerning minors' competence to make health care decisions is sound or ought to be revised and for managing the many cases in medical practice in which even current legal policy permits an assessment in the individual case of a particular minor's competence.

I note here that my concern throughout this paper is with minors' competence to decide about their medical therapy or treatment and not with their competence to participate in research. While much of the analysis presented here could be applied to minors' consent to research, research raises some distinct issues of its own that I do not address in this paper.[2]

There are both conceptual components and empirical components of the issue of children's competence. The conceptual components concern the nature of the concepts of competence and incompetence, and in turn of determinations about competence. The empirical components concern the degree to which children at particular ages possess the various decisionmaking capacities the conceptual analysis shows to be necessary for competence. I believe it is not unfair to say that much current legal and medical practice displays both considerable confusion about the conceptual issues concerning competence determinations, whether of children or adults, and an inadequate appreciation of the empirical evidence from developmental psychology about children's capacities. My discussion in this paper will be structured around the conceptual issues, integrating the relevant empirical data into the analysis at appropriate points. The argument and analysis are complex and I forego drawing conclusions for legal policy and medical practice until the analysis is completed. Readers who want a preview of these conclusions before setting out should consult the final section, Section VII of the paper.

There is one final preliminary point of considerable importance. I will be addressing the standards in particular cases and general policy for children's competence to decide about their health care *when they wish to do so*. Nothing in the analysis to follow implies that children should be pressured, forced, or even encouraged to decide about health care for themselves when they do not wish to do so but wish instead to have others, such as their parents, decide for them. If a child, for example, does not feel psychologically ready to take responsibility for a difficult treatment decision, then even if his or her other decisionmaking capacities appear well-developed it could be harmful to the child to have to take responsibility for the decision. Waiver by competent adults of their right to give informed consent and their transfer of that right to others is well recognized in medical practice, medical ethics, and the law, and deserves to be equally well recognized for competent minors.[3]

## I. THE CONCEPT OF COMPETENCE

Any person is competent to do some things and not other things. Thus, all judgments about a person's competence are task-specific. The task in question here is health care decisionmaking, but, for either children or adults, that still suggests too broad or global an understanding of competence. An adolescent under severe stress or suffering the effects of some medications may be incompetent, whereas when those stresses or effects are removed the adolescent may then be competent to make health care decisions. Competence thus varies over time with changes in the person's condition and so may be intermittent or fluctuating. Over longer time frames, children's competence will vary with developmental growth and change any judgment about a child's decisionmaking competence must be relative to both a particular time as well as the particular conditions in which the decision is to be made. The task in question must also be specified more precisely than decisionmaking, or even health care decisionmaking, since even a particular child at a specific time with specified decisionmaking capacities may be competent to make some health care decisions but not others.

Different health care decisions vary in the demands they make on the decisionmaker, for example, in the complexity of the information that must be understood or the balancing of risks and benefits of different alternatives that must be weighed. Where a number of closely competing and complex alternative treatments exist for a condition such as bone cancer, the understanding and reasoning required may be much greater than when there is only one standard relatively simple, low-risk, and clearly beneficial treatment for a less serious condition. Competence determinations, therefore, involve matching the capacities of a particular person at a particular time and under particular conditions with the demands of a particular decisionmaking task. Competence determinations are decision-specific because both the capacities of the person as well as the objective demands of the decision can vary substantially. Throughout this paper, when I speak of children's decisionmaking competence, I do not mean, of course, their capacity to make decisions on their own without the collaboration of health care professionals. What is in question, just as with adults, is their capacity, with the help of their physician, to understand the nature and consequences of alternative treatments sufficiently well to be able to give or withhold informed consent to a recommended treatment alternative.

Up to here I have appealed to the ordinary understanding of competence without elucidating it. For many cases this is adequate. With infants and very young children there is no question that they lack competence for all health care decisionmaking; perhaps, there might be no controversy as well that an exceptionally mature 17-year-old is competent to consent to a relatively simple and straightforward medical treatment imposing no significant risks. But adolescents and pre-adolescents constitute one of the largest and most important classes of patients of questionable or borderline competence for the health care decisions they commonly face. In order to clarify these cases of uncertain competence we require a more careful analysis of the competence determination.

## II. THE CAPACITIES NEEDED FOR COMPETENCE

What are the capacities needed, whether by a child or an adult, for competence in health care decisionmaking? There are many possible ways of breaking down what is needed and no one correct way, but at a quite general level I find it helpful to distinguish three broad sorts of capacities:

(1) capacities for communication and understanding of information;
(2) capacities for reasoning and deliberation;
(3) capacity to have and apply a set of values or conception of the good.

The first category of communication and understanding includes the various capacities that enable a person to become informed for a given decision. These include linguistic, conceptual, and cognitive capacities necessary for an understanding of the information relevant to the decision at hand. Combined with these must be sufficient life experience to appreciate the meaning of particular outcomes, what it would be or feel like to be in possible future states or to undergo particular experiences.

The second category of reasoning and deliberation, which obviously overlaps the first category to some extent, includes some capacity to engage in hypothetical or if-then reasoning necessary for drawing inferences about the consequences of making a particular choice and for comparing alternative outcomes. Some at least limited capacity to employ probabilistic reasoning about uncertain outcomes is commonly necessary, as well as the capacity to give appropriate weight to probable future outcomes in a present decision.

The third category – the possession of a set of values or a conception of one's good – supplies the standards by which to evaluate treatment alternatives and their various features as either benefits or risks, and to assign relative weights or importance to these features. The conception of one's good needs to be sufficiently consistent and stable at least to permit arriving at a treatment choice and maintaining that choice long enough for treatment to be carried out. The conception of one's good should also take account of one's future interests so that they will be given appropriate weight in one's present decisions, as well as reflect the predictable ways that one's values and goals will change over time. However, here as elsewhere, we must be careful not to demand more of children than we do of adults who are standardly deemed competent.

All of these capacities necessary for competent decisionmaking are possessed by different persons, as well as by a single person at different times, in different degrees. We will address shortly how the appropriate *levels* of capacities for competence should be determined, but it is fair to say that for most health care decisions normal adults generally possess these capacities in sufficient degree to be deemed competent. In particular cases, the effects of physical or mental illness, medications, alcohol, or other conditions may diminish or eliminate these capacities, either temporarily or permanently. Such conditions, however, represent impairments of otherwise normally sufficient capacities for decisionmaking competence in adults; this is, of course, the basis for the legal presumption of competence in adults. Children, on the other hand, begin life entirely lacking these capacities and then progressively develop them from the very earliest stages of childhood through adolescence. The presumption of decisionmaking incompetence of minors would seem to reflect the assumption that children generally lack these capacities in sufficient degree to be permitted to make their own medical treatment decisions. To what degree is that assumption justified?

## III. THE DEVELOPMENTAL EVIDENCE ABOUT CHILDREN'S DECISIONMAKING CAPACITIES

As a philosopher, and neither a developmental psychologist nor an empirical researcher, I have no new or original data or evidence to offer on children's decisionmaking capacities. What I shall do instead is to summarize very briefly a number of results from child development studies and theory that are relevant to general assessments of these capacities in children.[4] Though I will assign age ranges to the development of a number of capacities, I emphasize at the outset that these age references are only rough generalizations and that there is considerable variation among children, including variation outside the bounds of these age ranges, in the ages at which they reach particular stages in the development of the capacities needed for decisionmaking.

We can consider the capacities for understanding and reasoning together since the relevant empirical evidence and theory often does not distinguish clearly between them. At a minimum, the capacity for communication and understanding requires sufficient linguistic and conceptual development to enable the child to understand the semantic content of the information relevant to a particular treatment decision. As Grisso and Vierling conclude in their excellent review from a developmental perspective of minors' capacities to consent, "we have practically no systematic information regarding children's understanding of the meanings of terms likely to arise in situations in which consent to treatment is sought" ([10], p. 417). Physicians and other health care professionals often do use terms and concepts in decisionmaking contexts that are difficult or even impossible for many patients, whether children or adults, to understand. However, the question is whether it is possible to put information relevant to patient's treatment decisions in terms that children can understand. The principal understanding children need for most treatment decisions is not of technical medical data, but of the impact that treatment alternatives will have on their lives. While admittedly we lack systematic data on this, I believe there is little reason to hold that children who otherwise have the capacities to consent are barred from doing so by inadequate semantic understandings of necessary information. Even for difficult concepts like "death" there is probably an adequate understanding by early adolescence.

The major issues lie in the capacities I have lumped under reasoning and deliberation. As Grisso and Vierling note, among the capacities that may be needed for the reasoning process are the ability to sustain "one's attention to the task, ability to delay response in the process of reflecting on the issues, ability to think in a sufficiently differentiated manner (cognitive complexity) to weigh more than one treatment alternative and set of risks simultaneously, ability to abstract or hypothesize as yet nonexistent risks and alternatives, and ability to employ inductive and deductive forms of reasoning" ([10], p. 418). Developmental theorists posit that the degree to which children perceive the locus of control in a particular matter as internal and subject to their decision, or external and a subject of fate, will affect their attentiveness to the decision, awareness of its details, and the effort they will exert to gather information about it. Children below the ages of 12 to 13 are significantly more prone than older children or adults to see the locus of control as external to them. Lewis [16] found that children in the 6 to 9 age range often did not perceive themselves as deciding even when they were doing so. Role-taking skills are also thought to be necessary to enable a child to entertain a position presented to him by the physician as well as his own different position as both potentially valid, so that the alternatives can be weighed against each other. These skills are undergoing substantial development in the 8 to 11 age period, and are often quite well-developed by 12 to 14.

Perhaps most important for reasoning are several capacities Piaget [20, 21, 22] identified in what he called the formal operations stage of cognitive development. As characterized by Grisso and Vierling, "this stage includes the development of an increased cognitive capacity to bring certain operations to bear on abstract concepts in problem-solving situations." Included is the ability to "perform inductive and deductive operations or hypothetical reasoning at a level of abstraction that would be represented by many consent situations involving treatment alternatives and risks" ([10], p. 419). These capacities are related to children's understandings of the causation of disease and illness, which also appear to follow a developmental pattern. Beginning with quite magical views of causation of disease at around age 5, it is not until around age 12 to 13 that "most children begin to understand that there are multiple causes of illness, that the body may respond variably to any combination of agents, and that the host factors interact with the agent to cause and cure illness" [15]. Further, "emergence of the formal operations stage allows a child to become sufficiently flexible in thinking ... to attend to more than one aspect of a problem at once – for example, to entertain alternative treatments and risks simultaneously."[5] In the formal operations stage children's general problem-solving abilities increase markedly, as does their ability to consider novel data and to use logic in the solution of problems. With the usual qualifications about the variability with which different children reach a particular developmental stage, it is generally in the 11 to 13 period that the various skills and capacities of the formal operations stage appear in children.

What are some of the limitations of children related to their values or conception of their good? An important issue is whether their values adequately reflect their future interests. While children in the 7 to 13 age range have largely left the earlier magical stage of thinking and now view the world in concrete terms, they can have great difficulty anticipating their future ([11], p. 96). This can lead to two important problems: children may give inadequate weight in their valuing to the effects of decisions on their future interests and also fail to anticipate future changes in their values that may be predictable to others. Related is the instability in children's values in this period, particularly as they concern the child's future goals, and due in part simply to limitations in their experience, especially with adult rules. (Some features of this category of capacities for competence overlap the reasoning category, since limits in anticipating the future will result in prudential irrationality in reasoning when present effects receive disproportionate weight in comparison with future effects.) All of these factors, and others, underlie the concern about relying fully on the values or life plans of children below the age of 14. I reiterate again that the presence of these factors varies significantly among children of the same age, that, other things being equal, these limitations become progressively more prominent the younger the child, and that their impact can vary greatly depending on the nature of the treatment choice before the child.

In adolescents above the age of 13, a heightened concern with physical appearance in comparison with adults or children at earlier stages of development can have a substantial effect on treatment choices [15, 29]. It is important to avoid the mistake of judging such values mistaken or distorted simply because different from those of adults. Children quite properly will commonly have age appropriate values different from adults that reflect their current stages of development. Thus, for example, the effects of a treatment on the physical appearance of an adolescent may properly be given more weight than they might be given by an adult. The adolescent's incapacity may reside only in his or her failure to appreciate how the importance of that effect may often recede later as an adult.

As a very broad generalization, I believe the developmental evidence briefly summarized here supports the conclusion that children by age 14 or 15 usually have developed the various capacities necessary for competence in health care decisionmaking to a level roughly comparable to that attained by most adults. The policy implications of this are developed in the last section of this paper. However, there is an additional factor besides competence in decisionmaking that is necessary for valid informed consent – consent must be voluntary. Children's capacities to give voluntary consent also bear on whether they should have decisional authority about their treatment, and there is significant developmental evidence about these capacities as well. I have already noted above that whether children perceive the locus of control on a matter as internal to themselves or external in the environment affects their capacities for understanding and reasoning. This perception is relevant to voluntariness as well in its effect on children's understanding of whether the choice is really theirs or whether it rests with others, making opposition pointless. A related point is that "children (particularly young children) are unlikely to perceive of themselves as having rights ... children learn that they should 'obey thy father and mother – and anyone else bigger than they are'."[6] More generally, development theorists have observed that children below the age of 14 or 15 do not assert themselves well against authority figures, and of course the other parties involved in the consent process – parents and physicians – are likely to be perceived as strong authority figures. Indeed, children in the 11 to 13 age range have been found to be more conforming in their behavior than younger children of ages 7 to 9. On the other hand, for children in strong oppositional stages during later adolescence, the voluntariness of their *dissent* to the treatment recommendations of authority figures is in question. Grisso and Vierling conclude that below the age of 15 there is at least significant question about children's capacities to give voluntary consent.

## IV. DETERMINING A STANDARD OF COMPETENCE

As we have seen, in children even more commonly than in adults the capacities necessary for competence in health care decisionmaking are possessed by different persons, as well as by the same person at different times and stages of development, in varying degrees across a broad continuum. Does this mean that health care decisionmaking competence itself should be thought of as possessed in different degrees? In a great many contexts we do commonly speak of persons being more or less competent to perform any number of tasks. Nevertheless, because of the role the competence determination plays in the law, and in turn in medical practice, we should resist the notion that a person's competence to make a particular health care decision is a matter of degree, of more or less. What is that role or function of the competence determination in legal and medical practice?

The function of the competence determination can only be fully understood within the broader doctrine of informed consent of which it is a part. The doctrine of informed consent mandates that health care treatment only be given to a competent patient with that patient's voluntary and informed consent. The ideal underlying the informed consent doctrine is that of patients deciding, together with their physicians, what treatment, if any, will best serve their particular needs and aims. If the decision is not voluntary, but instead coerced or manipulated, it will likely serve others' ends, or others' views of the patient's good, not the patient's own view of his or her good. In that sense, an involuntary decision originates with the coercer or manipulator and not with the patient. If the decision is not informed, it will not reflect a sound understanding of how different alternatives might serve the patient's aims. Finally, if the patient is not competent the patient will be either unable to decide at all, or the patient's decisionmaking process will be seriously flawed and so less likely to result in a choice fitting the patient's aims and needs.

Within this very broad understanding of the informed consent doctrine, what more specifically is the function of the competence determination? For adults, who are presumed to be competent, the question is whether they should be declared incompetent for the decision at hand. The central consequence of the competence determination for adults is to sort them into two classes: (1) those whose informed and voluntary decisions about their health care treatment must be respected and accepted by others, and (2) those whose decisions may be set aside and for whom others will be designated as surrogate decisionmakers for them. This sorting function amounts to an "all or nothing" classification of persons regarding whether they are to have a right to make a particular treatment decision for themselves. The competence determination is *based on* matter of degree findings about a person's decisionmaking capacities and skills as brought to bear in a particular decision, but it is not *itself* a matter of degree determination. Competence is thus in this sense a threshold concept, not a comparative one.

Understood as a threshold concept performing this sorting function, the crucial question is *where* or at what *level* of decisionmaking abilities and performance on a particular decision this threshold separating the competent from the incompetent should be set. For adults, this is a question of how defective their decisionmaking capacities must be in a particular decision for the presumption of their competence to be overcome or rebutted, for them in turn to be found incompetent, and for another to be designated as surrogate decisionmaker for them. For children, because of the law's presumption of incompetence, the question is how good their decisionmaking abilities and performance must be on a particular decision to overcome or rebut the presumption of their incompetence and for them to be found competent to decide for themselves. How can the level of decisionmaking capacity required for competence be set in a non-arbitrary manner? The answer must be grounded in the values at stake in whether a person is permitted to decide about treatment for him or herself.

### The Values at Stake in the Competence Determination

For both adults and children there are two important values at stake: respecting their interest in self-determination while also protecting and promoting their health and well-being. As I will bring out shortly, both how a child's well-being is determined as well as the nature of a child's interest in self-determination are somewhat different from an adult's, but the values of both self-determination and well-being are at stake for adult and child. In each case these are patient-regarding values that are at issue – it is the *patient's* self-determination and well-being that are at stake in the determination of whether the patient will decide about his or her own treatment. I will discuss these values below, but I note first that the reason the competence determination can serve to sort adult patients into two groups – those whose decisions must be accepted as binding and those for whom others must decide – is that the patient's self-determination and well-being are the overriding values generally accepted to be at stake: there are no, or virtually no, substantial other-regarding values, that is, values serving the interests of others besides the patient, that are commonly asserted to be of sufficient weight to warrant lodging the decision with anyone other than the patient.

In decisions about health care treatment for children, however, there is a third substantial non-patient-regarding interest commonly asserted – the interest of *parents* in making important decisions about the welfare of their minor children. As a result, the determination of children's decisional authority must take account of this third interest as well, and so can only partially, but not completely, be brought within the competence determination. This is an important difference in the nature and role of the competence determination that has not been sufficiently appreciated in either the informed consent and competence literature or in health care practice. I shall examine briefly the two principal values at stake in setting a standard of competence for either children or adults, and then turn to the third value relevant to the allocation of decisional authority for the treatment of children, parents' concern to make important decisions about the welfare of their children.

## 1. The Child's Well-being

There is a long tradition in medicine that the physician's first and foremost commitment should be to serve the well-being of the patient, and this is no less so when the patient is a child instead of an adult. This tradition is compatible with the more recent informed consent doctrine if patients are assumed usually to make decisions they perceive will best promote their well-being and that, at least in general, competent individuals are better judges of their own well-being than are others. However, as is well-known and as physicians are often quick to point out, the complexity of many treatment decisions often strains patients' capacities for understanding. Moreover, the stresses of illness on patients that result in fear, anxiety, and dependency – not to mention the physical effects of illness itself – mean that patients' ordinary decisionmaking abilities are often significantly diminished in the setting of health care decisionmaking. Thus, patients' treatment choices may sometimes fail to serve their own good or well-being, even as they normally conceive their good. The protection and promotion of patients' well-being sometimes require patients to be protected from the harmful consequences of their own choices.

Besides the factors which often diminish the health care decisionmaking abilities of patients, whether adults or children, there are the more general developmental limitations discussed above in children's decisionmaking abilities. Other things being equal, and even in the best of circumstances for decisionmaking, the younger minors are, the less weight we give to their expressed preferences, to their views about their own good, in our judgment of what is best for them. We generally accept normal adults' settled aims and values as being ultimately determinative of their good or well-being, and so accept their choices as best for them except when their decisionmaking is defective in ways that result in their choices being incompatible with their settled aims and values.[7] The greater the limitations of judgment and experience of children and adolescents, the less weight we give to their expressed aims and values as ultimately determinative of their good. Thus, for example, a choice of a mature adult, based on an unusual but considered and settled conviction, such as a Jehovah's Witness' rejection of a blood transfusion, may be accepted as consistent with that person's overall well-being. A similar decision by a 9-year-old child, on the other hand, may not be considered consistent with the child's well-being because the child's aims, values, and commitments have not yet developed and endured to the point where we are prepared to accept them as the ultimate determinants of his or her well-being.

Unlike adults, a child's good is more fully determined by the developmental needs of children generally at that age than by his or her current but predictably transient goals and preferences. These developmental needs are based in significant part on the aim of preparing the child with the opportunities and capacities for judgment and choice necessary for exercising self-determination as an adult. A principal focus in judgments about children's well-being is on fostering these abilities and opportunities so that as adults they will be able to choose, revise over time, and pursue their own particular plans of life, or aims and values, now suited to the adults they have become.

Thus, in any particular society children need to develop certain general abilities and traits in order to be able, as adults, to enjoy a reasonable range of opportunities from among which to choose their own particular aims, pursuits, and life paths. This is why, for example, it is reasonable to require that all children receive some minimal level of education, as state laws generally do, that require children to attend school until they reach age 16; some minimal level of education is necessary to have a reasonable array of occupational opportunities and other pursuits to choose from as an adult. This is also why it would not be reasonable, however, to require that all persons pursue postgraduate education, since that is only necessary for the pursuit of a small subset of available alternative life plans; only those who choose particular pursuits like law or medicine need advanced education for them. Likewise, it may be reasonable to allow adults to assume a risk to health or even life because doing so is necessary to the particular life plan they have chosen for themselves. That may be life within a particular religion, as in the extreme case of the Jehovah's Witness who rejects blood transfusions thereby threatening his life or health; or life within a particular occupation such as a professional athlete who rejects the conservative treatment of amputation for bone cancer in the hopes of saving his limb and his career; or life defined by a particular social role such as the homemaker with rheumatoid arthritis who pursues long term steriod therapy with harmful consequences to her health because it is necessary for continued functioning in her homemaking role.

The general point is that children's well-being depends less on their current individual preferences and more on the objective conditions necessary to foster their development and opportunities than does the well-being of adults. One of the principal bases for involving adults in decisionmaking about their health care is to recognize this sense in which they are sovereign in the determination of their good. In the case of children, while this basis for involving them in decisionmaking is not absent entirely, it is substantially weaker than with adults.

One final point should be made about how children's well-being is at stake in whether they are permitted to participate in decisions about their own medical care. Involving patients, whether children or adults, in decisions about their treatment often promotes their well-being by increasing their level of compliance in cooperating with and carrying through the treatment.

## 2. *The Child's Self-determination*

The value of self-determination too constitutes a weaker basis for children's involvement in decisionmaking than it does for adults. By self-determination I mean people's interest in making important decisions about their life for themselves according to their own values and aims. It involves the capacities to form, revise over time, and pursue a plan of life or conception of the good. Having a conception of the good is more than merely having a set of desires, which young children possess as much as adults. Persons have a capacity for reflective self-evaluation, for considering what kinds of desires and character they want to have, what kinds of persons they want to be. There are, of course, limits to the extent to which persons can change their desires and character. In broad respects our natures are fixed and given to us by our biological nature and our environment, but within these broad limits we adopt particular values and create a unique self. In these ways people are capable of shaping their character and of taking responsibility for the kind of persons they are and will become. It is through the ongoing exercise of self-determination that persons become and are active and responsible agents shaping their lives and controlling their destinies.

From early childhood well into adolescence children are in the process both of developing the various capacities necessary for the exercise of self-determination and of beginning to exercise their imperfect capacities for it. An important developmental task of adolescence is the process of psychological separation from one's parents and the establishment of a sense of identity with some consistency and stability.[8] Moreover, the development of a conception of the good and of the capacity for responsible choice requires being allowed a certain freedom to choose for oneself; this freedom must include some freedom to choose badly, as the old adage that people learn best from their mistakes bears out. Nevertheless, the less well-developed the capacities for deliberation and choice required by self-determination are, the weaker are both children's interest in self-determination and the case for respecting either their desire to choose for themselves or their resultant choices.

In many cases children may have as strong a desire as adults to choose their medical treatment and to have the treatment they choose, but the basis for others' letting them decide and abiding by their decision is weaker. In part this is because one reason for respecting individual self-determination is that doing so tends indirectly to serve a person's well-being. If a person's well-being is subjective in the sense that it is ultimately determined by that person's subjective values and preferences, then in the exercise of self-determination in purposive, rational action persons pursue their good as they perceive it. Exercising self-determination will commonly result in the promotion of the person's well-being. This is the basis behind the common view that a person him or herself is the best judge of his or her own interests, though that is not, of course, to say that persons can never be mistaken about where their interests lie. Thus, when persons, whether children or adults, have seriously diminished or undeveloped decisionmaking capacities, the basis for believing that their choices accord with their well-being is weakened; there is less reason to believe that their well-being is indirectly served by respecting their self-determination.

The value many people ascribe to self-determination rests not just on its serving a person's well-being or interests, but also on a particular ideal of the person – we value being, and being recognized by others as, the kind of being capable of determining and taking responsibility for our own destinies. There is a dignity and integrity in being self-determining that is lost in even a satisfying subservience. Children who are in the process of acquiring rights of self-determination are often acutely sensitive to this point. It is this ideal of the person that is often expressed in patients' desires to make significant decisions about their lives for themselves, even if others (for example, physicians, parents, or even computers) might be able to decide for them at least as well if not better than they themselves could. Nevertheless, the plausibility of this ideal of the self-determining person assumes at least some minimal development of the various capacities necessary to its exercise.[9] The question of the current exercise of self-determination by beings or things that lack these capacities, such as cats and infants and stones, simply never arises. When these capacities for self-determination are possessed in substantially reduced degrees, as they are with developing children and adolescents, I believe the value of their self-determination and of their choosing for themselves is likewise diminished; more accurately, children are less able to realize the ideal and so the value it represents.

One other respect in which the importance of self-determination varies on different occasions is with the nature of the decision being made. For a particular person, whether child or adult, some decisions may bear directly on centrally important aspects of his or her life plan, for example choices that may force changes in a career path or violate strong religious commitments, while other decisions may have only minor or peripheral effects on the person's goals and values, whatever choice is made. The value or importance of one's self-determination being respected varies directly with how centrally and pervasively the individual's life plan will be affected by the choice being made.

An important part of children's and adolescents' interest in self-determination is not their interest while children in making decisions for themselves, but their interest as children and adolescents in developing the capacities later to be self-determining as young adults and thereafter. We have already noted this in discussing the application of the concept of well-being to children and adolescents. Looked at from this perspective, self-determination is principally, or at least significantly, to be achieved in the future by children and adolescents rather than exercised while still children and adolescents. At the same time, this process of the development of the capacities for self-determination requires that children be given progressively increased decisionmaking opportunities and responsibilities appropriate to their level of development and the seriousness of the choice.

We have now seen that though both the same values of well-being and self-determination are at stake in determinations of the competence of children as of adults, those values take different shape and importance with children. The very same trade-off for adults recurs with children – protecting their well-being from the harmful consequences of their choices when their decisionmaking capacities are defective must be balanced against respecting their interest in deciding for themselves when they are able. But what best serves children's well-being is determined less by their own preferences than for adults, while the value of their exercise of self-determination while children is also less than for adults. This means that, other things being equal, it is reasonable to employ a substantially lower threshold for intervention or setting aside a child's or adolescent's choice, as compared with an adult's, based on a conception of the child's well-being different from his or her own.

## 3. The Parents' Interest in Making Decisions Concerning Their Children

A third important value is commonly asserted in determinations of the decisionmaking authority to be accorded to children that is not asserted with adults, namely, the interest of parents in making important decisions about their children's welfare. What is the precise nature and basis of this purported interest? Several strains of reasoning can be identified in discussions of this issue.[10] One simply bases parents' decisionmaking role on the incapacity of children to decide for themselves, the consequent need for someone else to decide for them, and a general presumption that parents will usually do a better job of deciding than anyone else who could, as a general practice, be substituted for them. Because in most cases parents both care deeply about the welfare of their children and know them and their needs better than others do, they will be more concerned as well as better able than anyone else to ensure that decisions made serve their children's welfare. This argument accords to parents *no* independent interest or right to decide for their children and to enforce their choice even when it may *not* best serve their children's welfare. Instead, it makes the parents' claim to decide dependent on their deciding most in accordance with their children's welfare. This is essentially the standard argument for family members serving as surrogate decision-makers for incompetent adults; there too the surrogate has no independent right to make a choice that serves his or her interests or wishes as opposed to the wishes or interests of the incompetent patient.

A second line of argument appeals to the fact that parents must bear the consequences of treatment choices for their dependent children and so should have at least some control of those choices. Parents are held financially responsible for the costs of the treatment and bear as well some of the longer-term consequences, financial and other, of the particular choice made. As a result, it may be unfair to force them to bear the consequences of the treatment choice while denying them any input into it. Even if this argument is accepted, it would establish only an interest of the parents in having *some* input into and control over the treatment choice, but certainly not unlimited discretion. The child in nearly all instances bears the principal consequences of the choice, and so on this line of reasoning the child's interest should principally determine the choice.

A third line of argument claims a right of parents, at least within some limits, to raise their children according to the parents' own standards and values and to seek to transmit those standards and values to their children. Virtually no one today believes that children are simply their parents' property, to be done with as the parents wish. Rather, proponents of this line of argument claim that children begin life as a tabula rasa and only through a process of socialization and development will they acquire values, goals, and standards, together with sufficient experience and powers of judgment to warrant our respecting their choices. For this reason, the case for parents' discretion in treatment choices for their children who have not yet acquired stable and mature values of their own is more persuasive than the analogous case for family members deciding for now incompetent, but once competent, adults; previously competent adults, unlike children, once did have their own values and goals which can now guide others' choices for them. Someone must inevitably shape children's goals and values, and since we assign childrearing responsibilities in our society largely to the family, it seems reasonable to accord to the family as well some significant discretion in imparting its values to the children within it.

A related but distinct fourth line of argument appeals to various respects in which the family is a valuable social institution, such as its role in fostering intimacy; the family provides the most significant source of intimacy for many adults as well as the context in which children's own capacities for intimacy are developed. The family must have some significant freedom from oversight, control, and intrusion to achieve intimacy, and one aspect of this freedom or privacy can be the right, at least within some limits, to make important decisions about the welfare of its incompetent members.

## V. CHILDREN'S COMPETENCE AND CHILDREN'S DECISIONAL AUTHORITY

These various arguments for an independent right of parents to make health care treatment decisions for their children raise large and complex issues that cannot be pursued further here. That might seem to be no difficulty, in fact the reader might wonder why they have been taken up at all here, for what do they have to do with children's decisionmaking competence? I believe it is correct that these arguments do not directly bear on children's decisionmaking competence, but on the other hand they do bear directly on the issue of children's appropriate decisional authority. Consequently, the determination of children's decisionmaking competence cannot fully perform the function of settling their decisional authority, as the determination of adults' competence does. As already noted, when we set the threshold for competence of adults, and thereby decide whether their choices will be respected or instead a surrogate will decide for them, we balance protecting them from the harmful consequences of their choices against respecting their interest in making treatment decisions for themselves when they are able. Why does it make conceptual sense for this balancing to be part of the determination of whether the patient displays an appropriate level of decisionmaking competence?

Decisionmaking is a purposive activity in which people seek best to secure their overall aims. When a particular choice fails best to secure the decisionmaker's aims, given the knowledge and information available to the decisionmaker, this failure is itself evidence, in that instance, of reduced decisionmaking competence. One measure of how seriously defective a purposive process or activity is, is the degree to which it fails to secure its purpose or goal. Thus, the more the decisionmaker's overall aims or well-being seems to be jeopardized by his or her choice, the more serious the defectiveness or incompetence that can be presumed in his or her decisionmaking.[11] It is for this reason that the degree to which a choice is contrary to the decisionmaker's overall well-being, as determined by his own aims and values, can reasonably be taken to be a measure of the seriousness of overall decisionmaking incapacity in that instance. The threshold question of how seriously defective the decisionmaking and attendant threat to the decisionmaker's well-being should be to warrant protecting the decisionmaker for his or her own good must balance any other interest *the decisionmaker* has in whether this is done; the principal other interest is the patient's self-determination. The issue of paternalism – whether the decisionmaker should be protected from the potentially harmful consequences of his choice for his own good (as he determines it) – is settled in the assessment of his decisionmaking competence.[12] Essentially there are then two components of this assessment: first, the assessment of the seriousness of any incapacity in the decisionmaker's reasoning process, determined by the degree to which the decisionmaker's choice threatens his well-being; second, whether this threat is serious enough, in light also of the decisionmaker's self-determination interest in deciding for himself, to warrant for his own good transferring decisional authority to another by adjudicating him incompetent to decide for himself.

For adults, both legal and medical practice address this threshold question of whether decisional authority should remain with the patient, or be transferred for the patient's own good to another, as part of the question of the patient's competence to decide. Can the decisional authority of children be understood as a threshold determination of competence as well? In determining children's decisional authority, suppose not only their interests in exercising self-determination and protecting their well-being must be balanced, as with adults, but also the interests of their parents in determining the treatment choice according to the parents' standards. The parents' aims and values may be no part of the aims and values sought by the child, and so the failure of the child to decide in accordance with the parents' standards need bear *no relation at all* to any decisionmaking incapacity or incompetence of the child. This interest of the parents does not bear on the assessment of the *child's* decisionmaking competence, though it does bear on the distinct question of whether the child should have decisional authority. Only the values of protecting children's well-being and respecting their self-determination are relevant to a determination of their decisionmaking competence. If the parents' interest in determining the choice is to be given weight in whether to grant decisional authority to the child, then the assessment of the competence of the child can only be a part of, but cannot fully settle, that question of decisional authority. This is an important and inadequately recognized difference in the function or role of the competence determination in children and adults.

The child's competence will be even less determinative of decisional authority if some commentators are correct in claiming that there are still other substantial interests to be accounted for besides the three I have discussed, such as the state's interest in a healthy citizenry.[13] I assume, though I shall not argue the point here, that the interests of the child's well-being and self-determination, together with the interest of parental authority, form the predominant interests for virtually all cases concerning children's decisional authority about their medical care.

Despite the argument I have just made for distinguishing between determining a child's decisionmaking competence and determining his or her decisional authority, I will continue in what follows to address the question of the decisional authority about children's treatment in terms of setting the proper standard of competence. This way of framing the issue is too deeply entrenched in both medical and legal practice and policy to ignore, and for adults causes no difficulty. We must, however, not lose sight of the interests of parents in deciding for their children. Though these interests do not fit in a conceptually comfortable manner within assessments of children's competence, they are a proper part of that assessment when the competence determination is employed to settle children's decisional authority.

## VI. THE VARIABLE STANDARD OF COMPETENCE

We began discussion of the values at stake in the competence determination in the hope of clarifying the proper standard of competence. If in setting a standard of competence at least the three values of protecting the child's well-being, respecting his or her self-determination, and honoring legitimate parental authority are all at stake, and must often be balanced against each other, we can now draw several conclusions about the proper standard. First, there is no reason to believe there is one and only one objectively correct trade-off to be struck between these competing values, even for a particular decision under specified circumstances. Determining the proper trade-off goes beyond an empirical investigation of the child's decisionmaking capacities and is not simply a scientific or factual matter, but a value choice. The proper standard is thus an essentially controversial matter concerning the proper weight to be accorded these different values.

Second, while there may be no single, objectively correct balancing of these values, we can see that some standards of competence that have been proposed clearly fail to represent a reasonable balancing of them. At one extreme is a minimally paternalistic standard requiring only that the child be able to express a preference for some alternative. Since this standard would respect every expressed choice of the child, it fails to be any criterion of *competent* choice, and so fails either to provide any protection of the child's well-being against the harmful consequences of its decisionmaking incapacities or to give any weight to the parents' interest in deciding. By giving absolute weight to the child's present self-determination under all circumstances, it fails to balance the several values at stake. This standard is insufficiently paternalistic and fails to find incompetent some children who should be so judged.

At the other extreme are standards that are excessively paternalistic and that too readily and often find children to be incompetent. Perhaps the most important example of standards of this type are standards that look simply to the content or outcome of the decision and then apply some "objective" measure of the correct decision which is not based at all on the child's aims and values; for example, the standard that the choice be a reasonable one, what other rational persons would choose, or what the child's parents would choose. Using such standards, failure of the child's choice to match the allegedly objective standard of choice entails that it is an incompetent choice. Any such purportedly objective standard for the correct decision will ignore the child's own emerging and distinctive conception of the good and will substitute another's conception of what is best for the child; it will sanction so-called hard paternalism. Such standards maximally protect the child's well-being – according to the standard's conception of well-being – but fail adequately to respect the child's self-determination. Moreover, even such a standard's claim maximally to protect the child's well-being is only as strong as the objective account of the child's well-being on which the standard rests.

Objective standards of the "best" choice raise complex theoretical issues that cannot be explored here. For normal adults, I believe any standard of well-being that does not ultimately rest on their informed preferences is both problematic in theory and subject to intolerable abuse in practice. With children, however, there is more room in an account of their good for appeal to the objective conditions necessary for the development and preservation of the abilities and opportunities that will enable them later, as adults, to be self-determining agents with choices from among a reasonable array of life plans. The younger the child and the less well-developed and mature his or her values, the less weight others reasonably give to those values in their assessments of the child's good. Other things being equal, the younger the child, the stronger the basis for discounting his or her expressed values and substituting an opportunity-based objective conception of value. To the extent that children's good is defined in terms of developing the capacities and opportunities for determining the course of their own life as adults, their good also *cannot* be determined by the particular values and goals that their parents have chosen for themselves and might wish to apply to them.

An appropriate standard of competence, whether for adults or children, should focus on and address the *process* of their reasoning. The more mature the children's development of their values or conception of their good has become, the more the competence evaluation should seek to determine the extent to which their decisionmaking accurately assessed the available alternatives in terms of those values. Only by examining the process of children's reasoning can others determine whether and where children's choices may fail to secure *their own* aims and values, as opposed to fail to secure another's conception of what is best for them. The question remains for a process standard of competence: how defective should a child's decisionmaking be on a particular occasion for the child to be judged incompetent to make that decision? We can now see clearly the extremely important point that no single answer to that question is adequate. This is simply because both the effects on the child's well-being, as well as the importance of his or her exercise of self-determination, can vary greatly depending on the choice the child has made. The most important variable is how the expected consequences of the choice for the child's well-being may vary from clearly and substantially beneficial (for example, preventing serious, irreversible disability or loss of life) through trivial or negligible, to seriously harmful (for example, causing serious, irreversible disability or loss of life). The more adverse the expected consequences to the child of accepting his or her choice, the higher both the level of decisionmaking capacity and the certainty that level is attained that are reasonably required.

There is an important implication of this view that the level of competence required ought to vary with the expected benefits or harms to the child of acting in accordance with his or her choice – namely, that just because a child is competent to consent to a treatment, it does *not* follow that he or she is competent to refuse it, *and vice versa*. For example, consent to a low risk, life-saving procedure by an otherwise healthy child should require a minimal level of competence, while refusal of that same procedure by the child should require the very highest level of competence. This very same variability applies to adults as well.

Because the appropriate level of decisionmaking competence properly required for a particular choice must be adjusted to the consequences of acting on that choice, no single standard of decisionmaking competence is adequate for all cases. Instead, the level appropriately required for decisionmaking varies along a full range from low/minimal to high/maximal.[14] Table I illustrates this variation.

The presumed net balance of expected benefits and risks of the child's choice in comparison with other alternatives refers to the physician's assessment of the expected benefits in achieving the goals of prolonging life, preventing injury and disability, and relieving suffering from a particular treatment option as against its risks of harm. The table also makes clear that

TABLE I

| Presumed net balance of expected benefits and risks of child's choice in comparison with other alternatives. | Level of decisionmaking competence required. | Grounds for believing child's choice best promotes/protects own well-being. |
|---|---|---|
| Net balance of expected benefits and risks from child's choice is substantially better than for possible alternatives. | Low/minimal | Principally the benefit/risk assessment made by others. |
| Net balance of expected benefits and risks from child's choice is roughly comparable to that of other alternatives. | Moderate/median | Roughly equally from the benefit/risk assessment made by others and from the child's decision that the chosen alternative best fits child's conception of own good. |
| Net balance of expected benefits and risks from child's choice is substantially worse than for another alternative(s). | High/maximal | Principally from child's decision that the chosen alternative best fits own conception of own good. |

the relevant comparison is with other treatment alternatives, and the degree to which the net benefit/risk balance of the alternative chosen is better or worse than that for other treatment options. A choice might properly require only low/minimal competence, although its expected risks exceeded its benefits, because all other available alternatives had substantially worse expected benefit/risk ratios.

Table 1 also indicates, for each level of competence, the different grounds for believing that a child's own choice best promotes his or her own well-being. For *all* treatment choices of children, the persons responsible for deciding whether those choices will be respected should have grounds for believing that the choice, if it is to be honored, is reasonably in accord with the child's good and does reasonably protect or promote the child's well-being. When the child's level of decisionmaking competence is only at the low/minimal level, those grounds are only minimally that the child has chosen the option, but are principally others' positive assessment of the option's expected effects for the child's life and health. At the other extreme, when the expected effects of the child's choice for life and health appear to be substantially worse than an available alternative, the requirement of a high/maximal level of competence provides strong grounds for accepting the child's decision as establishing that the choice best fits the child's good (his or her own particular aims and ends). The very highest level of competence is required to rebut the presumption that if the choice seems not best to promote life and health, then the choice is not in fact reasonably related to the child's well-being. When the expected benefits for life and health of the child's choice are approximately comparable to those of the best alternative, a moderate/median level of competence is adequate to provide reasonable grounds that the choice promotes the child's good and adequately protects his or her well-being.

I have spoken above of levels of competence in a manner that suggests that it can be measured, perhaps on some single quantitative scale. In competence evaluations of adults the use of tests such as the Mini Mental Status Exam may further reinforce the notion that competence is formally testable and admits of a quantitative measure. To assume this, however, would be misleading if not outright mistaken. There are no simple tests in widespread use for children like the Mini Mental Status Exam for adults. But even if there were, such tests could not measure competence for a particular treatment decision. Instead, they could only serve as an initial crude screening of the patient's general capacities for understanding, memory, and reasoning. In any case of borderline or questionable competence the patient's understanding and reasoning in the decision at hand must be examined, and that will inevitably involve an informal process of discussion with the patient about that decision and his or her reasons for it. Any tests or exams for measuring general capacities for understanding and reasoning, whether of adults or of children, cannot measure how well a person's decisionmaking capacities are exercised in a particular decisionmaking situation, though they can sometimes serve as useful guides for defects to look for in the decision at hand.

This necessity for more specific and informal evaluation of the child's decision does not mean that competence does not admit of measurement at all. Competence evaluations are often criticized as intuitive, arbitrary, and lacking any sound theoretical basis because they are based on no single, uniform, and precise measure of competence. Moreover, lacking any such measure, they are particularly susceptible to evaluator bias and variability. The fact that there exists no single unified scale or numerical measure of competence applicable to every decision and decisionmaker does not imply, however, the competence evaluations are arbitrary and lack any sound theoretical basis. Instead, the lack of such a scale merely reflects the reality that competence involves too complex a meshing of various capacities and skills of each decisionmaker with the demands of a specific decision situation to yield any single, unified, formal summary. The potential situations are too numerous and the potential arrays of patient decisionmaking abilities and disabilities are too varied. In any particular case, however, the various capacities and skills of understanding and reasoning necessary to competence in decisionmaking could each be measured on a relative scale and reasons offered for the relative importance of specific deficiencies that limit a child's overall competence to make that decision. Thus, while competence evaluations are not amenable to any simple measure, they do have a sound theoretical basis and need not be merely intuitive, subjective, or arbitrary.

## VII. IMPLICATIONS FOR MEDICAL PRACTICE AND LEGAL POLICY REGARDING COMPETENCE

My principal concern to this point has been with the nature of the competence determination, with the relevant data, largely from developmental studies, concerning children's possession of the various capacities on which that determination should be based, and with how an appropriate level of competence required in a particular decisionmaking situation can be determined. I want now to consider some issues of general legal and medical policy regarding children's competence.[15] As noted at the outset of this paper, in most states the law presumes that minors (in most states, persons below the age of 18 years) are incompetent to give informed consent for their own health care. Most states also have statutes enabling children of specified ages, usually ranging between 12 and 17 years, to give informed consent for treatment for specific medical conditions. The conditions covered by such specialized consent statutes include alcoholism, drug abuse, emotional disturbance, mental illness, pregnancy, rape, sexual assault, transplantation, blood donation, and sexually transmitted diseases. Which of these conditions are covered by specialized consent statutes varies from state to state as does the age at which consent by a minor is permitted. These reductions in the age of consent, however, do not seem usually to be based on assumptions about the minor's competence. Instead, their rationale usually is to permit minors to seek and obtain treatment when requiring parental consent and/or notification for treatment would likely discourage many minors from seeking treatment important to their own and/or others' well-being.

Many states also have statutes specifying conditions in which minors are considered emancipated from their parents or guardians for the purposes of giving informed consent to medical treatment, including marriage, pregnancy, self-management, self support, being a parent, military service, and being a high school graduate. The conditions establishing emancipation vary considerably from state to state. Some states also have what have been called subjective rules permitting consent by minors of two general sorts: "(1) Conditional Minor – a minor may consent if he will be in serious danger unless health care services are provided: (2) Mature Minor – a minor may give consent if he understands the nature and purposes of the proposed treatment" ([6], p. 94). The conditional minor rule may be intended to be restricted to emergency or quasi-emergency conditions, and in any event in Alexander Capron's 1982 survey of state laws it had been adopted in only two states ([6], Appendix).

The mature minor rule has also been adopted in statutory law in only a small minority of states, though it probably has a broader basis in case law.[16] It retains the presumption that minors are incompetent to consent to their own treatment, and requires that their "maturity," that is, their understanding of the nature and purposes of the proposed treatment, be demonstrated in any particular instance to rebut that presumption. Moreover, the application of the mature minor rule has generally been restricted to minors very close to the age of majority, usually not less than 15, independent of the particular minor's general capacities to decide. Despite these limitations, the mature minor rule does come closest to rejecting the general legal policy that minors are not permitted to give informed consent to their own treatment. Does our analysis of competence and the available empirical data concerning children's decisionmaking capacities support further revision in legal policy and/or in medical practice, insofar as it is guided by legal policy?

Because I have been unable to assess here the merits of the various strands of argument identified earlier as commonly offered in support of parental authority to make treatment decisions for their minor children, I want first to address the policy question about children's competence independent of the claims and value of parental authority. Thus, if we look principally to children's interests both in having their well-being protected and in having their emerging self-determination respected, what ought legal and medical policy be? One relatively clear implication of our conceptual analysis of the nature of the competence determination together both with empirical studies of children's decisionmaking capacities and with development theory is that the general presumption that all minors are incompetent for health care decisionmaking is very difficult to defend.[17] As we saw above, there seem to be no differences of significance between adults and children of roughly ages 15 (some would say 14) to 17 years in their general capacities that are needed for health care treatment decisionmaking. This suggests that the presumption of competence for health care decisionmaking that holds for adults should be extended to minors in this age range as well. The most direct way of doing this would be through statutes that explicitly lower the age of majority for medical decisionmaking. At the time of Capron's 1982 survey, three states had enacted such statutes: Alabama, age of majority for medical decisionmaking set at 14; Oregon, at 15; South Carolina, at 16, with restrictions on consent for operations. It bears emphasis that a presumption of competence extended to 15- to 17-year-olds would be, just as with adults, only a presumption and so rebuttable in any particular cases by evidence concerning the child's decisionmaking in that case. This presumption of competence would cover both consenting to treatment without the need for parental consent, as well as refusing treatment to which the child's parents might or might not be prepared to consent.

Below age 15, there seems good reason to maintain the current presumption of incompetence because of the extent to which various general decisionmaking capacities then are usually still undergoing significant development and are appreciably more limited than for most adults. Nevertheless, in the 9 to 14 age period some children will demonstrate sufficient capacities to make particular decisions to be deemed competent to make them. Since the younger the child within this age period, the less often is competence likely to be demonstrated, the presumption of the children's incompetence for medical decisionmaking in the 9 to 14 age range should be understood to be increasingly strong the younger the child. Grisso and Vierling go somewhat further in concluding that "there may be no circumstances that would justify sanctioning independent consent by minors under 11 years of age, given the developmental psychological evidence for their diminished psychological capacities" ([10], p. 424). While this would be correct for the vast majority of cases, I believe the very great variability among children of this age in decisionmaking capacities, as well as the very great variability in the demands made by different decisions, justifies allowing for the possibility that the very strong presumption that 9- and 10-years-olds are incompetent to decide about treatment for themselves might on occasion be rebutted.

The principal implication of the fact that children in this 9 to 14 age range sometimes demonstrate competence may be for medical practice rather than legal policy. Physicians treating children in this age group should always explore treatment alternatives with children, as well as with their parents, in an effort to determine the children's preferences regarding treatment, to understand the reasoning on which their preferences are based, and to assess their competence for decisions. If parent and child are in agreement with the physician's treatment recommendation and the parents' competence is not itself in question, then the physician will have both the parents' consent and the child's assent and can proceed with treatment.

If the child is in disagreement with the physician's recommendation and/or the parents' choice, the first response, as always, should be to seek to resolve that disagreement through further discussion. When disagreement cannot be resolved, whether the child's disagreement is with the physician, the parents, or both, the physician should be obliged to seek to assess the child's competence in this decisionmaking situation. If the physician judges the child to be competent, it should be his responsibility to accede to the child's wishes if the parents are also in agreement with the child, or to serve as his patient's – the child's – advocate in pursuing further steps to enable the child's wishes to be heard and respected. Such a practice would insure that treatment of children in this age range would never proceed without either the child's assent to the treatment or an evaluation by the physician of the child's reasoning in the case in question that affirmed the presumption of the child's incompetence.

While the preceding suggestions concern medical practice, they would require some change in the law in virtually all states. What is needed is clearer legal authority, and in turn legal procedures, for rebutting the presumption of incompetence of children in this age group and for establishing their competence. I have no special suggestions about what those mechanisms should be, but, plainly, if they are desirable, it would not be impossible to develop them.

For children below the age of 9 it is probably reasonable to maintain the practice of treating the presumption of their incompetence as irrebuttable. Despite the significant variation in different children's decisionmaking capacities at any particular age, it is hard to imagine the circumstances in which this young a child's refusal of a treatment on which both the physician and the parents were agreed, or request for a treatment that both physician and parents opposed, evidenced sufficient decisionmaking capacities to be honored in the face of this opposition. However, just as with incompetent adult patients, such children should always have their treatment explained to them before they receive it in a manner appropriate to their abilities to understand. This is a way of respecting their dignity appropriate to their limited decisionmaking capacities.

Finally, while assessing the various complex arguments in support of parents' independent interest in deciding about their children's health care is beyond the scope of this paper, we can at least add one suggestion to remedy slightly this incompleteness. A minimal recognition of this interest, which is desirable as well on other grounds, would be provided by a policy for all dependent minors of encouraging joint decisionmaking between parent and child about the child's treatment. Even in the absence of a willingness of children to engage in joint decisionmaking with their parents, a general policy of parental notification before proceeding with treatment to which a competent child has consented is probably desirable except when treatment is needed under emergency conditions, when parental notification would seriously threaten to make the child's decisionmaking involuntary, or when parental notification would substantially discourage children seeking a particular kind of treatment. While this would certainly not satisfy some advocates of parental authority, it would at least go some small way in recognizing parents' quite legitimate concern about their children's health care.

*Brown University*
*Providence, Rhode Island*

## NOTES

[1] The analysis in this paper extends to the case of children the account of competence and incompetence of adults that I have developed elsewhere. While on the staff of the President's Commission for the Study of Ethical Problems in Medicine and Biomedical and Behavioral Research, I was the principal author of the account of competence and incompetence in the Commission reports, *Making Health Care Decisions* [23] and *Deciding to Forego Life-Sustaining Treatment* [25] and of its discussion paper, 'Patient Competence to Make Decisions About Medical Therapy' [24]. I was also the principal author of the chapter, 'Competence and Incompetence', in the longer study, 'Surrogate Decisionmaking for Elderly Individuals Who are Incompetent or of Questionable Competence' [5] and in [4]. I have drawn freely on both the ideas and, in some cases, the specific statement of them in these works in developing the basic framework of competence in the present paper. I am especially indebted to Allen Buchanan and to the other Commission staff members with whom I worked in what was truly a team effort on the two reports cited above: Alexander M. Capron, Joanne

Lynn, Marion Osterweis and Alan J. Weisbard. Edwin N. Forman, Rosalind Ladd, and Peter M. Smith generously provided written comments on an earlier draft of the paper, and Robert Holmes provided a very helpful written commentary at the conference at East Carolina University School of Medicine at which this paper was first delivered. Many others have provided helpful comments and criticisms of my views on this subject, though none can be held responsible for any misuse I may have made of their good influences.

[2] A good source of data and discussions on research involving children is the National Commission for the Protection of Human Subjects of Biomedical and Behavioral Research [19].

[3] See Alan Meisel, 'The Exceptions to the Informed Consent Doctrine: Striking a Balance Between Competing Values in Medical Decisionmaking' [17].

[4] For the account in this section of the developmental evidence about children's capacities, I rely heavily on the excellent review by T. Grisso and L. Vierling, 'Minors' Consent to Treatment: A Developmental Perspective' [10] and also on: G. B. Melton, 'Psychological Issues in Increasing Children's Competence' [18], C. E. Lewis, 'Decision Making Related to Health: When Could/Should Children Act Responsibly?' [16], M. A. Grodin and J. J. Alpert, 'Informed Consent and Pediatric Care' [11], L. A. Weithorn, 'Involving Children in Decisions Affecting Their Own Welfare: Guidelines for Professionals' [28], L.A. Weithorn and S. B. Campbell, 'The Competency of Children and Adolescents to Make Informed Treatment Decisions' [29], and S. Leiken, 'Minor's Assent or Dissent in Medical Treatment' [15].

[5] Grisso and Vierling ([10], p. 194). This is also consistent with the findings of Weithorn and Campbell [29] that 9-year-olds were able to reach reasonable conclusions, but were not able to attend to all relevant factors in complex cases and instead focused on a few prominent factors.

[6] Melton ([18], p. 24), quoting Keith-Spiegel [14].

[7] Discussion of the philosophical issues in conceptions of the good for persons can be found in Brandt, 'Two Concepts of Utility' [3], T. Schwartz, 'Human Welfare: What It Is Not' [27], and J. Rawls [26].

[8] See among many of his writings on this subject, Erikson [8].

[9] Strictly, the assumption is that the person either now or at some time in the past had the capacities necessary for self-determination. It is possible to respect a person's past choices for his present treatment even if the person currently lacks the capacities for self-determination.

[10] Representative philosophical discussions of this issue can be found in W. Aiken and H. La Follette (eds.), *Whose Child? Children's Rights, Parental Authority and State Power* [1], J. Blustein, *Parents and Children: The Ethics of the Family* [2] and L. Houlgate, *The Child and the State* [13].

[11] While serious decisionmaking incapacity may be implied by a choice that threatens the decisionmaker's overall well-being, a choice in accordance with the decisionmaker's well-being does not imply sound decisionmaking in that instance. Through mere chance or good fortune, a thoroughly disordered process of reasoning may nevertheless result in the best choice. (It is a simple fact of logic that unsound arguments may contain true conclusions.) This will then be a case in which incapacity in decisionmaking does not result in *serious* decisionmaking incapacity or in any need to protect the patient's well-being.

[12] It is worth noting that only the question of so-called soft paternalism, not hard paternalism, arises in this account. The difference is in whether the appeal to the subject's own good, in justification of interference with his doing as he wishes, makes reference only to his own conception of his good (soft) or to another's conception of what is best for him (hard).

[13] Both Grisso and Vierling [10] and Weithorn [28], for example, assert the relevance of this interest.

[14] Among other explicit defenses of a variable standard of competence are J. Drane, 'The Many Faces of Competency' [7] and W. Gaylin, 'Competence: No Longer All or None' [9].

[15] Capron's 'The Competence of Children as Self-Deciders in Biomedical Interventions' [6] contains an excellent review of the law regarding children's competence, as well as a very useful appendix detailing the standards in the different states.

[16] Cf. Holder [12], pp. 133–135.

[17] Capron [6] argues that as a matter of law this presumption is undergoing change and erosion.

## BIBLIOGRAPHY

1. Aiken, W. and La Follette, H. (eds.): 1980, *Whose Child? Children's Rights, Parental Authority, and State Power*, Littlefield, Adams, Totowa, NJ.
2. Blustein, J.: 1982, *Parents and Children: The Ethics of the Family*, Oxford University Press, New York.
3. Brandt, R.: 1982, 'Two Concepts of Utility', in H. Miller, and W. Williams, (eds.), *The Limits of Utilitarianism*, University of Minnesota Press, Minneapolis, pp. 169–185.
4. Buchanan, A. and Brock, D. W.: 1986, 'Deciding for Others', *Milbank Memorial Quarterly* **64**: Suppl. 2, 17–94.
5. Buchanan, A., with Gilfix, M. and Brock, D. W.: 'Surrogate Decisionmaking for Elderly Individuals Who are Incompetent or of Questionable Competence', National Technical Information Service, Washington, D.C. (forthcoming).
6. Capron, A.: 1982, 'The Competence of Children as Self-Deciders in Biomedical Interventions', in W. Gaylin, and R. Macklin, (eds.), *Who Speaks For the Child*, Plenum Press, New York, pp. 57–114.
7. Drane, J.: 1985, 'The Many Faces of Competency', *Hastings Center Report* **15**: 2, 17–21.
8. Erikson, E.: 1963, *Childhood and Society*, 2nd Ed., W. W. Norton, New York.
9. Gaylin, W.: 1982, 'Competence: No Longer All or None', in W. Gaylin, and R. Macklin, (eds.), *Who Speaks For the Child*, Plenum Press, New York, pp. 27–54.
10. Grisso, T. and Vierling, L.: 1978, 'Minors' Consent to Treatment: A Developmental Perspective', *Professional Psychology* **9**, 412–427.
11. Grodin, M. A. and Alpert, J. J.: 1983, 'Informed Consent and Pediatric Care', in G. B. Melton, G. P. Koocher, and M. J. Saks, (eds.), *Children's Competence to Consent*, Plenum Press, New York, pp. 93–110.
12. Holder, A.: 1985, *Legal Issues in Pediatric and Adolescent Medicine*, 2nd Ed.,

Yale University Press, New Haven.
13. Houlgate, L.: 1980, *The Child and the State*, Johns Hopkins, Baltimore.
14. Keith-Spiegel, P.: 1976, 'Children's Rights as Participants in Research', in G. P. Koocher, (ed.), *Children's Rights and the Mental Health Professions*, Wiley, New York, pp. 53–81.
15. Leiken, S.: 1982, 'Minor's Assent or Dissent in Medical Treatment', President's Commission for the Study of Ethical Problems in Medicine, *Making Health Care Decisions, Volume III: Appendices* U.S. Government Printing Office, Washington, D.C., pp. 175–191.
16. Lewis, C. E.: 1983, 'Decision Making Related to Health: When Could/Should Children Act Responsibly?' in G. B. Melton, G. P. Koocher, and M. J. Saks, (eds.), *Children's Competence to Consent*, Plenum Press New York, pp. 75–91.
17. Meisel, A.: 1979, 'The Exceptions to the Informed Consent Doctrine: Striking a Balance Between Competing Values in Medical Decisionmaking', *Wisconsin Law Review*, 413–488.
18. Melton, G. B.: 1983, 'Decision Making by Children: Psychological Risks and Benefits', in Melton, G. B., Koocher, G. P., and Saks, M. J. (eds.), *Children's Competence to Consent*, Plenum Press, New York, pp. 21–40.
19. National Commission for the Protection of Human Subjects of Biomedical and Behavioral Research: 1977, *Research Involving Children, Appendix Volume*, U.S. Government Printing Office, Washington, D.C.
20. Piaget, J.: 1965, *The Moral Judgment of the Child*, Free Press, New York.
21. Piaget, J.: 1972, *The Child's Conception of the World*, Littlefield, Adams, Totowa, N.J.
22. Piaget, J. and Inhelder, B.: 1969, *The Psychology of the Child*, Basic Books, New York.
23. President's Commission for the Study of Ethical Problems in Medicine and Biomedical and Bahavioral Research: 1982, *Making Health Care Decisions*, U.S. Government Printing Office, Washington, D.C.
24. President's Commission for the Study of Ethical Problems in Medicine and Biomedical and Behavioral Research: 1982, 'Patient Competence to Make Decisions About Medical Therapy', unpublished.
25. President's Commission for the Study of Ethical Problems in Medicine and Biomedical and Behavioral Research: 1983, *Deciding to Forego Life-Sustaining Treatment*, U.S. Government Printing Office, Washington, D.C.
26. Rawls, J.: 1971, *A Theory of Justice*, Harvard University Press, Cambridge, MA.
27. Schwartz, T.: 1982, 'Human Welfare: What It Is Not', in H. Miller, and W. Williams, (eds.): 1982, *The Limits of Utilitarianism*, University of Minnesota Press, Minneapolis, pp. 195–206.
28. Weithorn, L. A.: 1983, 'Involving Children in Decisions Affecting Their Own Welfare: Guidelines for Professionals', in G. B. Melton, G. P. Koocher, and M. J. Saks, (eds.), *Children's Competence to Consent*, Plenum Press, New York, pp. 235–260.
29. Weithorn, L. A. and Campbell, S. B.: 1982, 'The Competency of Children and Adolescents to Make Informed Treatment Decisions', *Child Development* 53, 1589–1598.

# CONSENT AND DECISIONAL AUTHORITY IN CHILDREN'S HEALTH CARE DECISIONMAKING: A REPLY TO DAN BROCK

Dan Brock argues persuasively that the general presumption of incompetence on the part of minors to decide about their medical care seems indefensible for children roughly above the age of 14 [1]. The presumption still exists and apparently remains strong in the law. But as Angela Holder points out [2], the courts are moving in the other direction. I am essentially in agreement with this, and find most of Brock's analysis of the notion of competence convincing. However, I am unconvinced that children at even earlier ages may not often have as much competence in this area as most adults. I stress this comparative judgment, because I think it is important that, in our concern for the well-being of children, we do not expect more from them than from adults in comparable situations. That having been said, I will argue that children nonetheless probably should not be given the authority to make such decisions, even though adults in similar situations should. That is, I will argue both that children have the necessary competence and that decisionmaking authority should not be extended to them.

Let me say at the outset that the issue is not one of competence to decide which medical treatment one should have, if by that we mean competence to decide on one's own what the treatment should be. Practically no one without medical training has such competence. Nor, obviously, is the question merely one of capacity to consent, since that is a minimal capacity, possessed by children almost from the time they learn to say 'yes' and 'no'. The question, rather, concerns competence to *share* in decisionmaking regarding medical care. This is the most that any of us are competent to do. Even if the final decision is ours about whether to undertake a particular line of treatment, we need the guidance of a competent physician to enable us to do this wisely.

But the idea of "decisional competence" seems to me to contain an ambiguity. There is a difference, that is, between being involved in a process in which you work out with a physician what will be the best course of treatment for you, and being presented with a judgment as to what others deem to be best for you and then being asked (perhaps after having it explained) whether you consent to the treatment. This is a matter of degree, and it may not always be possible to draw a sharp line between the two. But some people who might be marginally competent to engage in the former process are fully competent to consent in the second sense. I shall speak of the first competence as that in *decisionmaking*, the second as that to *consent*.

Now the doctrine of informed consent, as Brock states it, holds "that health care treatment only be given to a competent patient with that patient's voluntary and informed consent" ([1], p.189), i.e., that voluntary and informed consent is a necessary condition of treatment being administered. It is with the comparative competence of children younger (by an unspecified number of years) than 14 years of age *to consent* in this sense that I am concerned, even though I think sometimes they may be competent to be involved in decisionmaking in the former sense as well. I take it that "valid" consent, on this view, is consent by a competent person in an informed and voluntary manner.

Competence, Brock says, "varies over time with changes in the person's condition and so may be intermittent" ([1], p. 183). For this reason, "competence determinations therefore involve matching the capacities of a particular person at a particular time and under particular conditions with the demands of a particular decisionmaking task" ([1], p. 183). The abilities needed are:

"(1) capacities for communication and understanding of information

(2) capacities for reasoning and deliberation;

(3) capacity to have and apply a set of values or conception of the good". ([1], p. 184)

Reflecting on the abilities of children, Brock sees no particular problem with (1), some problems with (3), but the main problems with (2). Citing Grisso and Vierling, he says that among the capacities that may be needed for the reasoning process are "the ability to sustain one's attention to the task, ability to delay response in the process of reflecting upon the issue, ability to think in a sufficiently differentiated manner (cognitive complexity) to weigh more than one treatment alternative and set of risks simultaneously, ability to abstract or hypothesize as yet nonexistent risks and alternatives, and the ability to employ inductive and deductive forms of reasoning" ([1], p. 186). Insofar as I understand all of this, it sounds like a formidable set of abilities, and put in this way it is questionable whether children under the age of 14 (or perhaps at any age) have them. But these sound very much like the abilities necessary to *decisionmaking* rather than those necessary to *consent* meaningfully. A person, I suggest, might very well be competent (in the sense that he has normal intelligence and understanding, i.e., is not defective in any way) and be competent to consent but have decisionmaking abilities only to an imperfect degree.

Take a rough analogy. Decisionmaking competence in chess, as measured by one's ability to make the best moves in highly complex situations, varies with experience, study, native ability, and so on. Someone lacking these skills and this experience will possess such decisionmaking ability only to a limited degree. But even a rank amateur can understand highly complex maneuvers when they are explained to him. He is capable of understanding analyses of the games of grandmasters, even though he is incapable of going through those decisionmaking processes at such a level himself. Even with only a rudimentary knowledge of the game, a person can see and understand why certain moves at the highest level are the best. Now in the case of medical care, the capacity to consent validly, I suggest, should be understood as more akin to being able to understand why a move in chess is the best one than to being capable of going through the decisionmaking process by which some moves are arrived at. When such consent is voluntary and informed, I suggest it should be considered valid even though nothing approximating the full range of the other abilities is present.

A child can understand when he is sick and when he is well. He can understand that if he does $x$ he will get well and if he doesn't he won't. Or that if he does $x$ he will get well sooner than if he doesn't. To understand this doesn't require anything in the way of technical medical information about what precisely $x$ involves. He does need to know what it will involve in terms of his own experience; whether it's likely to hurt or cause discomfort or be incapacitating and for how long. But it is hard to see how there is much in the way of *essential* information that fairly young children (by which I now mean many of those between the ages of 7 and 14) cannot comprehend. Nor is it easy to see why highly sophisticated decisionmaking skills and reasoning processes are necessary. Not that sometimes they may not be important; if a complex set of alternatives is presented, with varying probabilities, and widely varying accessory effects to those desired, the basic understanding of the relevant facts will be correspondingly more difficult. But then it will be correspondingly more difficult for many adults as well. We don't require of adults that they have or even comprehend much in the way of medical information. And we certainly require little in the way of reasoning skills. Often people just ask their physician what he advises and then accept that without question. My point is that we should not exaggerate the abilities that are needed to consent validly.

Trust, I want to suggest, is a factor to consider in assessing competence to consent. If there is a sound basis for trust between a child and its parents (and, perhaps as an extension of that, between the child and the physician); trust that the parents have uppermost in their minds the well-being of the child; and trust that they will tell him what he needs to know to deal with the situation in the best way he can, a child may be fully capable of consenting to proposed treatment. Trust is important between patients and physician generally. The greater the trust, when it is well-founded, the more reliably and competently one can consent to proposed treatment. Well-founded trust, in this sense, can offset limitations in detailed knowledge about the particulars of treatment.

There are other factors, of course, which enter in. Children often are nervous or frightened when confronted with such situations. Often, as Brock points out, the degree to which children perceive the locus of control as external to them affects their attentiveness to the decision, to awareness of details, and the effort they will exert to gather information about it ([1], p. 186). But, I suggest, it is also true of many adults in health care situations. They are nervous, confused, ready to defer to the advice of the physician. Often, indeed, they are made to feel that it is an impertinence on their part to do otherwise. If we should be prepared to declare children incompetent on these grounds, why not adults as well?

Regarding the third element, a conception of the good, Professor Brock says that the children from 7 to 13 "can have great difficulty anticipating their future," and this can lead to "prudential irrationality in reasoning when present effects receive disproportionate weight in comparison with future effects" ([1], p. 187). But all of these things are also true of many adults. If by "prudential irrationality" is meant choosing courses of action or engaging in practices that are to one's long-run detriment even though they give some short-run rewards, people generally show a remarkable capacity for this, in everything from cigarette smoking and over-indulgence in alcohol to drug abuse and speeding on the highway. Many persons overeat, neglect their health, and allow medical problems to become acute before seeking treatment. Yet no one thinks they should be declared incompetent for these reasons. Either they do not have a clear and coherent conception of their own good and the means by which to realize it, or they do and are unwilling to pursue it or perhaps incapable of pursuing it. Children do not, as a rule, at a young age have a detailed conception of their good. But they are certainly capable of understanding many of its essential features – health, material well-being, happiness. And it is mainly only decisions pertaining to these values that parents or physicians or state authorities are competent to make, since they do not know any more than the child about the *particulars* of the life-plans the child will eventually choose. It is not as though there exists a clear conception of the child's good and the child is simply unable to perceive it. It is that this clear and determinate conception doesn't yet exist at all. For that reason adults are no better able to act on it on the child's behalf than is the child (though they are sometimes better able to see what will make realization of that good possible, whatever it turns out to be).

What if the child decides wrongly, consenting to treatment which, as it turns out, he might better not have consented to at all (doctors, after all, sometimes make mistakes); or which in the long run somehow redounds to his disadvantage in ways he didn't anticipate. Or what if he dissents from treatment that he really should have undergone and suffers some unanticipated detriment – up to and including death? Again, this happens to adults all the time. It no more counts against the competence of a child to consent than it does of an adult that these things may happen. It would count, of course, if it were known with any confidence that children of a given age either consented or dissented with significantly bad consequences for their own well-being to a significantly greater degree than adults. But since there is the presumption of incompetence on the part of children, and since they aren't generally accorded the right to be involved in decisionmaking in this way, there is unlikely to be much in the way of solid evidence to show that the results of their being so involved would be worse than they currently are for adults. Nor, it should be conceded, is there much solid evidence on the other side either.

Needless to say, below a certain age children lack such competence. The newborn not only do not have the capacity for informed and voluntary consent, they do not even have the capacity to consent or dissent. And in normal children by the age, say, of 18 they already do have the capacity. Where between the boundaries it emerges to a degree sufficient to warrant our saying that they have this competence I am unable to say, but I see no reason why it should not be at a younger age than 14. Again, I do not say that children at a younger age have a high degree of such competence. That I don't know; only that it very likely compares favorably with that displayed by adults.

For all of this, I don't believe that children in general should be given a final say in consenting to or dissenting from treatment, even though they should be involved in the process with parents and physicians to the full extent of their capacities. The reason why I do not, even though I suspect many children are competent to consent voluntarily in a relevantly informed way, is because the situation of children and adults is disanalogous in an important respect. In the case of children there normally are other persons (or a person), namely, parents, who have moral responsibilities toward them which they are obliged to honor, even if the child objects and even if it has the competence to consent or dissent in a valid way. That is, as Brock points out, we need to distinguish decisionmaking competence – by which I still mean the second of the competencies distinguished earlier, namely, that to give voluntary and informed consent and not necessarily to be involved in decisionmaking in a significantly fuller way – and what he calls decisional *authority*. Here I am not talking about legal authority. That is another matter. Rather, I mean authority in the sense of moral responsibility to look after the best interests of the child. Even demonstrated competence on the part of the child to consent or dissent from a particular course of treatment does not of itself suspend this responsibility on the part of parents.

This does not mean, however, that parental responsibilities include a right to insist on treatment for a child that is not in its best interests. As both Brock and Holder point out, at some point the parents' standards and aims may cease to be those of the child – as happens conspicuously among teenagers. Even there, however, I think parental standards are not without some moral weight, particularly vis-à-vis the state. The institution of the family carries with it some presumption that others should not intrude into it in a way that interferes with the effort on the part of parents to inculcate what they consider proper values in their children. This isn't to say that they may not sometimes be wrong in their judgments. Nor is it to say that parents do not sometimes act to the detriment of their children, as in the case of child abuse. It is to say only that it is unlikely *in general* that the state can do better, or sufficiently better to warrant extensive intrusion into the family.

If parental responsibility, morally speaking, is to do what is for the best of the child, then the concern of parents will be predominantly a consequential one. And as development of the capacities for self-determination will be in the interests of the child, this means that responsible parents will involve children in the decisionmaking process at as early an age as can reasonably be expected to benefit them. But although parental concern should properly be consequentially oriented in this sense, the *ground* of their moral responsibility will be deontological, arising from obligations bound up in the very relationship between parent and child.

This leaves a host of questions unanswered, of course, including what precisely the age is at which children generally have the competencies for which I have been arguing, and at what age decisional authority should pass from parents to children. On the former question, I would venture to say that 10-year-olds I have known seem to have that competence to a significant degree; and some 10-year-olds I have known have had it in a higher degree than most adults. Much younger than that I begin to have doubts. As for decisional *authority*, that I believe varies from situation to situation, depending both on the child and the parents and on the quality and nature of the relationship between them. My main concern is to suggest that these issues require the perspective of a moral and conceptual framework regarding the whole relationship between parents and children, the institution of the family and the legitimate role (if any) of the state in this area. We do not, it seems to me, now have such a framework.

*The University of Rochester*
*Rochester, New York*

## BIBLIOGRAPHY

1. Brock, D.: 1989, 'The Competence of Children for Health Care Decision-making', in this volume, pp. 181–212.
2. Holder, A.: 1989, 'Children and Adolescents: Their Right to Decide About Their Own Health Care', in this volume, pp. 161–172.

WILLIAM RUDDICK

## QUESTIONS PARENTS SHOULD RESIST

To take decisive part in a child's medical care, parents must know what questions to ask physicians – and themselves. I wish to examine three questions often asked and answered in pediatric decisions, namely,

(1) "What would I do if it were my child?" asked by physicians of themselves;

(2) "What would you do if it were your child?" asked of physicians by parents of ill children;

(3) "What is in the child's best interests?" asked by both physicians and parents.

These questions have various appeals but also serious defects. The first question ("A Physician's Question") may be easily answered, but lends itself to arrogant paternalism. The second question ("A Client's Question") may be used to test for various biases, but is more likely to impose them on anxious parents culturally different from their physicians. On balance, these two subjunctive questions are best avoided, or their answers accorded little weight in pediatric decisions.

The third question about a child's "best interests" is less readily answered. In trying to disentangle a child's interests from those of other family members, parents are prone to various self-deceptions and unrealistic self-denials. Physicians, social workers, and judges may find the question easier, given their "professional distance" from the family and their simplifying theories of child welfare. Indeed, this is "A Professional's Question," which parents should resist.

Parents should, I think, ask not about best interests, but about *acceptable lives*. Specifically, they should ask about the bearing of treatment options on a child's life, and on the lives which encompass it, including their own. ("Will the operation enable her to lead a tolerable life?," "Will the care she would require make our lives impossible?") Such "Biographical" questions make explicit factors which, although not always decisive, are morally relevant but likely to be obscured or distorted by the preceding three questions.

I apologize in advance to those who look to philosophy for answers and principles, not for questions. But attention to questions, their presuppositions and contexts, often simplifies the search for answers, or in some cases justifies the rejection of a troublesome question. For example, the common parental question, "What did I do wrong?" presupposes that a child's malady involves fault, perhaps even punishment – a presupposition which may distort parental solicitude. Out of guilt and a desire for penance, parents may approve heroic treatment, without asking about risks, benefits, or burden of convalescent care.

Of course, principles also may help us identify and formulate the right questions, as well as to answer and properly weight the answers to those questions. I once tried to formulate a principle ("the Prospect Provision Principle") that would guide pediatric decisionmaking by parents and judges [5], and I am tempted to revive it here. But I have become somewhat suspicious of principles, partly because they can discourage or suppress questions. In so doing, they limit the factors and options we ought to consider. For example, the Hippocratic principle, "Do no harm," leaves little if any room for questions about patient or parental rights.

Attending to questions is especially appropriate work for philosophers concerned with occupational ethics. Busy practitioners rarely have the leisure to listen closely and reflectively to the language in which their work is conducted. As spectators trained to detect and elaborate distinctions and inferences, philosophers can sometimes help to make clearer what practitioners are saying and, hence, what they are doing. Although presumptuous, the project of articulating practice is less so than formulating principles by which to assess work in which philosophers themselves do not actively engage.

## I. A PHYSICIAN'S QUESTION: *"WHAT WOULD I DO IF IT WERE MY CHILD?"*

In a symposium on infant euthanasia several years ago, a pediatrician declared that he never discussed treatment options with parents of severely impaired newborns. They were, he assured us, too distraught to make a rational decision about giving or withholding treatment. Rather, he decided the matter himself by asking himself, "What would I do if it were my child?"

Although initially stunned by his self-confidence, I came to see the appeal of this decision-procedure in the circumstances. The pediatrician was in charge of the neonatal unit in a hospital serving a small city and surrounding rural area in Vermont. Prenatal care was not routine, nor were amniocentesis and abortion. Accordingly, impaired and precariously premature infants were born to shocked parents unprepared for discussion of treatment options. Moreover, the physician wished to spare all parents, however stalwart, the burden of decision and whatever guilt or recrimination might follow. He himself seemingly suffered little indecision or guilt: he was able to answer his own question easily – a boon for an overworked clinician.

Despite these merits, I think there are serious objections to his decision-procedure. Acting in their place, the pediatrician prevents parents from taking part in decisions about their child's fate. This is surely usurpation of parental rights, and extreme paternalism.

Such paternalism is encouraged by a tendency among pediatricians and allied social workers to regard parents as secondary, incompetent patients. This view, I think, should be resisted. Pediatricians do have to deal with often highly anxious and distressed parents, some of whom may need psychiatric counselling, or even hospitalization. But unless the shock of an infant's birth (or older child's sudden illness or injury) produces mental collapse, parents should be regarded as *distressed clients*, not secondary patients. Accordingly, parents must be given time and aid to formulate and confront the pediatric options. It is not for pediatricians or other professionals to declare them incompetent because distraught, and make a decision which alters their lives as well as the infant's.

Paternalism is, however, not the pediatrician's basic flaw. His question and decision-procedure rest, as does much paternalism, on a deeper arrogance to which all professionals are prone in our culture. In asking and answering this question, the pediatrician must make one of two assumptions. He must assume that the parents share his parental values, or, if not, that his own values should be decisive. By "parental values" I mean to include what one wants, as a parent, for a child or children, as well as what one wants, as a parent, for oneself and one's co-parent (if there is a co-parent).

How likely is it that a pediatrician – an urban, male professional specialist and hospital officer – would have the same parental values as, for example, a rural Vermont farm woman? In assessing an infant's future quality of life, would he and she equally value verbal capacity and academic prospects, or physical appearance and social acceptability? Would she react similarly to a child's hemiplegia, or reduced lifespan? Would she be equally able to commit herself to long-term care of a child who would never fully mature? If overwhelmed by that care, would she have similar resources of aid?

Perhaps so, but a pediatrician cannot assume similar values. Nor can he assume equal commitment to saving life per se, regardless of quality. Studies of people who are "pro-life" and "pro-choice" in the abortion debate suggest that religion, class, education, and ambition make for strikingly different principles and practices [2]. And since contemporary pediatricians (unlike many midwives) *do* differ culturally from many of their parent-clients, our pediatrician would have to enquire about parents' parental values if the pediatrician thinks his question is a way of making a decision on their behalf. Needless to say, neonatologists have little time or occasion for such enquiries, and obstetricians may be of no help. An obstetrician providing regular prenatal care might know a woman's parental values; but unless there were troubling signs and symptoms, the topic of possible congenital defects would probably be avoided as unduly alarming. A hospital social worker or unhurried nurse might be able to determine parental values, but probably not in time to bear on urgent perinatal decisions and their consequent duties.

Alternatively, the pediatrician might not try to discover differences between his values and those of the parents; he might believe that his own parental values should decide matters. Such self-assurance – to put it charitably – might derive from self-righteous conviction or pride. Competition for professional school entry, long training, high income, the rhetoric of service to society, professional privileges, and occupational autonomy all contribute to the self-esteem of those who succeed. The question, "What would I do if it were my child?", may well express the view that professional status confers moral as well as technical expertise and authority.

## II. A CLIENT'S QUESTION:
### "WHAT WOULD YOU DO IF IT WERE YOUR CHILD?"

This professional arrogance (to put it *un*charitably) may be further fostered by parents themselves who ask, "What would you do if it were your child, doctor?" It has the form of a question which clients often put to professionals: "What would you do if you were in my situation?" Anxious and ignorant, clients want and are quick to accept direction.

There are, however, other uses for this question. Rather than a request for guidance, it might be a test of bias. Parents may be asking, in effect, "Would you treat *your* child in the very way you are proposing to treat mine?" They thereby may try to discover if their physician is proposing inferior treatment, in virtue of possible class or ethnic prejudice. Alternatively, the parents could be indirectly testing their own possible biases. Since physicians usually have far fewer worries about medical costs, parents might pose the question to see if they themselves are choosing lesser treatment on financial rather than therapeutic grounds, or on grounds of excessive parental solicitude and fear.

Perhaps a desire to spare their child pain and danger is inclining them to the simpler therapy. Were their pediatrician willing to put his or her own child through the more aggressive treatment, then (they might reason) the advantages must be clear.

Whether they are seeking guidance, reassurance, or self-knowledge by this question, parents should be slow to ask or act on the answer to this question. First, we rarely know how physicians feel about their own children: they may be distant, or even antagonistic parents. Or, physicians may become professionally callous to suffering, even in the case of their own beloved children. Moreover, it is only *imagined* suffering they are being asked to contemplate.

There is a second caveat. Physician's judgments of risk/benefit ratios may be distorted by professional optimism. As the very terms of that ratio suggest, physicians by profession are risk-takers and optimists. Otherwise, they would speak of *costs*, not risks, or, alternatively, of *possible* benefits, not simply of benefits (or "anticipated benefits"). By contrast, many parents tend to be cautious, protective, and pessimistic, especially during a child's illness.

The differences between optimists and pessimists run deeper than the celebrated glass of water suggests. Optimists see the glass as half full in the light of hopes that it is being filled, or at least, that there is enough water to go around. Pessimists entertain no such hopes; they are inclined to believe that the glass may be losing water already in short supply. The realist, by contrast, would demand evidence and probabilities for going beyond the obvious fact that the fluid is at the half-way mark.

There is a veneer of realism in medical practice: talk of risk/benefit ratios, statistical studies, and the like suggest that there are well-established probabilities and criteria for assigning patients to appropriate reference classes. But unlike the billiard balls and dice of probability theory, patients are not highly similar or easily classifiable. Hence, physicians insist that clinical judgment is an art, that each case is unique, and that outcomes are usually unpredictable.

Rather than cause for distress, this uncertainty allows physicians to press ahead, even in cases which are hopeless, statistically speaking. It is the *possibility* of a good outcome, not its probability, which drives much medical work. Parents may come to share a pediatrician's hopes and consent to risky or painful treatments for their children. On the other hand, they may refuse such therapies, dwelling pessimistically on the possibility of failure and its consequences. Or, they may as realists resist the physician's optimism. Whether pessimists or realists, should parents defer to a physician's optimism?

This is not the place to discuss these issues. But it should be clear that the anxious "Client's Question" cannot be the right question, if such differences between parents and physicians exist unexplored. We need other questions, preferably less tied to physicians' imagination and charisma. An obvious candidate is a question often asked by judges and social workers, when called into pediatric decisions, namely, "What is in the child's best interests?"

### III. A PROFESSIONAL'S QUESTION: *"WHAT IS IN THE CHILD'S BEST INTERESTS?"*

This question seems to escape the personal aspects of the questions we have been examining. It focuses our attention on the child, independent of what physicians can envisage for their own particular children. It also isolates a child's welfare from that of physicians, the hospital staff, siblings, and the parents themselves. As such, it would seem to protect a child's potentially competing interests against, for example, an overworked nurse who resists further admissions to her ward.

The notion of best interests has a revealing history. It came into play in custody disputes, but then was extended to the field of neglected children, including children so judged because of medical need. Lawyers and psychiatrists have attacked its use on several grounds, including its spurious objective impersonality [1]. No one, they say, knows very much about child development. When judges confront custody decisions, they often know too little about the particular child to make shrewd assessments of specific interests, long- or even short-term. The unacknowledged basis of their judgment tends to be their own intuitive conceptions of a good parent and "a normal life." Although ostensibly about a particular child, the best interests test becomes a masquerade for often highly personal, generalized, and culturally-prejudiced opinion. The vagueness of the notion gives judges license to impose their own values on others under the guise of "individualized justice." Parents have had their children placed in foster care for failing to maintain a clean house and nutritious diet, by the judge's standards.

Whatever a judge's shortcomings, we might suppose that parents are better able to answer the best interests question. They may have no greater knowledge of general child development, but they do tend to know their own children better than could even the most painstaking judge. Moreover, their cultural or class biases will be less distorting than will a judge's, for it is their biases which set the frame within which a child's interests must be fitted.

More precisely, their desires and beliefs partially define their child's interests, especially in the case of very dependent children.

Despite these advantages, parents are unpracticed in dissecting out the interests of a single family member. Their work is to weave various interests, to minimize or reconcile differences within a family. It is much easier for doctors, judges, and other officials to think of children as so separable from families. They see children, not in homes, but on professional turf, accompanied by parents who may be intimidated by such official and threatening settings. Professionals may even view their own children in this semi-detached way: the demands of professional training and work leave little time for family life or child care. Their children are likely to be raised by others; hence, they may underestimate the burden or disruption of extraordinary care for hopelessly ill or severely handicapped children.

In trying to answer the "best interests" question, parents must try to approximate this professional, or extra-familial view. But to do so is to simplify a child's interests and to neglect the interests on which a child's interests depend. Parents, in their knowledge and commitment to family, are likely to give answers subtly weighted in favor of others, or answers which imply standards of child care they are bound to fail. This can lead to decisions based on ideals of parental devotion and self-sacrifice which very few parents satisfy. The resulting sense of failure, self-reproach, and guilt which false ideals generate can serve neither parental interests nor a child's best interests, however determined.

Such overcommitment and unanticipated strains might be less if parents were allowed explicitly to consider and weigh their own interests, parental and otherwise, as well as the interests of siblings and other affected relatives, in making pediatric decisions. But the best interests question tends to suppress such interests as morally irrelevant by conceptually isolating a child from its familial context. In sum, physicians and parents may be able to answer the "Professional's Question," but should be slow to ask it.

Are there better questions parents should ask themselves and physicians – questions which do not distort and simplify?

## IV. A BIOGRAPHICAL QUESTION:
### *"HOW WILL TREATMENT AFFECT OUR LIVES?"*

As a start parents could ask: "Will the treatment improve our child's life?" "Will she have a life more worth living, a life we can endorse and foster?" Although not always answerable, such questions about lives are less abstract and artificially isolating than questions about interests. Unlike general notions of welfare and need, lives are defined by specific activities and relationships. For young children, those are primarily family activities and relationships. Dictionaries count 'child' and 'parent' as correlative terms, each defined in terms of the other – a semantic point which reflects biographical reality.

For this reason, it is a short and permissible step for parents to ask explicitly, "How will this treatment affect *our* lives?" But this is not to shift from a child-centered to a self-centered question. Since the lives of parent and child are so intertwined, in asking about a child's life, one is asking about parents' lives as well. (The topic of lives, in medical and other contexts, is explored at greater length in [3], [4], and [5].)

Of course, the lives of parents usually include more than children and child care, and increasingly so, as women reject a conception of motherhood as a semi-religious, all-encompassing vocation. The tension between parental and non-parental interests and responsibilities is a feature of most lives which pediatric decisions must recognize. To do so explicitly is to reduce the risks of self-deception noted above, as well as the false ideals of self-effacing self-sacrifice. A parent must be allowed to ask openly, "Will this operation save a child whose care will ruin my life and the lives of those to whom I am already committed?"

Of course, lives are often altered by events in unpredictable and rewarding ways. We may surprise ourselves and others by uncharacteristic responses to events as unprecedented in our lives as the birth of a severely impaired child. Optimists speak of "challenges" and trust that they can meet them and accommodate whatever the world presents. They extol general capacities for flexibility, endurance, courage, and loving care in the face of adversity. To such optimists, the desires and standards of less "open" people seem self-indulgent. For example, the desire for a healthy fetus or infant is dismissed by optimists as a selfish "desire for a perfect child." Such misrepresentation can be resisted if parents are allowed to think about their own lives.

It is very easy for even normally confident parents to lose their sense of themselves and their interests when faced with professional optimism, authority, and impatience. If parents are to exercise parental rights, however defined, or even to take effective part in pediatric decisions, they must know what questions to raise – and what questions to resist. These biographical questions will help them maintain the parental voice, by keeping parental caretakers clearly in view. They focus imagination on the joint lives of parents and children, however abnormal their duration or degree of dependency. Parents, as well as children, are liable to neglect in pediatric decisions.

Biographical questions do not, of course, preclude self-serving or negligent answers. Physicians, judges, and other people committed to children's welfare must be able to review and intervene in a parent's treatment decisions. My suggestion is that biographical questions would reduce the professional and cultural distance between parents and physicians in such reviews. And whatever pediatric decisions resulted would, I think, better reflect a daily, domestic reality, undistorted by abstract ideals of Life, Childhood, and Motherhood. Since their welfare is indivisible, both children and parents would benefit.

*New York University*
*New York*

## BIBLIOGRAPY

1. Goldstein, J., Freud, A., Solnit, A., and Goldstein, S.: 1986, *In the Best Interests of the Child*, Free Press, New York.
2. Luker, K.: 1984, *Abortion and the Politics of Motherhood*, University of California Press, Berkeley.
3. Rachels, J.: 1986, *The End of Life*, Oxford University Pres, New York and Oxford.
4. Rachels, J. and Ruddick, W.: 1989 'Lives and Liberty' in J. Christman (ed.), *The Inner Citadel: Essays on Individual Autonomy* (forthcoming).
5. Ruddick, W.: 1979, 'Parents and Life Prospects', in O. O'Neill and W. Ruddick (eds.), *Having Children: Philosophical and Legal Reflections on Parenthood*, Oxford University Press, New York, pp. 123–137.

H. TRISTRAM ENGELHARDT, JR.

## TAKING THE FAMILY SERIOUSLY:
## BEYOND BEST INTERESTS

William Ruddick leads us beyond a simple-minded appeal to the best interests of children in determining levels of treatment. Nothing ever happens to anyone only in terms of his best interests, nor should it. Treatment and care are provided to adults not just in terms of what would be best for them. Consideration is given as well, and correctly, to the financial, psychological, and social costs to others. Societies establish levels of treatment by also asking the question, how will particular policies of providing care affect the providers and those who pay for the care? Ruddick's "biographical" question for parents, "How will treatment affect our lives?", reminds us that families should be concerned about their own burdens as well. The best-interests standard as a single standard is unreasonable. It would suggest that choices about the best interests of children can be made apart from decisions regarding the best interests of their families and society. Taken in isolation, it would suggest that obligations to achieve the best interests of children are absolute and cannot be defeated by costs or by obligations to others.

This onesided focus on the best interests of children to the exclusion of psychological and financial burdens to the parents and others is encouraged by many medical codes of ethics concerning care for children. The American Academy of Pediatrics, for example, in its joint statement excludes consideration of future quality of life.

When medical care is clearly beneficial, it should always be provided. When appropriate medical care is not available, arrangements should be made to transfer the infant to an appropriate medical facility. Considerations such as anticipated or actual limited potential of an individual and present or future lack of available community resources are irrelevant and must not determine the decisions concerning medical care. The individual's medical condition should be the sole focus of the decision. These are very strict standards ([1], p. 559).

In contrast, the American Medical Association supports quality-of-life considerations. Though the AMA statement does not rule out considerations of cost, it argues that they should not be primary:

Quality of life: In the making of decisions for the treatment of seriously deformed newborns or persons who are severely deteriorated victims of injury, illness or advanced age, the primary consideration should be what is best for the individual patient and not the avoidance of a burden to the family or to society. Quality of life is a factor to be considered in determining what is best for the individual. Life should be cherished despite disabilities and handicaps, except when the prolongation would be inhumane and unconscionable. Under these circumstances, withholding or removing life-supporting means is ethical provided that the normal care given an individual who is ill is not discontinued ([5], p. 10).

Neither of these two statements suggests that the burdens of the parents could be sufficient to justify termination of care. But the AMA did recognize the preeminent role of parents in making such decisions.

In desperate situations involving newborns, the advice and judgment of the physician should be readily available, but the decision whether to exert maximal efforts to sustain life should be the choice of the parents. The parents should be told the options, expected benefits, risks and limits of any proposed care; how the potential for human relationships is affected by the infant's condition; and relevant information and answers to their questions. The presumption is that the love which parents usually have for their children will be dominant in the decisions which they make in determining what is in the best interest of their children. It is to be expected the parents will act unselfishly, particularly where life itself is at stake. Unless there is convincing evidence to the contrary, parental authority should be respected ([5], p. 11).

The AMA takes the position that the parents should usually remain in authority over their children, even when it comes to the issue of discontinuing life-saving treatment, on the premise that parents' love for their children will bring them to choose in a selfless fashion.

In many respects, the AMA position meets Ruddick's desiderata. It enjoins the physician to take parents seriously and to regard them as responsible and loving custodians of their children. In this respect, the AMA position contrasts starkly with the U.S. government Baby Doe/child abuse regulations [7]. Moreover, the benefit of doubt is given to the parents. The AMA statement suggests that others must show that the parents are choosing improperly, not that the parents must show they are choosing properly. But the role of cost to the parents in such decisionmaking is discounted: the physician is not asked to tell the parents about costs and burdens.

To find a clear and unambiguous affirmation of considerations of costs to the parents and society, one can turn to Catholic moral theology: the traditional distinction between ordinary and extraordinary care allowed parents to stop treatment when the parents' burdens of care were severe. There was no obligation to provide all medical care that was clearly beneficial. Moreover, concerns for the family as a whole could defeat obligations for further treatment. Consider, for example, the analysis provided by a Jesuit theologian in 1956 of a case involving a child requiring permanent tube-feeding to maintain life.

> If gavage is required over a long period of time, say six months, this care would clearly prove very burdensome for the parents, and therefore they need not undertake such care for the prolongation of the life of this child, just as they would under ordinary conditions have no obligation of prolonging their own lives by such means. ... It is one thing, however, to praise as noble and truly Christian the determination to preserve an infant's life at any cost, and quite another thing to declare that failure to use even extraordinary means will make the parents guilty of very serious wrongdoing in the eyes of God ([4], p. 89).

The parents were allowed to ask the question, How will treatment affect our lives?, and if the answer was that it would constitute a grave or serious burden, they were normally free to discontinue further treatment. Though best interests of the child were important, they were not the only consideration. Nor, given the author's belief in an afterlife, was preserving physical life at all costs seen to be a reasonable endeavor. The author's view reflected a well-established view in Catholic moral theology, one in fact rearticulated in 1957 by Pope Pius XII. The Pope argued that

> normally one is held to use only ordinary means – according to the circumstances of persons, places, times, and culture – that is to say, means that do not involve any grave burden [*aucune charge extraordinaire*] for oneself or another. A more strict obligation would be too burdensome [*trop lourde*] for most men and would render the attainment of the higher, more important good too difficult ([6], p. 1031).

This approach provided a protection against an idolatry of physical life or high technology that can extend life at great costs and often with little quality.

The theory of ordinary versus extraordinary care even provided a brake on the inclination of the wealthy to spend available resources on prolonging their lives. Edwin Healy not only argued that individuals were in general not obliged to expend more than two thousand dollars on care, but he argued as well that this limitation applied to the wealthy.

> The absolute norm, on the other hand, establishes a maximum amount beyond which no one need go in spending money to care for his health. This norm is based on that which *people in general* would find very costly. The average person would experience very grave inconvenience in paying for medical care which cost a great sum of

money. It is difficult to fix the amount exactly, but it seems that in normal times $2,000 or more would certainly constitute such a "great sum" for the average man. Hence if the treatment required for one's cure of a fatal disease would cost $2,000 or more, he would not be obliged to employ so costly a remedy.

Let us suppose that an individual whose health requires costly treatments is exceedingly wealthy. He could, without being caused any inconvenience by the expense, pay for such medical care. Despite his financial status, treatments costing $2,000 or more would be considered extraordinary means of preserving his life. He would, it is true, find those means easy which the average person would find hard, and therefore he is in a very unusual position. In his case the absolute rather than the relative norm should be applied ([4], p. 68).

Given the papal statement that one must consider burdens not only to one's self but others, it follows that these considerations would apply also to the recommendations by physicians to parents. Even given the considerable inflation since 1956, the absolute limit envisaged by Healy would make most neonatal treatment in intensive care units supererogatory, not obligatory.

This moral theological approach to drawing the line between ordinary and extraordinary care, between that care that is obligatory to provide and that that is not, apart from its special theological valences (and Father Healy's financial calculations) carries with it three important moral points:

(1) The saving of human life, even the life of a child, is not the only or not even the most important goal of individuals or families.

(2) Families may, just as states may, set limits to their investment in saving human life because of the costs and burdens involved.

(3) Families are communities with a moral integrity of their own within which children are produced and raised, and therefore under the prima facie authority of parents.

The first two points are perhaps the easiest to accept. It is implausible that the mere extension of life is the most important human goal. One remains alive not simply to live, but in order to achieve the goals that living makes possible. Physical life itself has only instrumental value. When the attempt to stay alive defeats the realization of ultimate goals, the endeavor loses its purpose and validity. The second point follows in part from the first. When the costs of preserving life undermine the goals one wishes to achieve by remaining alive, the duty to preserve life is defeated. Beyond that, no obligations of beneficence are absolute. Costs in and of themselves will at some point defeat all duties of beneficence. The final point, however, is probably the most problematic, for there is no longer a general acceptance of the family as a moral entity with authority and purposes of its own.

The recent focus on protecting members of the family from each other through child abuse and spouse abuse legislation has made the family less of a sovereignty in its own right. Some individuals find it implausible that parents may touch and punish their children without the child's explicit permission in a way that would under other circumstances constitute assault and battery. However, one must recall that the family has developed in the West on a robust presumption that children are under the sovereign authority of their parents, which authority includes the right and duty physically to reprove children (He that spareth his rod hateth his son; Proverbs 13:24) and to make decisions regarding their care. It is this understanding of familial integrity that the Baby Doe regulations and similar laws have brought into question.

This is not the place to give an account of parental authority and the status of infants, young children, and adolescents [3]. Still, one must at some time face the question of the latitude that parents have in making decisions at variance with general communal understanding or state policies regarding the best interests and proper raising of children. As one acknowledges some level of parental authority and familial integrity, the burden of proof shifts to the state and its agents and an enclave of self-determination opens for the family. In terms of such an enclave, one can recast the first two questions posed by Ruddick, so as to look for ways in which physicians can support parents in their choices. There is, after all, an important function served by asking the physician's perspective with regard to treatment. Physicans cannot tell parents what it would be like *for them* to raise a seriously handicapped child. But an experienced physician can sketch out in general terms the geography of pains, costs, burdens, possibilities, and rewards that particular approaches to treatment are likely to offer. Generally, parents will be facing the problem for the first time. Physicians, especially physicians experienced in neonatology or the treatment of serious childhood diseases, will have seen the problem before in many patients and within many families. They can *help* the parents place the issue in perspective. But Ruddick is right, the physician cannot put the issues in perspective at least without the parents. The physician cannot substitute for the parents' own perspective on the geography of possibilities.

The questions therefore cannot be abandoned but must be brought into proper context, both through institutional changes as well as the education of physicians. Physicians must be taught how to place their expertise and experience at the service of parents when they are confronted with difficult decisions regarding the level of treatment they should provide for their children. Often this will involve the use of social workers and other allied health professionals. These are all points made by Raymond Duff and Alexander Campbell in their 1973 recommendations concerning the provision of high-technology care to handicapped neonates [2]. These recommendations of Duff and Campbell reflect much of the sense of moderation found in the moral theological account of ordinary versus extraordinary care. But Duff and Campbell also placed an emphasis on the patients as decision-makers.

It has not been possible to carry Duff and Campbell's views through in the 1980s, given the Baby Doe regulations with their intent to set aside considerations of costs and burdens to the parents, as well as judgments regarding the future quality of life of the child. If physicians wish to aid parents in stopping treatment under circumstances that do not conform to the rigorous constraints of the Baby Doe regulations, it is unlikely to take place in an open and candid fashion, given the legal sanctions at stake. Though I agree with Ruddick that parents should be able to talk about acceptable lives rather than the best interests of their children, they are well advised not to do so within earshot of an Infant Care Review Committee which is charged with the enforcement of the final Baby Doe/child abuse regulations [7]. Though Duff and Campbell's recommendations are in agreement with Ruddick's suggestion that parents must be given time to formulate and confront their options, the current regulations do nothing to encourage this. The physician/parent relationship, which Ruddick appears to endorse, is unlikely to be available, at least with regard to neonates or other children covered by the new Baby Doe/child abuse regulations. Except for a very narrow set of cases, the regulations take no interest in and are in fact inimical to physician's aiding parents in making a decision regarding treatment based on their own values or how treatment will affect their lives.

Finally, one must not lose sight of the fact that it is often appropriate for physicians to tell parents, after they have made their decision, that "You are making the decision I would, if I had it to make." When parents have been helped in making a decision appropriate to their circumstances, values, and capacities, they may still be burdened by uncertainty and guilt. Though physicians must often make decisions about ending treatment, for most parents this may be the first time that they have confronted the agony of such choices. In affirming a reasonable decision by the parents, physicians can absolve parents of unwarranted feelings of guilt and uncertainty.

Ruddick's examination of the "Questions Parents Should Resist" leads, as it should, to a reexamination of current policy. To recognize the concrete web of moral meaning, which is realized through and nurtured by families, one will need to return to greater parental authority in making decisions regarding the treatment of children. If one withdraws pediatric decisionmaking from the context of the family, one will be pursuing, as Ruddick correctly suggests, an abstract goal of child-oriented best interests at the expense of the environment in which concrete understandings of the best interests of children can be developed and nurtured. The difficulty is not simply or primarily with medicine. The cardinal issues are the limits of state authority and the provision of opportunities for parents to work with physicians to determine which choices of treatment are acceptable and obligatory. Only by lodging such decisions within the family, supported by educated counselors, can one hope that such decisions will be made, not only with proper concern for children, but also without a false entrapment in the promises of costly medical interventions. The neonatal intensive care unit, combined with the new Baby Doe regulations, is the full-fledged realization of the technological imperative: Save for the cases of permanently comatose newborns, what is not futile or virtually futile must be done. At stake is our understanding not only of families but of our relationships to technology and to the various goals we achieve in living.

*Baylor College of Medicine*
*Houton, Texas*

## BIBLIOGRAPHY

1. American Academy of Pediatrics: 1984, 'Principles of Treatment for Disabled Infants', *Pediatrics* **73** (April 4), 559.
2. Duff, R. S. and Campbell, A. G.M.: 1973, 'Moral and Ethical Dilemmas in the Special Care Nursery,' *New England Journal of Medicine* **289**, 890–894.
3. Engelhardt, H. T., Jr.: 1986, *The Foundations of Bioethics*, Oxford University Press, New York.
4. Healy, E. F.: 1956, *Medical Ethics*, Loyola University Press, Chicago.
5. Judicial Council of the American Medical Association: 1984, *Current Opinions of the Judicial Council of the American Medical Association*, AMA, Chicago.
6. Pius XII, Pope: 1958, 'Address to an International Congress of Anesthesiologists', November 24, 1957, *The Pope Speaks* **4** (Spring), 395–396.
7. U.S. Department of Health and Human Services: 1985, 'Child Abuse and Neglect Prevention and Treatment Program; Final Rule', *Federal Register* **50** (April 15), 14878–14901.

# SECTION IV

# THE PEDIATRICIAN'S ROLE: THEORY AND PRACTICE

# INTRODUCTION

Pediatrics is so thoroughly integrated into contemporary culture that it is hard to believe it only emerged as a medical specialty in the late nineteenth century. Until the eighteenth century, it was simply the women's task to attend to childbirth and to the care of young children. When women needed tha advice of an expert, they turned to the midwife for help. Perhaps, then, we ought to discuss the nurse, midwife and mother as paradigmatic pediatric professionals. But if we may be forgiven for limiting our field, we will consider the pediatrician as a representative for a number of pediatric health care specialties: nurses, social workers, and therapists. In this final section, then, we examine the growth of the specialty of pediatrics and inquire what special qualities may be important in taking care of the children.

Peter English, in his paper 'Not Miniature Men and Women: Abraham Jacobi's Vision of a New Medical Specialty a Century Ago', explores the exciting time when the unique specialty of pediatrics emerged. What was unique about it, in comparison to other fledging specialties, was that these "practitioners carved out an age group in contrast to an organ system, instrument, or procedure; and nearly all disclaimed the title 'specialists,' priding themselves as 'generalists' in the care of children" (pp. 247–248). In addition, these early pediatricians were advocates for a special point of view. This point of view, argues English, was the special contribution of Jacobi (1830–1919) who "clearly pointed to the interrelatedness of health, disease, and social policies" (p. 261). Jacobi is considered to have had the first appointment as a pediatrics professor in the United States.

English points out how Jacobi's vision of interrelating public policy and children's health forged pediatrics into a professional group concerned with the welfare of children in all areas of their life. He sought to separate the expert in the care of the mother (the obstetrician) from the care of the infant, arguing that a different kind of expertise was required to attend to infants' and children's needs for adequate growth, development, and health care. This was a departure from the dual task of the midwife as caregiver to both mother and young child.

Jacobi's version was shared and bolstered by many others. English recounts the central role of preventive medicine and social policy in the views of many early pediatricians. Among the most influential was Henry Dwight Chapin (1857–1942), who pioneered the study of the social aspects of child health and argued that general practitioners needed better training in pediatrics. Good nutrition, sanitation, better food, and living conditions could prevent disease in childhood effectively. Their focus on normal development, good nutrition, and hygiene led pediatricians to be advocates for children's health. It is difficult to separate health care for children from good nutrition, social supervision and schooling. From the beginning, pediatricians recognized the importance of all of these for the self-fulfillment and the well-being of children.

In 'The Development of Pediatrics as a Specialty', Todd Savitt asks why pediatrics developed late in the nineteenth century. He answers that there are certain conditions that have to be present before medical specialties are likely to develop, and that by the end of the nineteenth century they had been fulfilled in the case of pediatrics. First, sufficient knowledge had developed to form a new specialty. Second, the population had become concentrated enough to support specialists in the care of children. Third, new institutional arrangements, like foundling homes and hospitals for children, made it possible to study children's diseases and train pediatric specialists. Like English, Savitt stresses Jacobi's special role in American pediatrics. He looks at Jacobi's European medical training in the care of children to help explain why Jacobi played such a key role at the time pediatrics developed in the United States.

In reviewing some of the important figures in the development of pediatrics, English makes the case that pediatricians were especially concerned about children's general well-being; they did not narrow their focus to technical medical questions. This tradition thrives today. Many pediatricians, like Peter English and Thomas Irons, actively protect children from abuse and neglect, and help to improve their lives and opportunities.

John Ladd's 'The Good Doctor and the Medical Care of Children' explores a fundamental philosophical question: Must a good doctor also be a morally good person? To answer that question, Ladd first articulates an account of moral goodness based on the concept of virtue and then defends his account against two objections. One objection claims that requiring doctors to be virtuous demands too much of them. Ladd replies that this view falsely separates minimal duties of persons from acts of supererogation and assimilates virtues to the latter. Instead, Ladd argues, virtues are for everybody; acting virtuously is morally necessary, though how and how fully we express the virtues may be optional. A second objection is that there is no clear logical connection between being a physician and having particular moral virtues. This objection, Ladd claims, relies on a view of the physician as a craftsman or technician whose sole purpose is to apply medical knowledge to the cure or alleviation of disease. As an alternative to this model, Ladd offers a professional model of the physician which emphasizes an open-ended, person-centered, and nontechnical interaction between physician and patient in addressing complex problems. If, however, the physician's role is understood in this broader way (and Ladd implies that professionalism is the more adequate model for the physician's role), then the virtues will be a natural and desirable part of that model. Because physicians' activities are open-ended, they have clear opportunities to be virtuous. Because those activities have a significant bearing on the well-being of others, we want physicians to act for the benefit of others and to do this primarily for its own sake, not for external rewards. The attributes of concern for others and intrinsic motivation are, Ladd argues, essential to the concept of virtue.

In his comment on J. Ladd, Stuart Spicker emphasizes Ladd's characterizations of virtue as essentially other-regarding, that is, concerned with the good of another or others. Spicker argues that this understanding of the virtues illustrates the importance of community and inter-relationships as preconditions for the very existence of persons. Viewing community as a fundamental ontological category, Spicker points out, provides additional grounds for rejecting both the craftsman model of the physician's role and a merely physicalistic account of human existence.

In 'Government by Case Anecdote or Case Advocacy: A Pediatrician's View,' Myron Genel argues that pediatricians have traditionally been children's advocates. They have done this effectively in many different ways, such as by providing primary care, doing research and influencing social policy. He focuses on "case" advocacy, the kind of advocacy where so much attention is focused on a particular case by the media, that this "national grand rounds" becomes a powerful tool for advocacy on behalf of groups of children with similar conditions. But case advocacy often generates poor policy. "To a great extent all of this media attention and public curiosity have understandable origins, but without concerted thoughtful analysis of the issues presented by these cases, we risk descent into what I term 'government by anecdote'" (p. 306). What may be a very appropriate decision, he says, when we focus on a particular case, may become less clearly so when other considerations are brought to bear, including the cost or allocation of limited resources. He illustrates this by discussing some of the complexities with regard to mechanisms for obtaining and distributng organs for transplantation, and the controversies over Baby Fae, Baby Jessie, and others. Policymakers, he says, are often eager to dispose of a single case, "while avoiding the more difficult and complex problems of class advocacy" (p. 309). That is, it is relatively easy to obtain care for a single child, but much harder to deal with the problems of a just distribution to children in similar circumstances.

With regard to the "Baby Doe" policies, Genel offers us a very important perspective. He was a fellow at the Institute of Medicine assigned to the subcommittee on investigations and oversights of the House Science and Technology Committee at the time the Baby Doe policies were being considered. Genel "was struck by the oversimplified and somewhat doctrinaire approach that was being considered by congressional staff wholly unfamiliar with the subtleties and nuances of medical decisionmaking" (p. 311).

In his commentary, H. Tristram Engelhardt, Jr., examines the idea of advocacy that Genel employs. Engelhardt argues that advocacy is a complex notion and that there is no one unambiguous sense of it used by Genel throughout his discussion. Engelhardt writes, "to be an advocate is to take a particular position for a particular person or group of persons over against other positions and other persons or groups of persons" (p. 321). To clarify the value of advocacy we have to examine to whom and on whose behalf the advocate speaks, which values or rights are appealed to, and the authority by which the advocate speaks. Engelhardt discusses in detail one of the examples cited by Genel. He argues that those who favor the Baby Doe Regulations advocate their vision of what is in the best interest of certain children. They succeeded in establishing their view, and it is a new and different way to make controversial treatment decisions for critically sick infants. It is a fundamental change because in the past parents had more authority to decide what, within a range of medically acceptable alternatives, was best for their child. Engelhardt urges us to reflect critically on the different ways in which one can advocate, and for the need to justify the position one adopts.

In 'Loving the Chronically Ill Child: A Pediatrician's Perspective', Thomas Irons examines the complex problems of providing comprehensive and compassionate care for chronically ill children. Irons presents the virtue of love as the key to revitalizing a doctor-patient relationship threatened by fragmentation and dehumanization. He understands love in this context as "a posture of humility, matched with genuine commitment of the physician's physical, mental, and emotional resources to the child and family he serves" (p. 323) This virtue was exemplified for Irons in the personal relationships and house calls of physicians he observed as a child. Irons recognizes that in the past thirty years significant advances have been made in the treatment of children with chronic illness, but notes that treatment programs for such children have also become more complex, demanding, and expensive. To avoid the pain of developing a full and personal commitment to the care of such patients, Irons notes, physicians use a number of defense mechanisms. These include paternalistic control over the patient's care, defensive medicine, retreat into the scientific aspects of care, and using consultants to distance oneself from the patient and family. In response to these temptations, Irons urges his fellow physicians to maintain strong personal relationships with their patients and families, relationships marked by genuine commitment and open communication.

In his comment on Irons, John Moskop explores Irons' application of the concept of love to the physician-patient relationship. Irons, he notes, understands love in this context as a personal and active commitment of one's physical, mental, and emotional resources in order to foster the good of another. While Irons focuses on obstacles to realizing this love, Moskop considers what the limits of the physician's commitment to patients should be. He cites warnings about the dangers of over-identification with the patient and questions how or whether love is a realistic ideal in the care of some patients. He reminds us, for example, of the so-called "hateful patient." Finally, Moskop notes that scarcity of resources may prevent well-meaning physicians from expressing love for their patients by providing the care they need.

PETER C. ENGLISH

## "NOT MINIATURE MEN AND WOMEN": ABRAHAM JACOBI'S VISION OF A NEW MEDICAL SPECIALTY A CENTURY AGO

It does not require special powers of observation to note when you move into the pediatric section of a hospital, clinic, or emergency room. Health providers have frequently abandoned ties and whites; they cover over their formal name tags with Smurf characters; and they decorate their instruments with attention-diverting animals.

To obtain necessary information, pediatricians learn a variety of verbal and non-verbal forms of communication. They will obtain historical information in various positions (on the floor, at the door). They usually abandon the time-honored sequence of examination that proceeds from "head to toe" (that derives its order from the Edwin Smith Surgical Papyrus), and they usually are quick to seek alternatives to painful manipulations and procedures.

Pediatricians forsake office prints of Van Gogh for Winnie the Pooh; they are willing to work with mobiles dangling in their hair; and they sacrifice valuable hospital or clinic space for playrooms or schoolrooms.

In off hours, we find pediatricians on boards of education, advising mental health departments, coaching athletic teams, and teaching Sunday school. They are more likely to tolerate school failure, delinquency, pregnancy, or acne (unless, of course, these conditions occur in their own families). They keep up with the latest fads: jams, spiked hair, Star Wars, GI Joe, Castle Greyskull. Among physicians, only pediatricians consider CABG a doll and not a coronary artery by-pass graft.

The pediatrician's research approach to medicine – at least on the surface – appears different, geared to patients who must be approached at different developmental levels. Where did this attitude come from? One traditional role of history is to seek origins, and in this paper the tack I will take is to investigate early pediatricians' own views of the origin of their specialty.

In the last decades of the nineteenth century, a new medical specialty emerged, one that focused on the particular health requirements of children. This new specialty was something of an oddity: its practitioners carved out an age group in contrast to an organ system, instrument, or procedure; and nearly all disclaimed the title of "specialist," priding themselves instead as "generalists" in the health care of children. This paper takes aim at these practitioners. What special health needs did they believe children required? What were the scope and definition of the field that eventually took the name "pediatrics"? What were the distinctive features and peculiarities that set this group apart?

My context will be the American scene, roughly from 1880, when the American Medical Association established a separate section on "Diseases of Children," until the early 1930s when the American Academy of Pediatrics, the American Board of Pediatrics, and the Second White House Conference on Child Health and Protection [78] had clearly recognized a distinctly separate field of medicine. One can find a number of narrative accounts of the history of pediatrics (see e.g. [1, 23]). What I will focus on is the *self-image* of early American pediatricians as expressed in their speeches and writings. My method of investigation was relatively simple. With the help of the *Index-Catalogue to the Surgeon General's Library* (five series) and the annual volumes of *Index-Medicus*, I searched for statements on the development of a separate discipline of child health. Here I am deliberately not using the term "pediatrics," for these most comprehensive indexes did not employ pediatrics as a key-word until after 1903, preferring "pediatry," "pedology," or "child health," among several others. I did not dwell on particular procedures, therapies, or discoveries, concentrating instead on the concerns of leaders and practitioners of child health about the formation of their discipline. My quest took me to a wide variety of sources, but I want to mention a few particularly rich ones at the outset: *Transactions* of the New York Academy of Medicine, Section on Obstetrics and Section on Pediatrics (frequently published in the *American Journal of Obstetrics* (1868–1890), The Chairman's Address of the AMA's Section on the Diseases of Children (1880–), the Presidential Addresses of the American Pediatric Society (1888–), and the *Transactions of the Association of American Teachers of Diseases of Children* (1906–28) [1, 4, 11, 23–25, 64, 65 72].

## I. THE DISEASES OF CHILDREN

Before addressing the several questions on the origins of a specialty devoted to child health, I want to begin by asking: What illnesses did these practitioners treat? In 1878, an attending physician at the Demilt Dispensary in New York published the diagnoses of two thousand children he had seen in the period of one year [55]. He excluded those children suffering from dermatological or surgical disorders who were seen by other physicians. His

## TABLE I
Diagnoses of 2,000 children, Demilt Dispensary, New York, 1878

| | | | |
|---|---|---|---|
| Adenitis, cervical | 15 | Icterus | 10 |
| Anaemia | 29 | Indigestion | 73 |
| Ascarides | 313 | Malnutrition | 25 |
| Bronchitis | 700 | Pertussis | 62 |
| Bronchitis chronic | 138 | Pneumonia | 28 |
| Catarrh, chronic nasal | 18 | Rachitis | 96 |
| Cephalalgia | 10 | Scrofulosis | 23 |
| Cholera infantum | 15 | Stomatitis, simple | 15 |
| Diarrhoea, simple | 187 | Stomatitis follicular | 21 |
| Eczema | 33 | Syphilis, congenital | 14 |
| Entero-colitis | 80 | Tonsillitis | 97 |
| Fever, intermittent | 14 | Tonsils, chronic enlargement of | 12 |
| Feverishness from dentition | 14 | Urticaria | 15 |
| Furuncles | 22 | | |

Source: [55].

## TABLE II
Diagnoses of 1,000 children, Bellevue Hospital, New York, 1888

| | | | |
|---|---|---|---|
| Eczema | 23 | Diarrhea | 64 |
| Whooping-cough | 72 | Constipation | 13 |
| Bronchitis | 255 | Dysentery | 17 |
| Broncho-pneumonia | 13 | Quotidian ague | 15 |
| Lobar pneumonia | 13 | Chronic malaria | 15 |
| Mitral insufficiency | 46 | Scrofulosis | 20 |
| Acute tonsillitis | 25 | Rachitis | 11 |
| Diphtheritic | 18 | Conjunctivitis | 73 |
| Gastric dyspepsia | 19 | Abscesses | 10 |
| Gastro-enteritis | 59 | | |

Source: [9].

## TABLE III
### Diagnoses of 1500 children, Durham, North Carolina, 1937

| Gastrointestinal disorders | | Total 236 or 15.73% | |
|---|---|---|---|
| Diarrhea | 109 | Gastritis | 12 |
| Bacillary dysentery | 24 | Constipation | 12 |
| Pylorospasm | 15 | Intestinal Influenza | 13 |

| Acute infections upper resp. tract | Total 656 or 43.73% |
|---|---|
| Upper respiratory infections (including rhinitis, pharyngitis, tonsillitis, catarrhal otitis) | 482 |
| Otitis media, purulent | 87 |
| Adenitis, cervical | 27 |
| Conjunctivitis | 16 |
| Sinus infection | 14 |
| Bronchosinusitis | 10 |
| Croup | 13 |

| Acute pulmonary infections | Total 222 or 14.80% |
|---|---|
| Pneumonia | 76 |
| Empyema | 10 |
| Bronchitis | 23 |
| Influenza | 79 |
| Pleurodynia | 14 |
| Hilum tuberculosis | 14 |

Source: [49].

## TABLE IV
Diagnoses of 80,000 children, Harriet Lane Home, Johns Hopkins Hospital, Baltimore, 1935

| | | | |
|---|---|---|---|
| abscesses | (11,176) | malformations | (2169) |
| adenoids | (2538) | malnutrition | (2820) |
| allergy | (584) | measles | (902) |
| anemia | (1064) | meningitis | (1124) |
| acute arthritis | (772) | mental retardation | (1927) |
| behavior problems | (2360) | neurosis | (766) |
| birth injury | (716) | otitis media | (9842) |
| bronchitis | (3542) | parasites | (585) |
| dental caries | (5458) | pertussis | (2293) |
| chorea | (764) | phimosis | (1933) |
| conjunctivitis | (1127) | pneumonia | (5291) |
| constipation | (763) | polio | (511) |
| dermatitis | (545) | prematurity | (877) |
| diet regulation | (9186) | pyuria | (1013) |
| diphtheria | (994) | rheumatic fever | (559) |
| dysentery | (1492) | pharyngitis | (8179) |
| eczema | (2484) | rickets | (4398) |
| enuresis | (1320) | scabies | (1241) |
| epilepsy | (691) | stomatitis | (1479) |
| eye abnormality | (1318) | syphilis | (1862) |
| gonorrhea | (622) | tonsillitis | (7928) |
| heart disease, acquired | (1556) | hypertrophied tonsils | (3789) |
| heart disease, congenital | (524) | tuberculosis | (4301) |

Source [3]

year of practice is shown in Table I. Only diagnoses made at least ten times are included here. Respiratory and gastrointestinal ailments accounted for more than half of all diagnoses. Of note was the high incidence of worm infestation. Surprisingly few children suffered from diphtheria, rheumatism, anemia, croup, mumps, or tuberculosis. A decade later, a physician at the Outdoor Bureau of Relief at Bellevue Hospital in New York, Dr. A. Brothers, published the diagnoses of one thousand consecutive children he had attended (Table II) [9]. Once more only diagnoses made ten or more times are listed here. Again, outpatient pediatrics primarily dealt with acute respiratory and gastrointestinal disorders. Of note was the relative rarity of scarlet fever, polio, and meningitis.

Fifty years later, the illnesses bringing children to Arthur London, a pediatrician in practice in Durham, North Carolina, were strikingly similar [49]. London's patients represented the first 1,500 children, with 3,830 diagnoses, under his care. Sixteen percent had gastrointestinal disorders, fifty-eight percent had respiratory diseases (Table III). Of note, 9 girls had vaginitis, 9 children had accidental poisonings, 25 had unexplained injuries, 12 had burns, and 3 had skull fractures. Thirty-eight children died. Many of these children were undoubted victims of child abuse.

In contrast to these purely outpatient surveys, Jay M. Arena and R. R. Harris collected the diagnoses of the initial 80,000 children seen in the wards and the dispensaries of the Harriet Lane Home at the Johns Hopkins Hospital [3]. Table IV contains a list of those diagnoses made more than 500 times. Some of these diagnoses we would group differently today. For example, most of the acquired heart disease, acute arthritis, and chorea resulted from rheumatic fever and would undoubtedly be added to that category.

## II. JACOBI'S VISION

Abraham Jacobi was the acknowledged American leader in the drive for pediatrics, its most consistently articulate spokesman, and an early historian [42, 43, 56, 57, 71]. I will argue that it was Jacobi's particular conception of pediatrics that became pivotal in the molding of American pediatrics. Jacobi (1830–1919) fled Germany in the aftermath of the 1848 revolution seeking refuge in Manchester, England. His novel practice in the diseases of children failed, so Jacobi emigrated to Boston where his efforts met a similar fate. He moved to New York City where he became in 1857 a lecturer in the Pathology of Infants and Children at the College of Physicians and Surgeons. In 1860, he was appointed to what is considered the first pediatric professorship in the United States, at the New York Medical College. In 1865, he moved to the University of the City of New York, and in 1870 he became the professor of pediatrics at the College of Physicians and Surgeons. Jacobi was a prolific writer on all aspects of pediatrics, and his papers published to 1909 filled eight thick volumes. In addition, he published *Diphtheria* (1880), *Intestinal Diseases of Infants and Children* (1887, 1890), and *Therapeutics of Infants and Children* (1896, 1903).

Of crucial importance in Jacobi's development as an advocate for children's health was his early experience as an attending physician at two foundling homes in New York City, Nursery and Child's Hospital and Infant's Hospital [29]. He witnessed the high mortality rate in foundling homes (often exceeding 75%) from malnutrition, acute epidemics, and psychological neglect in these well-meaning, but unhealthful, institutions. At the request of the Commissioners of Public Charities, Jacobi wrote a report in

1870 that condemned these institutions [44]. What emerged was Jacobi's recognition of the interrelatedness of public policy, institutions, the vulnerability of the developing child, nutrition, and psychological growth. Child health, then, needed to include far more than the study and treatment of disease in children.

When, a decade later, the American Medical Association (AMA) held its first separate session for the Section on Diseases of Children, Abraham Jacobi chaired the half-day meeting. In his address that opened the meeting, Jacobi claimed that the need for pediatrics stemmed from a general heightening of concern in American society over the welfare of children in all aspects of their lives (education, nutrition, hygiene, infant care, child labor, juvenile crime, mental health, etc.) and in particular from a recognition, spurred on by more accurate vital statistics that arranged illnesses by age, that infant mortality in the United States was unacceptably high [67]. Jacobi directed the aim of pediatrics at lowering the high infant mortality rate and at the special needs of the institutionalized child. The pediatrician also needed to know the pathology and therapeutics of children, which Jacobi believed differed from adults. For him the significance of embryology in particular separated pediatrics from general medicine. Even in this early statement, Jacobi believed that pediatrics was more than the diagnosis and treatment of diseases in children. The new discipline needed to address the promotion of health in the well child:

Nothing is more vital to the raising of the baby than its hygiene, which comprises more than feeding alone, and has to pay attention to dress, air, sleep, bath and exercise, both physical and mental ([41], p. 713; see also [70]).

Mary Corinna Putnam Jacobi (1842–1906), Professor of Materia Medica and Therapeutics at the Woman's Medical College of the New York Infirmary, helped to define the distinctness of pediatrics in her opening lecture on the "Diseases of Children" at the post-graduate Medical School in New York in 1883. She argued that the "basis of this specialization ... is the fact that the child represents an organism in a state of continuously progressive development" ([47], p. 121). For both Jacobis, part of the distinctiveness of pediatrics came from the role of embryology before birth and the steady development of children from infants to adults that occurred after delivery.

In 1889, the American Pediatric Society met in its first regular session in Washington, D.C. The Society's first president was Abraham Jacobi. In his address to the forty-three delegates (21 of whom came from New York City),

Jacobi tried to define the scope of pediatrics. He wanted, he said, to:

> prove that pediatrics does not deal with *miniature men and women*, with reduced doses and the small class of diseases in smaller bodies, but that it has its own independent range and horizon, and gives as much to general medicine as it has received from it ([46], p. 8).

Jacobi argued that tissues reacted differently at different ages; for example, tumors of children differed markedly from those of adults. There were anomalies and diseases that occurred primarily in children, such as diphtheria [27]. Those anomalies and diseases that occurred at all ages frequently manifested themselves differently in children. An example was acute articular rheumatism (later called rheumatic fever) in which children suffered more heart destruction but less joint disease than similarly stricken adults. Jacobi maintained that therapeutics for children contrasted significantly from therapy in adults. The surgery of children consisted primarily of repairing congenital malformations, such as imperforate anus or cleft lip. Diet played a more crucial role in infancy than at other ages. Of special significance, Jacobi believed that the health of the republic depended on the sound health of its children:

> Unless the education and training of the young is carried on according to the principles of a sound scientific physical and mental hygiene, neither the aim of our political institutions will ever be reached nor the United States fulfil its true manifest destiny ([46], p. 16).

And, in Jacobi's view, diarrhea and respiratory illnesses could not be eradicated until society dealt with the poverty of the crowded city.

## III. THE SCOPE OF PEDIATRICS

Some members of the new specialty were quick to explain why pediatrics was emerging in the last years of the century. Most explanations claimed that the general practitioner, the person who treated most sick infants, simply did not have the knowledge to do so. A doctor in Indiana, J. B. Casebeer, claimed that:

> Many of our most successful practitioners of medicine amongst the adult population have made signal failures when called upon to exhibit their skill in the treatment of tender children ([12], p. 327).

To another audience, Casebeer stated that the problem lay in the exclusion of children from the revolution in therapeutics that was engulfing medicine [13].

Henry Dwight Chapin (1857–1942), a New York pediatrician who pioneered in social aspects of child health, agreed that the major reason why pediatrics was emerging stemmed from the recognition of the ignorance of the general practitioner [75]. As an example, in 1898 Chapin described his examination of thirty-five recent medical graduates from ten leading medical schools who were applying to serve as house officers in general hospitals. All "stumbled badly" in pediatrics, and all failed to answer correctly his questions on the causes of death in diphtheria, the complications of scarlatina, and the symptoms of measles [18]. W. S. Christopher, a pediatrician from Chicago and an early president of the American Pediatric Society, argued in 1894 that children's health:

has always been handicapped by superstitious and nursery legends, which cling to it like barnacles to a ship's sides and impede its progress .... The subject has been treated in a step-motherly way by the medical colleges .... The new graduate is afraid of a baby, readily accepts the diagnosis of the grandmother and not infrequently follows her treatment ([20], pp. 49–50).

Christopher followed Abraham Jacobi's lead in the purpose of pediatrics. He argued that "it cannot be too often emphasized that children are not little men and women"[22] p. 779. Development made the central difference: "The child is an unstable human being, constantly changing, now developing this organ or system with great rapidity, [then others] ..." ([22], p. 779). Nutritive, sanitary, and psychic factors independently influenced the expression of diseases in developing organs. So important was development that Christopher called it the "keystone of pediatrics" in his presidential address to the American Pediatric Society in 1906. A human's development consumed one-third a lifetime. Important subsidiary sciences for child health, then, included genetics, sociology, developmental anatomy and physiology (for example, how the function and structure of the kidney changed with aging), and nutrition [21]. When Thomas Morgan Rotch, professor of pediatrics at Harvard, welcomed the Section of Diseases of Children of the AMA to Boston that same year, he specifically acknowledged Jacobi's emphasis in development as the heart of pediatrics. Rotch stated that prophylaxis and hygiene were equal partners in the discipline [61]. In 1911 Henry Chapin singled out the concern for development and hygiene as the difference between pediatrician and general practitioner [17]. Thirty-five years after Jacobi had inaugurated the Section on Diseases of Children of the AMA, in

1915, Lawrence T. Royster of Norfolk, Virginia, stressed to that same group, which now numbered more than 1500 members, that a pediatrician must be equally competent diagnosing and treating illnesses of all kinds of children, running a children's hospital, and advising government on children's hygiene, such as water supply, school life, and child labor [63]. Royster's broad definition of the pediatrician's role was squarely in the tradition pioneered by Abraham Jacobi.

Each discipline has its boundaries, and such concerns are particularly acute in the formation of a discipline. Just who should treat sick children? How should the pediatrician relate to the general practitioner, the obstetrician, or the surgeon? Many, like Abraham Jacobi, believed that most sick children went either to obstetricians or to general practitioners. Despite his collaboration with Emil Noeggerath in the founding of the *American Journal of Obstetrics* in 1868, Jacobi thought that there was little rational connection between obstetrics and pediatrics after delivery, ignoring the midwifery tradition of caring for both mother and infant. Others believed that maternal habits could beneficially or harmfully affect the developing embryo. Pediatrics, then, rightfully extended to conception [36]. In contrast, Jacobi argued that "the diseases of children must never be torn away from the general practitioner" ([41], p. 710). In his eyes, the role for the pediatrician was as teacher, consultant to general practitioners, leader of children's hospitals, advisor to governmental officials, and researcher. As the number of pediatricians grew, pediatricians took on the additional role of providing primary medical care to children, thus placing them in competition with the general practitioner. Should the pediatrician or the surgeon perform operations on children? Samuel Kelley, a surgeon from Cleveland, stated that the "medical side and the surgical side of pediatrics are very closely related" [48]. He predicted, correctly as it turned out, that surgeons would emerge as the operators but that pediatricians would be best suited to manage the special needs of children in the pre- and post-operative periods.

Many leaders in pediatrics accepted Jacobi's charge that their special role was as teachers to medical students and to recent medical graduates. In the first years of the century, Harvard's Thomas Morgan Rotch recounted the progress of inserting material on pediatrics into the curriculum for medical students. Rotch recalled the evolution: From an occasional lecture on a pediatric topic included in a lecture in general medicine, to an optional elective for senior students, to a separate department with a course and examination required for graduation [62]. In 1898, J. P. Crozer Griffith, professor of pediatrics at the University of Pennsylvania, had surveyed the growth of the pediatric curriculum in 130 American medical schools. Of the 117 schools who replied, sixty-four (55%) had a chair in pediatrics; 43 (37%) had a chair that included pediatrics with another discipline; and three had a lectureship. Only seven schools had no training in pediatrics. Of the 64 chairs, 49 were held by full members of the medical faculty; 15 were "clinical" professors [34].

John Lovett Morse, who became the second professor of pediatrics at Harvard, described the required course for third year students at Harvard in 1905. It consisted of 30 lectures, most devoted to infant feeding, and specific diseases of the gastrointestinal, respiratory, and cardiac systems. Fourth year students could select an elective that took them entirely into the pediatric clinics and wards for a period of a month [53]. The value of such instruction is that the student brought an informed knowledge of specific illnesses into general practice. Clemens F. von Pirquet, the first professor of pediatrics at Johns Hopkins, believed this was particularly important for the study of acute infectious diseases – common illnesses that comprised so much of everyday practice. Before the insertion of pediatric courses into the curriculum, medical students missed this training because, Pirquet claimed, teaching hospitals failed to admit either children or patients of any age with infectious diseases [54].

General practitioners also desired additional training in the diseases of children. A number of medical schools developed "post-graduate" courses to satisfy this demand. Table V shows how post-graduate students spent their month at Harvard's course in 1915. As the schedule demonstrates, time was divided among the Boston City Hospital, Boston Dispensary, Children's Hospital, Infant's Hospital, Massachusetts General Hospital, and the outpatient department [52].

What were the special problems of the hospitalized child? It is certainly beyond the scope of this paper to investigate the history of children's hospitals (a greatly neglected topic) [68]. Nevertheless, I would like to offer a few reflections. One measure of the child welfare movement was the creation of separate children's institutions of all kinds. The most prominent of these were orphanages or foundling homes that proved very successful in gathering children. The U.S. Census in 1890 listed 780 such institutions housing 100,000 children. In New York State, 140 foundling homes cared for one of every 100 children, in New York City for one in 35 [35]. Although it was not their purpose, these institutions became medical facilities because of the biological and psychological vulnerabilities of developing children [20]. Most had infirmaries and dispensaries that served as training sites for future

TABLE V

Schedule of a postgraduate course in pediatrics, Harvard Medical School, 1915

July

| Hours | M | Tu | W | Th | F | S |
|---|---|---|---|---|---|---|
| 9–10:30 | Infants' Hospital Wards | Infants' Hospital Wards | Infants' Hospital Wards | Infants' Hospital Wards | Infants' Hospital Wards | Infants' Hospital Wards |
| 11–12:30 | Mass. Gen. Hosp. | Boston Dispensary | Mass. Gen. Hosp. | Boston Dispensary | Mass. Gen. Hosp. | Boston Dispensary |
| 1:30–3 | Children's Hosp. Wards | Children's Hosp. Wards | Children's Hosp. Wards | Children's Hosp. Wards | Children's Hosp. Wards | Children's Hosp. Wards |
| 3:30–5:30 | Boston City Hosp. | Children's Hosp. Lab | | Boston City Hosp. | Children's Hosp. Lab | |
| 4–6 | | | Children's Hosp. Outpatient | | | Children's Hosp. Outpatient |

Source: [54].

pediatricians such as L. Emmett Holt, John Howland, Joseph O'Dwyer, Job Lewis Smith, Edwards A. Park, and William P. Northrop. Even hospitals created specifically for the medical care of infants and children, such as Babies Hospital in New York, could not escape the broader social ramifications of treating poor children. Luther Emmett Holt explained that the reason forty-five percent of children admitted to Babies Hospital in 1898 died resulted from the fact that most were abused, starved, or drugged in addition to suffering from some illness. Holt realized that just being within the institution undercut the precarious health of these children. Many developed "hospital marasmus." Holt lamented that "the disease can be cured provided the child is sufficiently strong and old enough to stand the strain of life in a hospital" ([39]; see also [28]). In 1917, New York City claimed sixty-four hospitals with 2469 beds for sick children. With 800 beds, the New York Foundling Hospital was by far the largest [14].

Of the discoveries that pediatricians made during these years, two stand out so far above the others that they deserve mention, if not analysis. By far the most widely discussed topic was infant nutrition [2]. Pediatricians believed that the high infant mortality rate resulted primarily from diarrhea that they attributed to improper diet and to the inability of poor mothers to nurse their infants. A paramount goal was a safe, nutritious diet for the developing child. As Henry Dwight Chapin exclaimed in 1920, "it is a truism to state that proper nourishment forms the foundation of satisfactory growth and enduring health" ([16], p. 364). The wide dissemination of diphtheria anti-toxin was the other advance receiving close attention. Diphtheria anti-toxin assumed almost symbolic proportions as one of the first specific therapies to combat a relatively common children's disease. Although diphtheria was not as common as bronchitis or diarrhea, as the lists from dispensary visits showed, it was common enough to extend into every physician's practice. Pediatricians hoped that anti-toxin would be the harbinger of pediatrics of the future. Just a few years after the introduction of anti-toxin in the United States in 1897, Edwin Rosenthal, a pediatrician from Philadelphia, claimed that "the history of anti-toxin is inseparably connected with pediatrics" ([58], p. 284). Henry Chapin believed that the successful rise of pediatrics was intimately tied to the introduction of diphtheria anti-toxin, "the greatest triumph of medical therapeutics" ([19], p. 1086). Abraham Jacobi wrote well-received textbooks on both nutrition and diphtheria.

## IV. CHILD DEVELOPMENT AND NUTRITION

As we have seen, Jacobi's concern for the child's vulnerable development was a common thread that ran through nutrition, diseases unique to infancy, and the dangers of hospitalization. Thomas Morgan Rotch was a pediatrician who attempted to quantify this concern. Rotch is normally remembered for his "percentage" system of artificial feeding in which he advocated continual and precise changes in the protein and carbohydrate content of formula. In a sense, Rotch attempted to give infant feeding the mathematical precision that "formula" implied. What should be remembered is that the principle behind this complicated system was that infant digestion was developmental: babies require and are able to digest different nutrients than toddlers are able to digest, and so forth. Less well known was Rotch's pioneering use of X-rays to measure development. Rotch understood that a child's size alone did not signify maturity; there was simply too much variation in size in the population. Rotch found that radiographs revealed a distinct pattern of development in the order of appearance of epiphyses and centers of ossification that more precisely reflected development. The X-ray then could serve as a measure of growth and development [60]. In a similar vein, Arnold Gesell perfected means of measuring mental growth and development. As Gesell phrased it,

> Growth is itself a unifying concept wihch depolarizes undue distinction between mind and body, between heredity and environment, between health and disease ([31], p. 1055).

In his initial address to the AMA's Section on Diseases of Children, Abraham Jacobi exhorted his listeners that "nothing is more vital to the raising of the baby than its hygiene, which comprises more than feeding alone ..." ([41], p. 713). In addition to treating and diagnosing illnesses, the pediatrician had to advise parents on infant hygiene in order to facilitate healthy development. What were the contents of this advice that became central to "the well baby visit"? One way to answer this question is to examine the commonly used American pediatric textbooks. These textbooks had a distinctive style, an initial general section devoted to infant hygiene followed by a far longer section that detailed specific diseases. Job Lewis Smith's *A Treatise on the Diseases of Infancy and Childhood* (1869) spent 70 of its 600 pages in matters of hygiene. Smith wrote on the care of the mother during her pregnancy, prevention of mortality in early life, lactation, selection of the wet nurse, course of lactation and weaning, artificial feeding (only 3 pages), and the normal physical examination [69]. Infant hygiene to Smith meant successful breastfeeding. Thomas Morgan Rotch devoted 284 of 1200 pages of his *Pediatrics, The Hygiene, and Medical Treatment of Children* (1869) to infant hygiene. Specifically, Rotch addressed the normal infant at term, normal development, breastfeeding, artificial feeding, special care of the premature infant, and general principles of examination and treatment. By far the largest section (130 of 284 pages) dealt with artificial feeding, but Rotch's textbook clearly reflected concerns for development and hygiene [59]. Luther Emmett Holt spent 140 of 1100 pages of his textbook, *The Diseases of Infancy and Childhood* (1897), on various issues of infant hygiene. Specifically, Holt noted that his purpose was to provide his readers with information to instruct mothers. Holt singled out bathing, clothing, preparation of the nursery, care of the eyes, and ventilation as special areas of concern. Physical growth and development, nutrition (breast was still best but Holt gave the most space to artificial feeding), and vaccination rounded out his section on hygiene [38]. J. P. Crozer Griffith spent 250 of 1500 pages on infant hygiene in *The Diseases of Infants and Children* (1920). Griffith discussed bathing, clothing, sleep, exercise, amusement, breastfeeding (26 pages), artificial feeding (42 pages), foods other than milk, diet after the first year, diet in illness, peculiar characteristics of diseases in infancy and childhood, mortality and morbidity, and guidelines for therapeutics [33]. Two of these textbooks, many editors and editions later, continue to serve the American pediatrician. Holt's textbook has had John Howland, Luther Emmett Holt, Jr., Henry Barnett, and currently Abraham Rudolph as editors. Griffith's textbook was subsequently edited by A. Greene Mitchell, Waldo E. Nelson, and eventually Victor C. Vaughn, R. James McKay, and Richard Behrman.

## V. "SOCIAL PEDIATRICS"

Jacobi clearly pointed to the interrelatedness of health, disease, and social policy. Henry Chapin was an early advocate of this particular thread of Jacobi's conception of pediatrics. The appreciation of a social approach came to Chapin early in his career when he noted that children discharged from the hospital frequently returned in worse shape because life at home was so detrimental to health. In 1890, Chapin organized a group of women volunteers who made home visits. A paid woman physician assumed these responsibilities in 1894, but Chapin found that a nurse "social visitor" actually performed the tasks of supervising nutrition and hygiene better than the physician. The "social visitor" always had enough funds to buy food when necessary. In 1915, Chapin's social visitors made 2270 visits; in 1916, 2308 visits [15].

In 1913, Ira S. Wiles exhorted the Medical Society of the State of New York that pediatricians must be "leaders in the preventive medical work" ([76], p. 1257). Mortality and morbidity were intimately linked to sociology, and the pediatrician had to get involved with day nurseries, child labor laws, foster care, milk depots, and school health. Wiles coined the term "social pediatrics," by which he meant "pediatrics interpreted in the light of social origins and effects of the diseases of infancy and childhood" [77]. Henry J. Gerstemberger, a pediatrician from Cleveland, supported this approach and claimed that his medical students spent time making home visits and talking with religious, municipal, and philanthropic organizations as well as the more traditional work on the ward, in the clinics, and in the milk laboratory [30].

William N. Bradley spoke of the close relationship between pediatrics and child welfare groups in an address to the Philadelphia Pediatric Society in 1915 [5]. Bradley identified at least eighteen groups that had to work together for effective child health, including physicians, mothers, maternity nurses, schools, day nurseries, boards of health, and pediatric societies [5, 26].

Concern for a social approach to pediatrics led to a reorganization of the course for medical students at Albany Medical College in 1920. The novel syllabus [66] is reproduced in an appendix to this paper. Not all medical schools were willing to go this far toward "social pediatrics." John Lovett Morse published the class plan for Harvard's course in Pediatrics also in 1920 (Table VI) [5]. At Harvard, medical students learned of the social

TABLE VI

Arrangement for a course of lectures, Harvard Medical School, 1920

| | |
|---|---|
| 1. Diphtheria. | 20. Acid intoxication. |
| 2. Exanthems. | 21. Diseases of the blood. |
| 3. Exanthems. | 22. Congenital heart disease. |
| 4. Infant feeding – metabolism. | 23. Acquired heart disease. |
| 5. Breast feeding. | 24. Bronchitis and broncho-pneumonia. |
| 6. Artificial feeding – general principles. | 25. Pneumonia and empyema. |
| 7. Calculations in home modification. | 26. Diseases of thymus and status lymphaticus. |
| 8. Indigestion in infancy. | 27. Diseases of the kidney. |
| 9. Indigestion in infancy. | 28. Pyelitis and enuresis. |
| 10. Infectious diarrhea. | 29. Tuberculosis in infancy. |
| 11. Other diseases of digestive tract. | 30. Tuberculosis in childhood. |
| 12. Feeding of older children. | 31. Congenital syphilis. |
| 13. Premature infants. | 32. Infantile paralysis. |
| 14. Congenital malformations. | 33. Cerebral paralysis. |
| 15. Diseases of the new-born. | 34. Feeblemindedness. |
| 16. Diseases of the new-born. | 35. Meningitis. |
| 17. Scurvy. | 36. Infant mortality. |
| 18. Rickets. | 37. Infant welfare. |
| 19. Spasmophilia. | 38. Infant welfare. |

Source: [51].

dimension of pediatrics in only two or three of the 38 lectures, and those were clustered at the end of the course.

## VI. TAKING STOCK

What should this specialty in child health be called? By 1915 "pediatrics" had supplanted "diseases of children," "pediatry," and "pediology" among others. What was the fuss? When Juliet asked, "What's in a name," her speech tried to persuade that names should make no difference. In the hope of not trivializing Shakespeare's tragedy, the upshot of *Romeo and Juliet* is that names can make a great deal of difference. Some physicians, concerned about their developing specialty, thought "diseases of children" was too narrow and did not encompass growth and development, hygiene, and the social aspects of pediatrics. Nevertheless, the American Medical Association (1880) and the British Medical Association (1883) both labeled their child health arms "Section on Diseases of Children." Some pediatric journals also reflected this narrower view: *American Journal of Diseases in Children* (1911–), *Archives of Disease in Childhood* (1926–), and the *British Journal of Children's Diseases* (1904–41). The new term pediatrics was taken to be much broader than its rivals. As Harry McClanahan argued, "pediatrics" was more inclusive, taking into account prenatal care, intrapartum transit, heredity, nutrition, environment, and development [50]. Pediatrics became the generally used term, but did it always represent more than "diseases of children"? Certainly medical students at Harvard came away from their course with a different viewpoint from their counterparts in Albany. Henry Shaw in his Presidential Address to the American Pediatric Society argued that the leaders in the field gave more emphasis to the broader concerns of child health than did the rank and file. As proof, he cited that 25 of 40 presidential addresses to the society had dealt with hygiene, development, history, literature, social issues, and development, but only 68 of 1196 papers delivered at the scientific sessions covered similar topics. Shaw worried that pediatrics was hardly more than "diseases of children" [67].

What about the fact of Jacobi's broader definition of pediatrics in the hands of the practitioner? To answer this question, let's return to Arthur London's analysis of the initial 1500 children under his care. Table VII indicates that routine care, education of mothers, immunizations, and general examinations made up 39% of his practice [49]. The broader concerns of pediatrics reached into London's everyday practice.

## TABLE VII
### Diagnoses of 1500 children, Durham, North Carolina, 1937

| Summary | Total 3830 | |
|---|---|---|
| Newborn and congenital defects | 129 | 3% |
| Routine care | 469 | 12% |
| Immunizations | 604 | 16% |
| Routine examinations and miscellaneous | 430 | 11% |
| Acute upper respiratory infections | 656 | 17% |
| Acute pulmonary infections | 222 | 6% |
| Gastrointestinal disorders | 236 | 6% |
| Acute infectious diseases | 278 | 7% |
| Accidents | 63 | 1% |
| Allergic conditions | 99 | 3% |
| Chronic conditions | 43 | 1% |
| Nutritional and glandular conditions | 41 | 1% |
| Disorders of the nervous system | 108 | 3% |
| Cutaneous lesions | 243 | 6% |
| Genitourinary disorders | 92 | 2% |
| Surgical conditions | 117 | 3% |

Source: [49].

The post-war years were a period of taking stock. L. Emmett Holt, looking back upon his career, believed that pediatricians should be proud in their role of lowering the rate of infant mortality [37]. Between 1896 and 1922, infant mortality under five years of age in New York City fell to 9.6% from 14%. Holt attributed this striking decline, much of it in a reduction of diarrheal deaths, to pasteurization and the improvement of the supply of safe milk to more infants, to diphtheria anti-toxin, and to better hygiene – in other words, all primary concerns of pediatricians. Where pediatrics needed still more work, Holt contended, was in mental hygiene, prevention, periodic physical examinations, and health teaching. After examining the organization of the specialty, Holt argued that three types of pediatricians were required: the researcher who would develop new ideas, the clinician who would translate these ideas in the wards and clinics, and the public health pediatrician who would extend these experiences to the general population. Holt clearly conceived the relationship among the three groups as hierarchical, with the most vital group, academic research pediatricians, housed in departments of pediatrics within medical schools and children's hospitals.

Borden S. Veeder, who followed John Howland as professor of pediatrics at Washington University, St. Louis, was not as satisfied with the state of pediatrics in the 1920s. Veeder concurred that pediatrics, with 1900 practitioners in the AMA's Section on the Diseases of Children, was thriving, but he worried that the specialty was losing one of its distinctive characteristics, what he called "the child angle." Pediatricians knew much about diseases in children, but not enough about the normal growth and development of children [73]. Veeder was particularly critical of Holt's "trickle-down" hierarchical organization for the future of pediatrics. Veeder believed that public health and child hygiene were the most important for the well-being of children, but Holt had accorded these concerns the least prestige in his organizational chart. Veeder claimed that these efforts required the best brains and the most rigorous research, and he hoped the pediatricians in medical schools would tackle them [74]. If pediatricians remained in the laboratories, many prominent issues of pediatrics would be ignored.

J. P. Crozer Griffith worried about the ascension of academic pediatrics for other reasons. He believed that the pediatrics as practiced in the 1920s placed too much reliance on the laboratory and that the rare and unusual were awarded recognition over the prominent and common. As proof, he cited the tables of contents of several current issues of pediatric journals that contained discussions of obscure conditions [32]. Joseph Brennemann, a pediatrician of Chicago, lamented that the primary site of teaching and research had shifted from the dispensary to the children's inpatient service. He claimed that the more realistic place for teaching, diagnosis, therapeutics, and decision making should remain in the outpatient department. But this area, which Brennemann believed was also a fertile ground for creative research, was quickly becoming a second class citizen:

Dispensary work has long been, and in many quarters still is, considered a sort of necessary drudgery or apprenticeship that the younger man must serve for many years, partly to supply material for the older men on the wards .... ([7], p. 1704).

In addition, Brennemann was critical of a trend in research-oriented pediatrics that encouraged overly-zealous therapeutic intervention. He believed that close study of children revealed that the healing power of nature was particularly curative and that pediatricians simply had to ignore "The [perceived] necessity of doing something [that] is deeply rooted in human psychology ..." ([8] p. 3).

This account has stressed the importance of Abraham Jacobi, intellectually and organizationally, to the rise of pediatrics in the United States. Contemporary pediatricians shared this view. For example, Samuel S. Adams in his Presidential Address to the American Pediatric Society in 1897 reviewed the evolution of pediatric literature. In his account, Adams divided the history of the American pediatrics into a pre- and post-Jacobi period. He claimed that there had been only 24 works devoted to pediatrics published in America before Jacobi [2]. Abraham Jacobi, in his own historical accounts, found many additional precedents. Indeed, he pointed out that Ludwig Meissner's *Grundlage der Literatur der Paediatrik* contained 7000 titles pertaining to children's diseases prior to 1849. But Jacobi had a broader vision of pediatrics than those who went before him. In Jacobi's opinion what distinguished American pediatrics after 1860 were its humanitarian origins in religious and child welfare groups, its devotion to development, and its concern for social policy issues [10, 40, 45].

Criticism and self-analysis in the post-war years were strengths. Despite concern from several fronts that pediatrics had drifted away from Jacobi's conception of growth and development, hygiene, social policy, the vulnerability of children, and diseases of children, it is clear that this conception of pediatrics remained alive in 1930. Even Borden Veeder, the critic, claimed that the aims and objectives of pediatrics continued to be:

Knowledge that enables a child to be well born, to come into the world strong and healthy, to thrive lustily without halt in infancy, to develop and grow in a normal way during childhood, to be able physically and mentally to attain a sound education, to acquire good habits of living and avoid bad ones, to avoid psychological abnormalities so that the child adjusts himself to his social environment, to pass through the difficult period of puberty and adolescence in a sound normal maturity, to prevent by scientific measures so far as we are able to infectious diseases of childhood ... ([67], p. 7).

Abraham Jacobi would have been pleased.

*Duke University School of Arts and Sciences*
*Duke University School of Medicine*
*Durham, North Carolina*

## APPENDIX: INSTRUCTION IN SOCIAL PEDIATRICS AT ALBANY MEDICAL COLLEGE, 1920

The Child in Health:
    Anatomy and physiology of infancy and childhood.
    Difference from adults.
    Growth and development.
    Periodic physical examination.

Vital Statistics and Demography:
    Birth registration. Stillbirths. Illegitimacy.
    Methods of improving.
    Mortality statistics:
        Rates at different ages and seasons.
        Effect of season and climate.

Mortality During Childhood:
    Definition and significance.
    Distribution in the United States and other countries.
    General causes: Prenatal; natal; postnatal; preventable; nonpreventable.
    Causes by age periods.
    Effect of poverty and ignorance.
    Influence of domestic and social conditions:
        Age and nationality of mother.
        Effects of alcohol and venereal disease.
        Food; nursing; milk; proprietary foods; diet.
    Preventive methods:
        The mother (maternal work; number of children; age of mother).
        The child.
        The surroundings.
        Social conditions: housing, sanitation, etc.

Prenatal and Maternity Care:
    Childbirth statistics.
    Causes of death: Baby: fetal and congenital; mother.
        Instruction of expectant mothers.
        Systematic examinations.
        The mother in industry.
        Regulation of midwives.

    Prenatal clinics and maternity centers.
    Care of mother during pregnancy.
    Care of mother: at confinement; hospital.
    Prevention of blindness.
    The prenatal nurse.

Infant Hygiene:
    Foundling asylums; baby farms; boarding out and adoption; infant hospitals.
    Instruction of mothers: breast-feeding, etc.; proprietary food.
    Importance of pure, clean and safe milk.
    Infant welfare stations: municipal; private.
    Day nurseries: objects: regulations and inspection; the child welfare nurse.

Child Hygiene:
    Preschool period: diet and nutrition; physical examination; correction of defects; posture, teeth, adenoids, tonsils, rickets; nursing schools.

School age:
    Medical school inspection; physician-nurses.
    Periodic examinations: weight, height, etc.
    Early treatment of defects: vision, hearing, nose and throat, skin, etc.
    Dental clinics; mouth hygiene.
    Treatment of dental defects.
    Prevention of infectious diseases.
    Diet and nutrition; school lunches.
    Mental examination; special classes.
    Physical training; open air classes.
    Supervised play; recreation and playgrounds.
    Health education of teachers and pupils.
    Little mothers' leagues; Junior Red Cross; Crusaders.
    School sanitation, ventilation, fumigation, lighting, cleaning.
    Adjustable seats and desks.
    The school nurse.

Care and Education of Abnormal Children:
    Backward and mentally deficient children.
    Institutional care; commitment.
    The crippled child.

The blind child.
The deaf and dumb child.
The delinquent child; juvenile courts.

The Child in Industry:
    State and national legislation.
    Approved standards of child labor.
    Employment certificates.
    Educational and physical requirements.
    Supervision and periodic examinations.
    Widows' and mothers' pensions.

Tuberculosis in Children:
    Physical examinations.
    Protection of exposed children.
    Home supervision.
    Preventoriums; sanatoriums; day camps, etc.
    Follow-up work.

Child Welfare Propaganda:
    Extension and educational work.
    Exhibits, posters, moving pictures, newspaper publicity, pamphlets, etc.
    Lectures and demonstrations.
    Administration of Child Welfare centers.
    National Child Welfare Organizations:
        American Child Hygiene Association.
        Child Health Organization.
        Child Labor Committee.
        American Public Health Association.
        Parent-Teachers Association.
    State and local child welfare organizations.

Health Agencies:
    Federal:
        Children's Bureau.
        U.S. Public Health Service.
        Department of Education.
    State:
        State department of health.

Division of child hygiene.
Vital statistics.
Public health nursing.
State board of charities.
State department of education.

Municipal:
Health department.
Board of education.
Child welfare stations.

Private:
Day nurseries.
Maternity centers.
Playground associations, etc.

Source: [66].

## BIBLIOGRAPHY

1. Abt, A. F.: 1965, *Abt-Garrison History of Pediatrics*, W. B. Saunders, Philadelphia.
2. Adams, S.: 1897, 'The Evolution of Pediatric Literature in the United States', *Archives of Pediatrics* 14, 401–29.
3. Arena, J. and Harris, R.: 1935, 'The Frequency and Distribution of Diseases in Children', *Southern Medicine and Surgery* 97, 520, 22.
4. Beaven, P. W.: 1955, *For the Welfare of Children. The Addresses of the First Twenty-Five Presidents of the American Academy of Pediatrics*, Charles C. Thomas, Springfield, Illinois.
5. Bradley, W. N.: 1915, 'The Relations of the Pediatrist to the Community', *Archives of Pediatrics* 32, 106–14.
6. Bremner, R. (ed.): 1971, *Children and Youth in America, A Documentary History*, Harvard University Press, Cambridge, Vol. 2, Parts 7–8, 958–82.
7. Brennemann, J.: 1926, 'The Outpatient Department in the Teaching of Pediatrics', *Journal of the American Medical Association* 87, 1704–08.
8. Brennemann, J.: 1930, 'Vis Medicatrix Naturae in Pediatrics', *American Journal of Diseases of Children* 40, 1–17.
9. Brothers, A.: 1889, 'Report of 1000 Consecutive Cases of Diseases of Children, with Special Reference to Children', *American Journal of Obstetrics* 22, 386–98.
10. Caulfield, E.: 1957, 'General State of American Pediatrics in 1855, With Particular Reference to Philadelphia', *Pediatrics* 19, 456–61.
11. Caulfield, E. J.: 1931, *The Infant Welfare Movement in the Eighteenth Century*, P. B. Hoeber, Inc., New York.
12. Casebeer, J. B.: 1883, 'Paediatric Medicine and its Relation to General

Medicine', *Journal of the American Medical Association* **1**, 327–30, 361–64.
13. Casebeer, J. B.: 1884, 'Who Shall Treat the Children', *Archives of Pediatrics* **1**, 737–743.
14. Chapin, H. D.: 1917, 'A Pediatric Center in New York City', *Archives of Pediatrics* **7**, 874–81.
15. Chapin, H. D.: 1917, 'Hospital Social Service for Children', *Medical Record* **91**, 353–55.
16. Chapin, H. D.: 1920, 'How the Pediatric Teaching of Nutrition May Affect the Nation's Welfare', *Journal of the American Medical Association* **75**, 364–67.
17. Chapin, H. D.: 1911, 'The Fundamental Principles of Pediatrics', *Journal of the American Medical Association* **57**, 599–601.
18. Chapin, H. D.: 1898, 'The Question of Instruction in Diseases of Children', *Medical Record* **53**, 658.
19. Chapin, H. D.: 1917, 'The Relation Between Pediatrics and General Medicine', *Journal of the American Medical Association* **68**, 1085–88.
20. Christopher, W. S.: 1894, 'A Plea for the Study of Pediatrics', *American Journal of Obstetrics* **29**, 49–56.
21. Christopher, W. S.: 1902, 'Development, the Keystone of Pediatrics', *Archives of Pediatrics* **19**, 481–88.
22. Christopher, W. S.: 1894, 'Pediatrics as a Specialty', *Journal of the American Medical Association* **23**, 779–81.
23. Cone, T. E., Jr.: 1979, *History of American Pediatrics*, Little, Brown, Boston.
24. Cone, T. E., Jr.: 1985, *History of the Care and Feeding of the Premature Infant*, Little, Brown, Boston.
25. Cone, T. E., Jr.: 1976, *200 Years of Feeding Infants in America*, Ross Laboratories, Columbus, Ohio.
26. Eastman, A. C.: 1916, 'Pediatrics as a Specialty in the Smaller Cities', *Boston Medical & Surgical Journal* **174**, 739–40.
27. English, P.: 1985, 'Diphtheria and Theories of Infectious Disease: Centennial Appreciation of the Critical Role of Diphtheria in the History of Medicine', *Pediatrics* **76**, 1–9.
28. English, P.: 1978, 'Failure-To-Thrive Without Organic Reason', *Pediatric Annals* **7**, 774–81.
29. English, P.: 1984, 'Pediatrics and the Unwanted Child in History: Foundling Homes, Disease, and the Origins of Foster Care in New York City, 1860 to 1920', *Pediatrics* **73**, 699–711.
30. Gerstemberger, H.: 1914, 'Shall a Department of Pediatrics Include in Its Curriculum the Theoretical and Practical Training of Medical Students in Social-Medical Work Among Infants and Children', *American Journal of Obstetrics* **70**, 855–59.
31. Gesell, A.: 1929, 'Infant Behavior in Relation to Pediatrics', *American Journal of Diseases of Children* **37**, 1055–75.
32. Griffith, J. P. C.: 1929, 'Simplicity in Teaching', *Archives in Pediatrics* **46**, 312–17.
33. Griffith, J. P. C.: 1920, *The Diseases of Infants and Children*, W. B. Saunders, Philadelphia.
34. Griffith, J. P. C.: 1898, 'The Rise, Progress, and Present Needs of Pediatrics',

*Journal of the American Medical Association* **31**, 947–51.
35. Hart, H.: 1893, 'President's Address: The Relation of the National Conference of Charities and Correction to the Progress of the Past Twenty Years', *Proceedings of the National Conference of Charities and Correction*, 1–32.
36. Hoag, W. B.: 1912, 'Why Our Obstetricians Should Either Extend Their Line of Endeavors or Confer Earlier with the Pedotrophist, with Apologies for the Coinage of a New Word', *American Journal of Obstetrics* **65**, 355–58.
37. Holt, L. E.: 1923, 'American Pediatrics, a Retrospect and a Forecast', *Journal of the American Medical Association* **81**, 1157–60.
38. Holt, L. E.: 1897, *The Diseases of Infancy and Childhood*, D. Appleton & Co., New York.
39. Holt, L. E.: 1898, 'The Scope and Limitations of Hospitals for Infants', *Archives of Pediatrics* **15**, 801–14.
40. Jacobi, A.: 1917, 'A History of Pediatrics in New York City', *American Journal of Obstetrics* **75**, 334–37.
41. Jacobi, A.: 1880, 'An Address on the Claims of Pediatric Medicine', *Transactions of the American Medical Association* **32**, 709–14.
42. Jacobi, A.: 1902, 'History of American Pediatrics Before 1800', *Janus* **7**, 460–65.
43. Jacobi, A.: 1917, 'History of Pediatrics in New York', *Archives of Pediatrics* **23**, 1–11, 144–49, 191–98.
44. Jacobi, A.: 1872, Inaugural Address, Including a Paper on Infant Asylum Foundlings (delivered before the Medical Society of the County of New York), in *Collectanea Jacobi*, Vol. 6, pp. 217–318.
45. Jacobi, A.: 1904, 'The History of Pediatrics and Its Relation to Other Sciences and Arts', *Archives of Pediatrics* **21**, 301–33.
46. Jacobi, A.: 1890, 'The Relations of Pediatrics to General Medicine', *Transactions of the American Pediatric Society* **1**, 6–17.
47. Jacobi, M.: 1883, 'Opening Lecture of Diseases of Children, at the Postgraduate Medical School, New York', *Boston Medical and Surgical Journal* **108**, 121–23, 145–48.
48. Kelley, S.: 1910, 'In What Department and by Whom Should the Surgical Diseases of Children be Taught', *American Journal of Obstetrics* **62**, 941–43.
49. London, A. H., Jr.: 1937, 'The Composition of an Average Pediatric Practice', *Journal of Pediatrics* **10**, 762–71.
50. McClanahan, H. M.: 1910, 'The Advantages of Specializing in Pediatric Practice', *American Journal of Obstetrics* **62**, 935–39.
51. Morse, J. L.: 1920, 'A Plan for a Graduate Course in Pediatrics for General Practitioners', *Journal of the American Medical Association* **75**, 363–64.
52. Morse, J. L.: 1915, 'Boston as a Pediatric Center', *Archives of Pediatrics* **32**, 401–08.
53. Morse, J. L.: 1905, 'The Teaching of Pediatrics', *Journal of the American Medical Association* **45**, 570–10.
54. Von Pirquet, C. F.: 1910, 'The Importance of a Thorough Teaching of Infectious Diseases of Childhood in the Medical Curriculum', *Johns Hopkins Hospital Bulletin* **21**, 316–20.
55. Porter, R. B.: 1878, 'Report of 2000 Cases of Disease in Children, Treated at

the Demilt Dispensary, New York, with Notes', *American Journal of Obstetrics* **11**, 278–91, 544–58.
56. Preble, E.: 1932, 'Abraham Jacobi', *Dictionary of American Biography*, Charles Scribner's Sons, New York, Vol. 5, pp. 563–64.
57. Robinson, W. (ed.): 1909, *Collectanea Jacobi*, The Critic & Guide Co., New York.
58. Rosenthal, E.: 1900, 'The Value of this Section's Work and How to Advance It [Section on Diseases of Children, AMA]', *Journal of the American Medical Association* **35**, 281–87.
59. Rotch, T. M.: 1896, *Pediatrics, the Hygienic and Medical Treatment of Children*, J. P. Lippincott, Philadelphia.
60. Rotch, T. M.: 1910, 'The Conditions Pertaining to the Safeguarding of Early Life from a Paediatric Point of View', *New York Medical Journal* **91**, 1269–79.
61. Rotch, T. M.: 1906, 'The Opportunity of the Pediatrician', *Journal of the American Medical Association* **47**, 729–30.
62. Rotch, T. M.: 1903, 'The Study of Pediatrics in its Relation to Medical Education', *Journal of the American Medical Association* **40**, 949–53.
63. Royster, L. T.: 1915, 'The Pediatrician and the Section on Diseases of Children', *Journal of the American Medical Association* **65**, 567–69.
64. Ruhrah, J.: 1925, *Pediatrics of the Past*, P. B. Hoeber, New York.
65. *Semi-Centennial Volume of the American Pediatric Society, 1888–1938*, American Pediatric Society, Menasha, Wisconsin, 1938.
66. Shaw, H. L. K.: 1920, 'Social Pediatrics', *Journal of the American Medical Association* **74**, 1275–76.
67. Shaw, H. L. K.: 1929, 'The American Pediatric Society and Preventive Pediatrics', *American Journal of Diseases of Children* **38**, 1–9.
68. Smith, C.: 1983, *The Children's Hospital of Boston: Built Better Than They Knew*, Little, Brown, Boston.
69. Smith, J. L.: 1869, *A Treatise on the Diseases of Infancy and Childhood*, Henry C. Lea, Philadelphia.
70. Thompson, M.: 1886, 'Why Disease of Children Should Be Made a Special Study', *Journal of the American Medical Association* **7**, 399–402.
71. Truax, R.: 1952, *The Doctors Jacobi*, Little, Brown, Boston.
72. Veeder, B. S.: 1957, *Pediatric Profiles*, C. V. Mosby, St. Louis.
73. Veeder B. S.: 1923, 'Pediatrics and the Child', *Journal of the American Medical Association* **81**, 517–18.
74. Veeder, B. S.: 1924, 'The Trend of Pediatrics', *Archives of Pediatrics* **41**, 5–12.
75. Ward, P.: 1973, 'Henry Dwight Chapin', *Dictionary of American Biography*, 3rd supplement, Charles Scribner's Sons, New York, pp. 159–60.
76. Wiles, I. S.: 1913, 'Social Pediatrics', *American Journal of Obstetrics* **67**, 1255–57.
77. Wiles, I. S.: 1923: 'Social Pediatrics', *Archives of Pediatrics* **40**, 174–88.
78. *White House Conference on Child Health and Protection*, The Century Co., New York, 1931.

TODD L. SAVITT

# THE DEVELOPMENT OF PEDIATRICS AS A SPECIALTY

Peter English's "'Not Miniature Men and Women': Abraham Jacobi's Vision of a New Medical Specialty a Century Ago" [1] provides important insights into the history of pediatrics. Using the writings and speeches of early pioneers in American pediatrics, he has described some of the arguments these leaders used to convince fellow physicians and the public of the need for a medical specialty whose sole concern was children. In recounting how early leaders in pediatrics demonstrated that children were not "miniature men and women," Dr. English necessarily covers a number of topics in a short space. He gives a very close-up view, from the perspective of the participants themselves. In this commentary, I will take a few steps back and survey more of the scene to gain a sense of how the arguments English presents fit into the larger picture of turn-of-the-century medicine and society.

I include society here because the development of pediatrics as a specialty in the United States and Europe at the turn of the century stemmed in part from the recognition of a severe social problem (the care of needy and unwanted children) and from its resolution as a public policy issue (e.g., through the establishment of foundling homes). Unlike other medical specialties that focused on body systems or on approaches to therapy, pediatrics singled out people of one age group and considered all aspects of children's mental and physical development in society as being within its purview. So consistently and well did pediatrics fulfill this goal of general concern for children's well-being that until the 1930s, at least one respected medical historian has argued, "the specialty had a social rather than a scientific rationale" for its existence ([6], p. 219). Such an assessment probably exaggerates reality, and would likely have caused Abraham Jacobi to respond heatedly and defensively. English uses the words and ideas of pediatrics' founders to demonstrate that early practitioners did offer more than social policy justifications for the introduction of a children's medical specialty in late nineteenth and early twentieth century America, though these policy issues certainly played a major role. What were the medical and social contexts?

Actually, interest in pediatric problems was growing in both Europe and the United States throughout the nineteenth century ([3], pp. 84–124). By 1850, for example, Friedrich Ludwig Meissner could publish, in Leipzig, a huge bibliography of pediatrics covering the period 1472–1849. The founding of children's hospitals, the publishing of books and articles on various aspects of children's diseases, and a growing awareness of the social and medical plight of children further attests to rising concern for the well-being of children. Many nineteenth-century European medical schools even taught pediatrics along with other aspects of medical practice. No doubt Jacobi's medical training in Europe included courses on and clinical work with children. So he brought with him to an already interested United States that special training and strong personal concern for the health problems of children that allowed him to become the first full-time instructor in pediatrics in this country.

Why did this new discipline arise and take root in the United States after Jacobi took those initial steps to establish it? Both children and physicians had been in the United States for a long time, yet few healers, until the late nineteenth century, devoted their full energies to treating just that patient population. Why was that? One can ask the questions negatively: Did general physicians not care about children? Did they not perceive children's problems as unique? Were they afraid they would not earn enough money treating only children? Was medical knowledge not far enough advanced to establish and maintain such a specialty? Though there were, in these and earlier times, physicians interested in limiting their practices to children, conditions were not proper for support of pediatricians in the United States. We can phrase the question more neutrally: What was occurring in late nineteenth-century America that allowed, even fostered, the development of pediatrics?

Pediatrics did not develop alone as a medical specialty in the late nineteenth century. Other groups of physicians were beginning to limit their practices to, for example, women, otolaryngology, surgery, neurology, dermatology, genito-urinary problems, and gastro-enterology, and to form specialty societies in order to keep up with new knowledge and with colleagues in these fields. William Rothstein, a sociologist writing about the nineteenth century American medical profession, has proposed several necessary conditions for the formation of specialties:

Specialization could not develop in medicine until a number of conditions were fulfilled: (1) a medically valid body of medical knowledge and techniques had to develop in a given specialty; (2) urban population aggregates had to become sufficiently large to support a specialist in the practice of his specialty; and (3) institutions and arrangements within the profession had to make it financially rewarding for a physician to restrict his practice to a specialty ([5], p. 207).

The development of pediatrics fulfilled these conditions for medical specialization, but it also contained some wrinkles not present in the growth of other specialties. Let us look at each of Rothstein's points as they relate to pediatrics.

(1) "A medically valid body of medical knowledge and techniques had to develop in a given specialty." Pediatrics' pioneers struggled long and hard to convince general practitioners and their other medical peers that caring for a child's health required special training and knowledge different from that then offered in medical school. Medicine in general was, at this time, developing a new body of knowledge and so was passing through an exciting period. As physicians adopted the germ theory and found ways to slow or prevent bacterial growth in humans (immunology), as surgeons took advantage of anesthesia and antiseptic techniques to develop and perfect new corrective procedures, as new diagnostic tools like the stethoscope, ophthalmoscope, laryngoscope, and X-ray allowed physicians to "see" inside the human body without opening it, public confidence in the medical profession rose. What principles and practices distinguished what came to be known as pediatrics from those of general medicine? Jacobi and his colleagues, Dr. English shows, explained how childhood development, nutrition, hygiene, diseases, and treatments like diphtheria antitoxin, for example, made pediatrics different. I will not repeat those points here.

(2) One aspect of that new body of knowledge relates closely to Rothstein's second condition for the development of medical specialization: "Urban population aggregates had to become sufficiently large to support a specialist in the practice of his specialty." As I mentioned earlier, children and physicians had been around for a long time before pediatrics emerged as a specialty. But Irish and German immigrants of the early nineteenth century and southern and eastern European immigrants of the late nineteenth century swelled the populations of most cities, especially in the Northeast. It was not only the now increased concentration of children that suddenly caused physicians to sit up and take notice of children's diseases, however. It was, as Dr. English points out in an earlier article [5], the extremely poor health and living conditions of these children, whose parents often earned little money, left them to care for themselves during working hours, fed and clothed and housed them poorly, and barely cared for them in times of sickness, that spurred physicians and social reformers to respond to the special needs of

children. Proponents of child welfare, primarily in cities like New York, Philadelphia, and Boston, then established institutions, Rothstein's third condition for medical specialization.

(3) "Institutions and arrangements within the profession had to make it financially rewarding for a physician to restrict his practice to a specialty." Foundling homes and hospitals for children provided general practitioners and specialists in, for instance, surgery, neurology, dermatology, and ear-nose-and-throat, a chance to observe patterns and differences in children's social and psychological development, and in their responses to diseases and treatments. Here also were places where those physicians concerned with children could do their medical training and become pediatricians. They could further develop the unique body of knowledge that the Jacobis, Luther Emmett Holt, Job Lewis Smith, and fellow pediatricians would then use in their campaigns to convince other physicians of the need for a children's medical specialty. In meeting a social need with medical components (caring for poor neglected children who often had health problems), sensitive physicians founded pediatrics ([4], pp. 79–80).

As Dr. English indicates, other physicians did not always respond to the new specialty with warm enthusiasm. General practitioners had to be convinced, partly because their pocketbooks were at stake. Less threatening at first because originally pediatrics was to be a consultative specialty from which the general practitioner simply took instruction in treating sick children, pediatrics soon became a "primary care specialty" that took patients from the generalist. Pediatrics became a profitable specialty for physicians to enter as medical knowledge, population dynamics, and social and medical institutions joined together in the late 19th century. The time was right for the emergence of a new profession based primarily, but not solely, on social needs.

*East Carolina University School of Medicine*
*Greenville, North Carolina*

## BIBLIOGRAPHY

1. English, P. C.: 1989, '"Not Miniature Men and Women": Abraham Jacobi's Vision of a New Medical Specialty a Century Ago', in this volume, pp. 247–273.
2. English, P. C.: 1984, 'Pediatrics and the Unwanted Child in History: Foundling Homes, Disease, and the Origin of Foster Care in New York City, 1860 to 1920', *Pediatrics* 73, 699–711.

3. Garrison, F. H.: 1965, 'History of Pediatrics', in *Abt-Garrison History of Pediatrics*, W. B. Saunders, Philadelphia, pp. 1–170.
4. Rosen, G.: 1944, *The Specialization of Medicine with Particular Reference to Ophthalmology*, Froben Press, New York.
5. Rothstein, W. G.: 1972, *American Physicians in the Nineteenth Century, From Sects to Science*, Johns Hopkins University Press, Baltimore.
6. Stevens, R.: 1971, *American Medicine and the Public Interest*, Yale University Press, New Haven, Connecticut.

JOHN LADD

# THE GOOD DOCTOR AND THE MEDICAL CARE OF CHILDREN

This essay will be concerned with the question: must a good doctor also be a good person? More precisely, the question is: does the conception of what it takes to be a good doctor include as part of its meaning the requirement that he or she also be a morally good person? One can imagine asking the same sort of question about other roles or relationships: for example, does one need to be a morally good person to be a good friend, a good parent, a good child, a good judge, a good minister, or a good counselor? In the case of, say, a good chess player the answer would, of course, be NO. A person could be a good chess player but a thoroughly immoral person. As far as the others are concerned, the answer is more problematic. Such is the case of the good doctor.

The main question of this essay is a philosophical question of the traditional kind that Plato and Aristotle might have asked. For it has interesting philosophical ramifications, relating, for example, to general problems about the nature of goodness as well as more specific problems about the meaning of "good of a kind," of "good doctor" and of "good person." As I explore the basic question of this essay, I shall draw on a long tradition of philosophical analysis, including particularly some of the things that Aristotle has to say about these subjects. I mention this right at the outset because I want to emphasize that the question of this paper is of considerable theoretical interest to philosophers quite apart from its practical implications.

But the question, involving as it does questions about the criteria of a good doctor, obviously also has significant practical implications. For it would seem to be a necessary prerequisite for medical education, including the planning of the curriculum and the selection of students, that we start with some conception of what it is to be a good doctor. In general, we need to know this for the evaluation of doctors and their performance in many of the contexts where such evaluation is required. If only as an ideal to strive for, it would be useful to have some idea of what makes a good doctor. Indeed, our question is a timely one, for the medical establishment itself, medical associations as well as medical schools, have begun to concern themselves with the moral side of the selection, training, and certification of physicians.

Even laymen, e.g., parents as they go about selecting a doctor for their children, might find it useful to have an idea of what a good doctor is. For particularly in the care of children one might think that it would be important to have a doctor who is a morally good person – in some sense or other – in addition to being a good medical technologist.

In sum, there are both practical as well as philosophical reasons for asking what special moral qualities, if any, are needed for someone to be a good physician. Indeed, I should add that it should come as no surprise to find both kinds of reasons together because I am prepared to argue that, in the final analysis, the philosophical and practical cannot be separated from each other when one is doing ethics.

Obviously, it is impossible in a short essay to do justice to such a complex subject, much less to provide a complete or definitive answer to the question. The best that I can hope to do here is to suggest a program for further study by presenting some of the issues and indicating some ways in which they might be resolved.

Before continuing, I should point out that the question that we are concerned with here is different from many of the questions that are discussed in medical ethics, which more often than not belong under what has been called *quandary ethics*. Quandary ethics is concerned with "situations in which it is difficult to know what one should do" ([17], p. 14). It assumes that it is the business of ethics to provide guidance in such situations and perhaps even come up with an authoritative answer. Baby Doe quandaries come to mind. The present question is not so much concerned with quandaries of this sort but with what Pincoffs calls "moral enlightenment, education and the good for man," that is, with general issues relating to conceptions of good people practicing medicine in a good society: what we should value and what we should strive for [17].

Generally speaking, the underlying issue is whether or not doctors in their capacity as physicians have special moral duties and responsibilities over and above those of providing a medical service as a medical technologist, as an applied medical scientist, or as a "medical craftsman," as I shall call him. Here the plumber or auto-mechanic model of the physician comes to mind; do the moral requirements incumbent on doctors encompass anything more than "doing a good job" or "fulfilling their side of the contract" as might be expected of plumbers or auto-mechanics? [3] After all, even though they do not take the Hippocratic Oath, plumbers and auto-mechanics, like ordinary persons, are supposed to be honest and "to do no harm"! A more old-fashioned way of stating the issue is to ask: is medicine a calling or is it simply a business? (Under the latter we might also include pursuing a career as a course of lucrative employment "that offers advancement or honor.")

It should be observed that this essay is concerned with only one particular aspect of a doctor's duties and responsibilities, namely, what is to be expected of him morally in his capacity of physician in contrast to that of employee, contractee, gatekeeper, or arbiter of rights. Elsewhere, I have distinguished between the external and the internal morality of medicine [14]. The requirements of external morality arise from considerations coming from outside medicine such as patients' rights, for example, the rights of patients to refuse treatment even when treatment is medically warranted, or requirements stemming from a doctor's contractual obligations as the employee of a hospital. The internal morality of medicine, on the other hand, refers to norms governing medical practice as such that determine what is good clinical medicine. Here I intend to focus on the latter.

As far as the internal morality of medicine is concerned, it is clear that questions about what makes a good doctor depend on a number of other basic issues, such as: what are the aims of medicine? What is the purpose of clinical practice? What kind of service is the physician expected to provide? What kinds of problems is a clinician supposed to handle? How are the problems defined? And, most importantly, what kind of relationship should hold between the physician, the patient, and other persons whom he serves in connection with the medical treatment, such as parents, guardians, and family? Since these basic questions can be quite complicated, especially when it comes to pediatrics, I can only hint at the answers here.[1]

The situation is further complicated by the fact that today's (and tomorrow's) physician is practicing in a changing medical world, which has transformed his old roles and has thrust new roles on him. Thus, it is no longer always certain whom the doctor is working for and to whom he is responsible.

In pediatrics, this kind of question often presents a real dilemma: does the doctor's responsibility lie with the parents (the clients), or with the child (the patient)? Clearly the issues here may be extremely complicated indeed. Besides the question of who should decide about treatment, there are questions of what to tell the child and who should do it, and other aspects involving other members of the family or the community.

Sometimes the problem is stated in the terminology of rights: who has the right to decide? who has the right to know? who has the right to withhold information? and so on. Or more broadly, where do the rights lie? Who is to determine what they are and who is to enforce them? and how? These unresolved issues often make it difficult for the doctor to decide what to do, not because the rational medical course of action is uncertain, but because children, parents, public officials, health care providers and health care management – not to mention lawyers – are at odds over what should be done and who should decide. Unfortunately, as I have argued elsewhere, the solution to these quandaries is sought by invoking rights. Baby Doe-like situations are only the more dramatic instances of quandaries that are translated into conflicts of right, partly no doubt because our society has become a litigious society.

Interpreting and attempting to resolve these dilemmas as conflicts of rights is a kind of legalism, which, I argue, often distorts the underlying ethical issues [13]. I have suggested that the concept of responsibility is often a more satisfactory conceptual tool for understanding and evaluating these problematic situations. In this essay, I shall avoid the concept of rights altogether, for the reasons mentioned and also because my concern is not with quandaries but with more common, everyday situations in which the doctor is called on to be a good doctor, as when he has to deal, say, with children, in a primary care situation. Being a good doctor involves much more, I submit, than simply being a respecter of rights or even a good arbiter of rights who, like a judge, is supposed to decide fairly and wisely between conflicting interests, claims, and rights in a problematic health care situation.

It should be noted at this juncture that the emergence of new relationships and new requirements in the new medical world to which I have called attention makes it no longer plausible to equate a good doctor with a paternalistic one, one who is a moral dictator or a moral entrepreneur (to use Eliot Freidson's phrase) [8]. Crass paternalism, although regrettably still often practiced, is now passé. "Shared decisionmaking" is increasingly the order of the day and this essay will discuss the notion of a good doctor in that setting.

## I. GOODNESS AND VIRTUE

Let us now return to our original question: must a good doctor also be a good person?[2] Assuming that by "good person" we mean someone who is "morally good," our first question is: what do we mean by "moral goodness"? The traditional answer is that it means moral virtue and since there are many virtues, there are many ways of being morally good. Before turning to the concept of virtue, however, I shall briefly examine two other views of moral goodness. Although I regard neither of them as adequate, a discussion of them will bring out certain facets of the problem that we need to consider.[3]

To begin with, there is the utilitarian answer, which says that a good person is one who performs or has a tendency to perform good acts, that is, acts that produce happiness or reduce suffering. The classical objection to this position is that it makes success a necessary, perhaps even a sufficient, condition of being morally good. Success of this kind depends on favorable and often fortuitous circumstances, that is, good luck. So construed, it follows that utilitarianism would have to admit that a good man also has to be a lucky man and that a scoundrel might, on this definition, on occasion turn out to be a good person! This is because, according to utilitarianism, internal motivation is pertinent only as it relates to the actual production of utility or to the tendency to do so. This stance about moral goodness is thought to be morally objectionable by many philosophers, including Kant. I agree with them.

It should be noted, however, that even if from one point of view the utilitarian conclusion about moral goodness is unsatisfactory, it might still be used in another way to answer our original question about what makes someone a good doctor, namely, the results. For probability of success of this kind, i.e., tending to produce good things, is precisely what we have in mind when we talk of good plumbers or good auto-mechanics, that is, good craftsmen: none of these would be any good if he did not usually *succeed* at his chosen task. Accordingly, if we opt for the craftsman view of doctoring, a Dr. Fixit, we can accept a utilitarian answer to the question: what is a good doctor? (I shall examine the craftsman approach in more detail later.)

Another theory of moral goodness, which I also find unsatisfactory, identifies a good person as a person who always does his duty, does what he morally ought to do, and scrupulously obeys the moral rules, fulfills his obligations and respects the rights of others. Moral goodness, in other words, means being *conscientious*. Again, somewhat the same objection brought against utilitarianism applies here, too, namely, that mere compliance with or conformity to rules does not seem enough to qualify a person as morally good. A scoundrel might find it to his advantage to "be moral" (i.e., conscientious) in this sense. He might find it profitable to do what is right; being "moral" is good for business – it pays off. Against this, we feel quite properly that to be really good the action somehow has to come from within the person, "from the heart," as it were; otherwise it is hypocritical.[4] (I shall return to this point presently.)

In addition to the difficulties already mentioned, it should be pointed out that equating moral goodness with either of the foregoing conceptions trivializes the question about a good doctor, because it makes the moral evaluation of doctors and of their performance in principle no different from what it is for anyone else, e.g., plumbers and auto-mechanics. The question is begged from the outset, namely, that there is no internal morality of medicine.

Let us now turn to the traditional concept of moral virtue as an alternative way to go about trying to answer the question: what makes someone, e.g., a doctor, a morally good person? The virtue concept identifies being morally good with being virtuous, that is, having moral virtues of one sort or another.[5]

One immediate advantage for our purposes of using the concept of virtue is that there are many different virtues, that is, many ways of being morally good. This makes it possible to articulate the concept of moral goodness so as to be more specific about how and why a particular person, e.g., a doctor, is morally good.

The list of virtues has varied over the centuries. For the present discussion we may include such qualities as: kindness, considerateness, compassion, sympathy, benevolence, generosity, trustworthiness, honesty, truthfulness, loyalty, responsibility, courage, and, of course, wisdom. Others would add conscientiousness, humility, fairness, and justice. As far as we are concerned here, it is really immaterial which specific virtues are included in our list. An analogous list could be compiled of the opposites of the virtues, namely, vices; but I shall not bother to give a list of them, because it is more or less self-evident.[6]

## II. THE TRADITIONAL CONCEPT OF VIRTUE

A few general remarks about the traditional conception of virtue will help prepare the way for answering our question about the good doctor, because, as I shall argue, a good doctor characteristically exemplifies certain virtues.

First, virtues are considered moral qualities of the person himself and not simply properties of what he does or does not do, or of what he succeeds or fails to bring about. For, as I have already suggested, moral virtues are in some important sense "internal to the moral agent." Aristotle says that a virtuous action must proceed from "a firm and unchangeable character" ([2], 1105a35). In a somewhat different tradition, Kant says that virtue means to have a certain kind of maxim or end ([12], Introduction). They both imply that there is something inside the person that makes him virtuous over and above, or beyond, his compliance with moral rules or his successes.

For such reasons, virtues are often said to be character traits that a person has over a period of time and that are manifested in a variety of situations. As a more or less permanent trait a person is presumed to have a virtue independently of the circumstances in which it is exercised. Later in the essay, I shall argue that this aspect of virtues has to be qualified.

A second attribute of virtues is that to ascribe a moral virtue to someone, say, kindness or courage, is automatically to praise him. By the same token, to ascribe the opposite to someone, a vice, is to condemn him morally. Virtues are characteristically praiseworthy and vices are blameworthy. It is often maintained that to hold a person virtuous is to single him out for praise – as against other people.[7] I shall argue against this bias in favor of selectivity later in this essay.

In the meantime, it should be observed that the concept of virtue is essentially a social concept in the sense that it is used by others to evaluate a person-actor and his conduct. In that sense, ascriptions of virtue are intended to reflect the attitudes of outsiders–recipients, third parties, even impartial spectators, towards a person-actor, his character, his motives, and his actions. The social side gives rise to certain logical peculiarities of virtue attributions that will be discussed later and that form the basis of my eventual conclusion that virtues are essentially other-regarding qualities of persons and their conduct.

One further point about virtues needs to be mentioned here. A more specific explanation of what it means to say that virtuous conduct relates to something "internal" in the actor is that virtue implies a certain special kind of motivation. As Wallace points out, having a virtue means that one is *intrinsically motivated* to do certain things as contrasted with being "instrumentally motivated" to do them ([20], pp. 117–8). If one is honest, for example, one will act honestly "for its own sake," as it were, and not because it pays off. Sanctions and incentives motivate instrumentally (or externally) and if one needs them to act, then one does not have the virtue in question.

The requirement of intrinsic motivation explains why, if one believes (or knows) that in a certain case Dr. Brown had an external incentive (e.g., monetary profit) for acting morally, e.g., taking good care of his patients, it would be incorrect or a mistake to say on that account that he is virtuous, e.g., a compassionate or thoughtful person. In other words, an attribution of virtue can be quashed if it can be shown that the person in question was only acting for personal gain, because intrinsic motivation is implied whenever we ascribe a virtue to a person. Another corollary is that virtue, as depicted here, (conceptually) excludes basing virtue on self-interest, at least of the mundane variety.[8]

To sum up. These three attributes are part of the traditional repertoire of philosophical ideas about the virtues: virtues are character traits, they are praiseworthy, and they imply intrinsic motivation. Later in the essay, I shall propose two "emendations" to the traditional concept, which I believe will round out the concept of virtue and make it applicable to doctors in their capacity as physicians. The emended concept of virtue will provide the basis for answering our original question.

## III. TWO OBJECTIONS TO THE USE OF THE CONCEPT OF VIRTUE

At this point, we need to consider two common objections to using the ethical category of moral virtue to clarify the concept of the good doctor. The first objection is that bringing in virtue as a requirement for a good doctor is to demand too much of doctors; it requires them to be saints or heroes. The second objection is that it does not make any sense to include virtues among the criteria of a good doctor because there is no clear logical connection between being a doctor and having particular moral virtues. Professionalism, it is contended, lays down requirements about the kinds of qualifications that physicians ought to have and the kinds of services they ought to provide, but not what kind of person they ought to be from the moral point of view. I shall discuss these objections in the next two sections and shall try to show how one might go about answering them.

I shall begin with the first objection, which is that requiring doctors to be virtuous demands that they be morally better people than most others, that is, that they be exceptionally good. After all, the objection goes, when all is said and done doctors are just ordinary people; it would be "nice" if they had these virtues, but it would also be "nice" if everybody had them. To claim that doctors are, or ought to be, "paragons of virtue," if it comes from the doctors themselves simply shows how arrogant they can be or if it comes from their patients or a child's parents simply expresses a fond hope or is just a piece of extravagant flattery! In any case, it is not to be taken seriously.

Although this objection has plausibility when stated in this way, when examined closely it can be seen to rest on a basic misconception of the nature and role of the virtues, namely, that virtues are supererogatory and only for the select few. This view of virtue comes from a thoroughly mistaken and mischievous view of ethics, a view that I shall call a "minimalist ethics."

## IV. MINIMALIST ETHICS AND ACTS OF SUPEREROGATION

A minimalist ethics of the sort I have in mind maintains that ethics consists of a limited set of precepts setting forth what is morally necessary. These precepts relate to such things as rights and obligations flowing from them or to moral rules (or strict duties) of one sort or another. In other words, they relate to acts that are obligatory, forbidden, or permitted. The kind of ethics that explicitly or implicitly sets limits to duty could be called a "minimalist ethics" or "legalism." Whatever particular version it takes, this kind of ethics conceives of morality as a body of strict moral requirements of a quasi-legal sort, which, although limited in scope, are strong in stringency and represent what individual actors are said to be "obliged" to do. (The analogy with legal obligation is not coincidental.)

The logical consequence of a minimalist ethics of this kind is that acts over and above the call of duty (e.g., the observance of rights) belong in the category of *acts of supererogation* (= extras, options).[9] Such acts are optional in that, although they are good to do, they are not wrong not to do, that is, it is permissible not to do them. They might be regarded as a sort of "moral luxury." Those who perform them are singled out as especially praiseworthy; they are like saints and heroes [19]. Among the examples that are generally cited of acts of supererogation are saintly and heroic acts, acts of charity, generosity and giving, acts of kindness and consideration, and various kinds of volunteering.[10] The general assumption is that such acts are for *gratis*, as it were, and exceptional.[11]

It follows from a minimalist ethics that being a good doctor might mean two different things. Either (1) that he is good because he does everything that is morally necessary, that is, he does everything that is a duty, such as fulfilling all his obligations to his patient and respecting his rights; or (2) that he is good because he gives "extras," he performs acts of supererogation for his patients such as giving them extra time and attention or treating them with extra kindness or compassion, that is, he performs gratuitous acts for which he deserves special praise and that qualifies him to be included in the class of saints and heroes. This is the kind of exaggeration that is taken for granted in the objection being considered here.

Quite apart from philosophical objections to this dichotomization, it is difficult to see how to apply it to the actual conduct of doctors in their practice. If a pediatrician takes special pains to explain to a parent or child what is going to happen to him or consents ("volunteers") to perform a particularly painful bone marrow aspiration instead of letting a resident do it, are these "morally necessary acts" or are they "acts of supererogation"? Is he just doing his duty or is he acting like a saint or hero? Surely it is difficult to say which it is. For it becomes obvious that if one considers the array of things that doctors actually do for their patients it is impossible to draw the line between what they do "in the line of duty" and what they do "over and beyond the call of duty." The reason for this is simply that there is no such line. The dichotomy is a false dichotomy.[12]

The factitious character of the dichotomy is a result of bypassing the ethical category of virtue, which, I argue, does not fit under either the morally necessary or the morally optional (supererogatory). It should be evident from the list of virtues given earlier, that none of them is entirely supererogatory in the sense that they represent an optional and exceptional moral value. Nor are they as mandatory and stringent as are, say, strict obligations. Yet they represent an ordinary and common requirement of morality – in between the other categories, if you wish. Virtues are, as it were, for everybody; they are good for everybody and they are good for everybody to have. Soldiers do not have to be heroes, but they are expected to be courageous. Bankers and politicians do not have to be saints, but they are expected to be honest. Being considerate, kind and thoughtful are not extras – luxuries or frosting on the moral cake, so to speak – but something that is everyday and normal.

One of the attributes of virtue in general is that there are degrees of virtue and of particular virtues, so that some individuals may in certain respects be more virtuous than others. Complete absence, however, of a virtue where it is fitting (choose any from the list) quickly degenerates into its opposite, a vice. Thus, a person who is never generous may be said to be mean: for the opposite of generosity is meanness. The same goes for other pairs of virtues and vices. (It would be a mistake, however, to be too precise or formalistic about these categories.) In any case, it is important to note that, although one may be more or less virtuous, the complete absence of a virtue is a vice and in that sense virtue is not entirely optional.

Virtuous acts are, however, optional in another sense, namely, that there is latitude in the way and degree to which a virtue is exercised. As Kant says, the *when* and the *how* is up to the free choice of the individual ([12], p. 390). For example, how and to whom to be generous, kind, or thoughtful is up to the individual and different individuals can choose to be virtuous in different ways. Unlike stringent moral requirements such as obligations and duties that are usually set forth in rules, virtues are not rule-governed but depend on the actor's perception of what is required in a larger sense. (This will be explained later.) But as I have pointed out, even though virtue is optional in this sense, it is not absolutely optional; for otherwise, morality would condemn us to live in a world mostly dominated by vice, except when we are lucky enough to have saints and heroes to assist us in overcoming our problems – gratuitously, as it were!

## V. THE GOOD DOCTOR AND THE GOOD CRAFTSMAN

The second objection to the use of the concept of virtue in the analysis of what it means to be a good doctor is that there is no clear logical connection between being a physician and having one or another of the virtues. It is obvious that this objection is closely connected with the minimalist ethics just discussed. It is based on what might by analogy be called a "minimalist" conception of the doctor's role, that is, one in which the role of a physician is restricted to the role of a technologist. Following Aristotle, I shall call this restricted conception *the craftsman model*. In order to understand what is at issue here, it will be helpful to examine briefly what Aristotle says about craftsmanship (*techne*) and its relevance to our question about the good doctor.[13]

Among the various activities involving reasoning that can be said to be "excellent," Aristotle identified one particular kind that he called *techne*, a term that has been variously translated as "art," "skill," or "craft." He includes medicine under this category, along with shipbuilding, architecture, military strategy and household economics. The distinctive feature of crafts in this sense is that they involve making something and what is made, the product, can be distinguished and separated from the craftsman's activity. The craftsman's activity is evaluated by reference to the value of the product and not by reference to anything intrinsic to the activity or to him as a person. In terms that I used earlier, a good craftsman is a person who succeeds in bringing into existence something good – a good artifact or a good state of affairs. (For medicine this would be health.) Success is all that counts and, as Aristotle concedes, good craftsmanship is often a matter of luck (chance) ([2], 1140a18).

Two corollaries to Aristotle's analysis may be noted. First, the motivation of a craftsman is irrelevant to the value of the product *per se*. The craftsman may pursue his work for all sorts of incentives: for money, for status, or for fame – or just for fun. His motivation, in Wallace's words, is "instrumental."

Second, the craftsman's moral obligations are no different from those of other persons, i.e., he ought to respect the rights of others, to fulfill his contracts, to provide good service for the fee, etc. In terms I have used elsewhere, craftsmanship as such involves only "external morality."

The craftsman model as I have described it here corresponds quite well to one fashionable view of what it means to be a good doctor. According to this view, the aim of medicine, as it is for Aristotle, is health, and the role of the physician is to restore patients to health, that is, to cure or heal them. Depending on how far one is willing to go in medical reductionism, say, by defining health as the absence of disease, the aim of medicine and the doctor's role may be said to be the application of medical knowledge to the riddance, reduction, or alleviation of the patient's disease. If we couple the idea of health as the absence of disease with what has come to be known as the biochemical model of disease, we have a strong case for the craftsman or technological interpretation of medicine according to which if the physician succeeds he is a good doctor and if he does not, then he is not a good doctor. In this respect, the "modern doctor" is like a good auto-mechanic, he is a successful repairman.

We now need to ask: does the craftsman model, brought up to date, provide a satisfactory account of the aims of medicine and of the doctor's role? There are a lot of people who think that it does; the model is an increasingly popular one. Others, however, find the model unacceptable on a number of counts. One would expect that pediatricians, for example, would find it to be unsatisfactory as a model for the medical care of children, which, at least at the primary care level, could hardly be reduced to a technological operation. For it seems unwarranted in this context to rule out the human and social factors that are interwoven with a child's medical problems as beyond the scope of the physician's proper concern. Furthermore, with children there are always nagging problems connected with growing up, not to mention those connected with child-parent relations, that make the physician-patient encounter more than a simple repair operation. There are numerous facets of the treatment of children that cannot be pigeonholed as neatly as the craftsman model requires.

## VI. PROFESSIONALISM AND THE AIMS OF MEDICINE

As an alternative to the craftsman model, let us examine another model of the doctor's role and of the physician-patient relationship, which I shall simply call "professionalism" [15]. It might be called: "medicine as a vocation" ([4], pp. 119–20). Unlike the craftsman model, which allows for an interpersonal relationship between doctor and patient (and parents) only incidentally, the interpersonal side is of the essence for professionalism (in the sense intended here). In L. J. Henderson's terms, the physician-patient relationship constitutes a "social system" [9].

I cannot describe this alternative concept of the role of the doctor in detail here. So I shall simply mention a few important points.[14]

First, the professionalism (vocation) model conceives the goal of medicine less narrowly, less specifically and in a less well-defined way than it is conceived by the craftsman model. Considered as a profession, medicine undertakes to treat less tangible, fuzzier, and more intractable kinds of problems, many of which are not purely technological in nature. Indeed, one of the principal tasks in a client's first encounter with a professional is the identification of the problem itself. It is sometimes said that professions are oriented to the process (procedures) rather than to the product (success); for a lawyer can be a good lawyer even if his client loses in court and a doctor can be a good physician even if his patient dies. The aim of the service provided by a professional is to assist his clients with their problems, the assumption being that he can (almost) always do something to help a client even when he cannot save him. There is a lot that a humane physician can do for a dying child and his parents.

The professional relationship, in theory at least, focuses on persons rather than on things (e.g., bodies) or cases. That means that the service may and usually does include advice, teaching, consultation, and conversation. As I have said, the problems that a person or a family brings to a professional generally come without neat labels indicating what is wrong; rather, the situations are often confused, open, and in need of identification and clarification.[15] Decisionmaking resulting from a professional/client-patient relationship should, according to this model, not be imposed but shared. The doctor's role is not to dictate a decision but to assist in its being made; the optimum is a decision that reflects the consensus of all parties – one that they "can live with." That is not because a better (i.e., wiser) decision will be made that way, but because the integrity of the interpersonal relationship depends on trustful sharing. It should be clear that to provide the type of service intended here, everyone – physicians, nurses, patients and families – needs a change of orientation from the craftsman mentality to a sharing interactive model of some kind.

Second, at this point we can enlarge the scope of our inquiry beyond physicians to include other health care professionals such as nurses and social workers, and, since we are particularly interested in pediatrics, their counterparts, the recipients of the service, will include not only the child but also the parents and the rest of the family. In the new medical world and in pediatrics in particular, we are dealing with interlocking social systems, where relationships have to be established and conflicts resolved. In this context, medical diagnosis and decisionmaking must be looked at as a group undertaking rather than as an engineering project (or managerial task). In a sense, then, the moral job for a doctor becomes how best to create and sustain what Ray Duff calls a "moral community" [7].

A lot more needs to be said about this approach to the doctor's role, but what has been said here should suffice to make it clear that a larger view of this role will lead to an entirely different conception of what a good doctor is from the conception based on the craftsman model. Therefore, in order to answer the question about the good doctor we need to look for an ethical concept of the good doctor that makes sense when it is applied to the interpersonal, social process model of the doctor/client-patient relationship that I have just sketched. This is where virtue comes in.

## VII. VIRTUE RE-EXAMINED: SOME LOGICAL CONSIDERATIONS

Let us return to the concept of virtue once more. Right at the start, we should note that the concept itself has a complicated structure and has certain peculiar logical attributes that need to be taken into account in our analysis. These special attributes are rarely discussed in the literature, but they are nevertheless relevant for our inquiry because they provide the background and the arguments for the emendations to the traditional concept of virtue that I shall offer shortly.

The first thing to observe about virtue is a sort of paradox, namely, that it cannot serve as an end-in-view, that is, as something to be directly aimed at or to strive for. In this regard it is not an end like winning a race, learning French, or becoming a doctor. Unlike the latter, there is something odd about adopting the aim of becoming a generous, kind, or courageous person.[16] Virtues as such cannot directly become goals; on that account they might be said to be *transparent*.[17]

Since virtue as such cannot be an end or goal, it follows that performing virtuous acts cannot be means to the end of being virtuous in the sense that running part of the way is a means to winning a race, learning vocabulary is a means to learning French, or going to medical school is a means to becoming a doctor. Accordingly, one cannot give to charity in order to be generous; rather, charitable acts are part of being generous. By the same token, a person cannot (and should not) use people (e.g., the poor) or even use money as a means to being generous; persons and things cannot be used as a means to virtue simply because virtue cannot itself be an end. All this is part of the "logic," so to speak, of virtue.

Another example may help to show the pertinency of this point about transparency for clinical medicine. The nature of virtue is such that patients and their suffering cannot be used as a means to a doctor's becoming a compassionate doctor as they *can* be and sometimes, unfortunately, are used as a means to making money or advancing a career. Inasmuch as virtue cannot serve as a direct end, the child and his suffering serve not as the means but as the occasion for a doctor to be compassionate. They are the objects of his compassion and his end, a virtuous one, is the relief of the child from suffering. These constitute his aims, what he has in mind, so to speak, rather than the evaluation of his conduct as virtuous. The evidence for this is that he can determine for and by himself whether or not he has attained the end aimed at (e.g., relief of suffering), but he cannot determine for and by himself whether or not he is virtuous. Self-testing for virtue is not practicable.

Closely related to this paradoxical attribute of virtue is a peculiar asymmetry that exists between the way that the actor, the performer of a virtuous act, and others, such as the beneficiaries and third parties, perceive the act in question. The actor does what is considered to be virtuous by others simply because he regards it as something that needs to be done, that is, he perceives it as a requirement of the objective situation. Thus, a kind person acts kindly in response to another person's need and not for the sake of being kind. In the eyes of others, however, what the actor does is virtuous, say, a kind act. In general, then, the concept of virtue is such that it requires looking at one and the same act from two different points of view, under two different descriptions, from the point of view of the actor and from the point of view of others, recipients and observers.

The transparency and the asymmetry just pointed out suggest that we might distinguish between the *formal* and the *material* sides (or descriptions) of a virtuous action. The formal side is its being virtuous, which, I would claim, is a function of how it affects and is judged by others and of its role in producing and sustaining moral relationships between persons. The material side is the act or activity relating to the objective intended by the actor, that is, the state of affairs that the actor has committed himself to, such as the health or relief from suffering of the patient. In deliberating about what to do, the actor thinks only of the material side, e.g., what is to be done and the end-in-view; that is because the formal side is, as it were, transparent to him. On the other hand, others, beneficiaries and third parties, can think of the situation also on its formal side, i.e., that, how, and why it is a virtuous thing to do.

The validity of the distinction should be apparent when we think of typical illustrations of virtuous conduct. All we have to do is to think of how a courageous, compassionate, or generous act looks to the person performing it: how he would describe it and how he would explain and justify it. For such a person, the material side is the reality. For others, as I have pointed out, the formal side may be more salient. These "logical" points about the concept of virtue provide support for one of the proposed emendations of the traditional concept of virtue, namely, that virtues are essentially other-regarding.

Another but equally important logical attribute of the concept of virtue is its adverbial character, that is, the way that it functions as a "modifier." For it is a significant fact for the analysis of the concept of virtue that grammatically virtue-words and words for virtue and for particular virtues and their cognates are typically used as *adverbs*. That is, we say that a person acts virtuously: he speaks courageously, he gives generously, and so on. Indeed, we use a person's acting virtuously as evidence of his being virtuous. Dr. Jones is a compassionate doctor in that he treats his patient compassionately or with compassion, and so on.

Although the virtues are also predicated of persons, motives, character traits, and actions, we could, perhaps, conclude that they primarily represent ways of acting – and, by extension, qualities of relationships between persons in the context of their conduct towards each other. Inasmuch as they function as modifiers (like adverbs), I propose, then, that virtues be regarded, in their primary sense, as *modes of conduct* rather than as properties of particular kinds of persons or of particular kinds of acts. I shall call this the "modal interpretation of virtue."

## VIII. TWO EMENDATIONS OF THE CONCEPT OF VIRTUE

The logical attributes of the concept of virtue just mentioned suggest that the traditional concept of virtue might be emended in two ways: first, that virtues are other-regarding and second, that virtues are situation-dependent. These emendations will make it easier to use the concept of virtue to answer the question what makes a person a good doctor. Let us examine the proposed emendations more closely, starting with the first.

The strands of the argument thus far can be gathered together to reach the conclusion that, as modal qualities, moral virtue in general and specific virtues like those mentioned reflect qualities of interpersonal relationships, where one person (the actor) does something for another (the recipient). In other words, virtues, as a kind of moral value, characterize (or supervene upon) actions and activities as they flow from and embody relationships between two parties, which might be called: Ego and Alter, Self and Other, or actor and recipient. The actions and activities reflect and determine the quality of the relationship, which includes the attitudes and objectives (motivations) of the actor. The resulting quality of the conduct is what *makes* a person, as an actor, virtuous.

The duality of actor and recipient, the two poles of the relationship, explains both the transparency and the asymmetry mentioned earlier that led to the distinction between the material and the formal sides of a virtuous act. The interpersonal relationship itself may often be between unequals, in which case it would not be reciprocal. In the case of children, especially infants, we have unequals; for they are the receivers and doctors and parents are the givers. The interpersonal relationship does not, however, have to be between unequals: it may be between equals and reciprocal, a situation that is implied in the contractual model, where benefits are supposed to be mutual.[18]

In any case, virtuous action always implies another person, a recipient or beneficiary, who may be a child or an adult, one or many, and real or imaginary.[19] For, to be honest implies being honest to someone; to be kind implies being kind to someone; the same goes for all the other virtues on the list. It is unnecessary to go into further detail. The point is that virtues cannot exist in a vacuum and so a solitary person on a desert island cannot have virtues – or vices – for that matter. (Unless one thinks of oneself as consisting of two persons, as a Kantian might.)

Generally speaking, a virtuous action on the part of one person represents materially a commitment to the well-being (safety, health, integrity, etc.) of another. In different words, Kant says that virtue implies a commitment to the Idea of Humanity in other people (and oneself) [12]. The application of these notions to doctors should be self-evident; for doctors (and other health care professionals) can be and are virtuous insofar as they are committed to the welfare ("Humanity") of the patients, children, and families under their care.

Arguments like these can be used to support the first emendation of the traditional concept of virtue, viz. that virtues are other-regarding.[20]

The second emendation that I propose is that, being modal, virtues are not personal properties of individuals considered in abstraction from their roles and relationships with others. Rather, they inhere in particular acts and activities, in particular situations and in relation to particular people. Thus, they are part of the social context in which people live, work, and relate to others. The modal or adverbial side of virtuous action shows very clearly that virtues characterize ways of doing things and of relating to others.

The traditional account of the virtues errs in assuming that they are primarily properties of persons. It thereby implies that an individual possesses a virtue independently of its being exercised and takes it with him wherever he goes. In other words, one is virtuous (courageous, generous, kind, fair) in the abstract, *tout court*, and without reference to a context. As personal properties, virtues are comparable to other personal properties such as eye color or mathematical ability, which one possesses without reference to their being exercised. Aristotle suggested that a man can be virtuous while asleep ([2], 1095b33).

This account of the virtues usually qualifies the property view of virtues by holding that, unlike eye color, virtues are dispositions, e.g., dispositions to perform certain kinds of acts. Be this as it may, what I want to emphasize is that we need to focus on what the disposition is a disposition to do. From the point of view of evaluating dispositions, one could compare virtues in this regard with the operating properties of an automobile; when parked in the garage, it has the 'disposition' to run (like the virtuous man asleep). But the disposition is worthless if it is never exercised. It has no social reality. The same could be said of virtues.

The crucial point is that the property model ordinarily is oriented towards the person rather than towards his conduct. The modal model does the opposite. The difference between the orientations is evident if we ask: do we praise an action because it comes from a virtuous man or do we praise a man, i.e., call him virtuous, because he performs virtuous actions? The two different answers stress two entirely different sources of the moral value of virtuous conduct. The second emendation of the traditional concept that I propose here pays more attention to the *what, how*, and *what-for* of an action than to the *who*.

The modal interpretation of the concept of virtue also explains why a man who is kind or heroic in one situation can be exactly the opposite in another situation. A man can be a kind doctor and an unkind father, or the other way around. Accounts report that, although saintly in many regards, Gandhi treated his wife and his family abominably.[21] If one examines the list of virtues given earlier, it is clear that most if not all virtues are context-dependent in the sense intended in the second emendation and therefore serve as additional evidence for it.

## IX. THE GOOD DOCTOR AS A VIRTUOUS DOCTOR

Much of this essay has been concerned with breaking down the barriers to the use of the concept of virtue in the elucidation of the idea of a morally good doctor. I have sought to dispose of a few misconceptions concerning the physician's role and have offered two emendations of the traditional concept of virtue with a view to showing how the concept of virtue might be applicable to doctors in their role of physician.

The main thrust of my argument is that virtuous action always relates to kinds of actions or activities that take place in particular contexts involving particular interpersonal relationships, where one person does something for another. Therefore, it is not strange at all to find that the concept of virtue can be applied to individuals acting in specific roles such as the role of a physician, of a health care provider, or of a parent or friend. There is nothing amiss in speaking of someone as a compassionate or kind doctor, a courageous patient, a thoughtful parent, a considerate child, a responsible manager, or a fair judge, meaning thereby that he is morally good in one of these particular capacities. That means that the persons in question are *pro tanto* good, that is, good on account of the way in which they act in the contexts of the roles and relationships in which they live and work. But, as I have pointed out, being virtuous in one of these regards does not mean that the same person will be virtuous in other contexts and situations. It is possible, for example, for an individual to be kind and considerate as a doctor but not as a neighbor.[22]

Summarizing several points made earlier, there are three conditions relating to the attribution of virtues that are relevant to their application to doctors, to pediatricians in particular. First, the actors in question must have an opportunity to be virtuous. That means that there must be some latitude for them in choosing what to do, how to do it, and whom to do something for. For being virtuous implies that there are morally better and morally worse ways of doing something, e.g., a particular task. Second, the actors' actions and activities must have a significant bearing on the well-being of other persons, e.g., recipients such as patients and families. Finally, in order to be virtuous, the actors must be intrinsically motivated to do what they do for the recipients of their actions and activities.

Considering these conditions as well as other aspects of the concept of virtue that have been discussed, it seems reasonable to conclude that the doctor's role, like other socially significant roles, definitely provides the kind of opportunity in which attributions of virtue are feasible and appropriate.

Indeed, it seems obvious that it is both possible and desirable to have doctors who are kind, compassionate, generous, courageous, and wise in their capacity *as doctors*. Virtues like these are elements of what it means to be a good doctor.

In the absence of these virtues, a doctor might qualify only as a craftsman, perhaps a good one like a good plumber, because one or other of the conditions of virtue just mentioned is lacking. It is much worse, however, if a doctor fails to be virtuous, not through lack of opportunity, but because he has made doctoring into something else, where the patient and his suffering have become the means for his private ends, as when the doctor turns into a businessman whose chief object is to make money or into an organization man whose chief object is to advance up the escalator to success. Insofar as a doctor adopts on principle as his primary goal an object like one of these, the absence of virtues looked for in a physician degenerates into a vice, that is, moral corruption, which may be defined as a principled reversal of moral priorities.[23] That sort of corruption may appear to be banal; but then, as Hannah Arendt points out, most evil is banal [1].

*Brown University*
*Providence, Rhode Island*

## NOTES

[1] I have discussed these questions in other articles. See, for example, [14] and the bibliography therein.

[2] It is important to note certain typical logical characteristics of the term 'good' that provide the parameters of possible answers. Goodness is a peculiar kind of property; in fact, philosophers have often denied that it is a property at all; it is said to be "dependent" or, in technical jargon, "supervenient." For a thing, whatever it is, is always good for some reason; it makes no sense to say that something is good for no reason at all. Hence, when someone says that X is good, it is germane to ask: why? what makes it good? Accordingly, right at the beginning we have to ask: what *makes* a person a good doctor? and what *makes* a person a morally good person or a virtuous person?

[3] For label-lovers, the two conceptions might be called "teleological" and "deontological" respectively.

[4] Here I should warn against a frequent misinterpretation of Kant's moral philosophy, which attributes to him the view that what he calls a "good will" is identical with a virtuous will. Unfortunately, most writers who comment (ignorantly) on Kant have not taken the trouble to read his work on the virtues. See [12], especially p. 395.

[5] A good general discussion may be found in [17], pp. 75–100.

[6] See [17] for further details.

[7] "We can think of the language of virtue as providing the set of categories in terms of which we can justify our choices of persons" [17], p. 78.

[8] That is not to say, of course, that virtue and self-interest might not be compatible given some sort of fancy metaphysical or theological concept of self-interest; but that sort of concept is not at issue here. Theories of salvation or of perfectionism might also be accommodated in some way.

[9] For a comprehensive survey of the concept of supererogation, including a historical account and a bibliography, see [10].

[10] See [10] for details and a similar list.

[11] For a critique along the lines advanced here, see [17]. I have borrowed the term "moral luxury" from this book.

[12] The dichotomy is a logical consequence of legalism in ethics. See [13].

[13] Aristotle discusses *techne* in a number of places. The principal discussion is to be found in [2], 1140a1–23.

[14] Some of the points made here are elaborated in greater detail in [15] and [18].

[15] For more on the identification of problems, see [14] and [18].

[16] Of course, we might try to copy the conduct of someone we believe to have virtues like these in the hope that we will become virtuous. But here it is the person or his conduct that we strive to emulate, not the virtue itself. We cannot directly aim at that.

[17] Ordinary ends of the sort that one can aim at might, in contrast, be said to be opaque.

[18] This illustrates one of the difficulties of the contractual model of the physician-patient relationship, for in the medical encounter physician and patients are rarely equal and the benefits are lopsided.

[19] The recipient might be a god, an imaginary person, or a personified entity such as the State. Religious virtues such as faith, sanctity, or purity might be accounted for in this way.

[20] I think that Hume might agree with this [11]. Pincoffs, using the example of Gandhi, argues against it. See [17], especially chapter 7.

[21] Pincoffs summarizes some of these facts about Gandhi in [17], chapter 7.

[22] Here one might ask the casuistical question: can a doctor who cheats on his income tax nevertheless be a morally good *doctor*? The answer I propose here would be Yes.

[23] Kant defines a vice as "an action contrary to duty that has been adopted as a basic principle (maxim)" [12], p. 390.

## BIBLIOGRAPHY

1. Arendt, H.: 1963, *Eichmann in Jerusalem: a Report on the Banality of Evil*, Viking Press, New York.
2. Aristotle. *Nicomachean Ethics*.
3. Bayles, M. D.: 1981, 'Physicians as Body Mechanics', in A. L. Caplan *et al.* (eds.), *Concepts of Health and Disease*, Addison-Wesley Publishing Co., Reading, MA, pp. 665–675.
4. Bellah, R. N. *et al.*: 1986, *Habits of the Heart*, Harper & Row, New York.
5. Blois, M. S.: 1984, *Information and Medicine*, University of California Press, Berkeley, CA.

6. Bursztajn, M. D. et al.: 1981, *Medical Choices, Medical Chances*, Dell Publishing Company, New York.
7. Duff, R. S.: 1980, 'Moral and Ethical Dilemmas: Seven Years into the Debate about Human Ambiguity', *Annals of the American Academy of Political and Social Science* **447**, 19–28.
8. Freidson, E.: 1970, *Profession of Medicine*, Dodd, Mead & Company, New York.
9. Henderson, L. J.: 1935, 'Physician and Patient as Social System', *New England Journal of Medicine* **213** (2 May), 819–23.
10. Heyd, D.: 1982, *Supererogation: Its Status in Ethical Theory*, Cambridge University Press, Cambridge.
11. Hume, D.: 1789, *A Treatise of Human Nature*.
12. Kant, I.: 1797, *Metaphysik der Sitten, Zweiter Teil, Metaphysische Anfangsgruende der Tugendlehre* Friedrich Nicolovius, Koenigsberg. English Edition: *The Doctrine of Virtue*, M. J. Gregor (trans.), 1964, Harper & Row, New York. (Numbers refer to the standard Prussian Academy edition.)
13. Ladd, J.: 1979, 'Legalism and Medical Ethics', *Journal of Medicine and Philosophy* **4**, 70–80.
14. Ladd, J.: 1983, 'The Internal Morality of Medicine: An Essential Dimension of the Patient-Physician Relationship', in E. Shelp (ed.), *The Clinical Encounter*, Kluwer Academic Publishers, Dordrecht, pp. 209–231.
15. Ladd, J.: 1984, 'Philosophy and the Moral Professions', in J. Swazey and S. Scher (eds.), *Social Controls and the Professions*, Oelgeschlager, Gunn and Hain, Cambridge, Mass, pp. 11–30.
16. MacIntyre, A.: 1981, *After Virtue*, University of Notre Dame Press, Notre Dame, Indiana.
17. Pincoffs, E.: 1986, *Quandaries and Virtues*, University of Kansas Press, Lawrence, Kansas.
18. Schon, D. A.: 1983, *The Reflective Practitioner*, Basic Books, New York.
19. Urmson, J. O.: 1958, 'Saints and Heroes', in A. I. Melden (ed.), *Essays in Moral Philosophy*, University of Washington Press, Seattle WA, pp. 198–216.
20. Wallace, J. D.: 1978, *Virtues and Vices*, Cornell University Press, Ithaca NY.

STUART F. SPICKER

## COMMENTS ON JOHN LADD'S 'THE GOOD DOCTOR AND THE MEDICAL CARE OF CHILDREN'

John Ladd's paper explores two important questions: What properly constitutes the principal aim of medicine and of the good doctor? Is there something about the medical care of patients, especially children, that requires special non-technical qualities? Frequently, answers to these questions focus on medical traditions. Ladd, however, turns to the domain of moral philosophy. I shall not iterate Ladd's analyses and conclusions, but merely discuss a few issues which, I hope, will stimulate further reflections.

First, it would be interesting to ask Ladd why he finds it so useful to return to, or resurrect, virtue theory from the history of moral philosophy or ethics. He clearly rejects Aristotle's theory of virtue ethics, on the grounds that (1) the craftsman model of medicine and doctoring is far too narrow and would equate the actions of the auto-mechanic with the actions of physicians, and that (2) Aristotle's virtues tend to be descriptions of the characteristics of individual persons and do not adequately signal social or communal roles or enterprises. Ladd suggests that the traditional concept of virtue be emended to take account of the other-regarding and context-dependent character of virtue. Given these emendations, however, virtue ethics is able to avoid the false dichotomy between duty and supererogation implied by a minimalistic ethics. Ladd's conclusion that virtue requires interpersonal relationships suggests the importance of the notion of a moral community in health care as proposed by Yale pediatrician Raymond Duff. It is what Ladd tersely expresses with the statement "Virtues are for everybody." Yet they are not abstract; they are conceptually clear only if a self and an other pre-exist in this essentially relational character.

The emendation of the traditional conception of the virtues (and vices) which Ladd advocates as a fruitful line of approach in addressing the meaning of the 'good physician' and the 'proper task of practicing medicine' has, it is important to observe, been addressed by a regrettably little-known Scottish philosopher who died not long ago – John Macmurray. In his important books, *The Self as Agent* [1] and *Persons in Relation* [2], Macmurray construes persons (or persons in relation) as irreducibly social; that is, an individual exists when he or she is in dynamic relation with an Other. One's being is confirmed, so to speak, only in this relationship. I exist, then, only as one element in the complex,"You and I." On this dyadic interpretation it follows that a person is clearly a human being, a biologically and genetically endowed individual of a specific sort, but such a human being, completely alone in the world (unless defining himself or herself in terms of his/her relations to other persons), could not be a person.

One corollary that follows from Macmurray's conceptual scheme, and that bears strongly on Ladd's emendation of the traditional theory of virtue ethics, is that if the "I-You" is the unit of the personal or the communal, community becomes a special characteristic of being a person. So whereas every human being is an individual, the individual is not the unit of personal existence, and hence traditional virtue language which tends to characterize individuals is virtually meaningless.

Ladd's work relies on traditional virtue ethics while pointing us beyond it. It seems to me that this approach is useful and Ladd offers us a framework which would enable us to clarify many important matters such as: what is the relation that should obtain between being *a good doctor* and being *a good person*? When should physicians and patients assume formal roles and duties to each other, and when should they not? Ladd is especially to be congratulated for clarifying why referring to physicians as auto-mechanics is both horrendous and bewitching. This also exposes the corollary and equally mischievous and muddled notion, that our living bodies are nothing more than complex machines. Such reductionism and physicalism leave only estimates of technical excellence as a criterion for what a good physician should be. They do not leave a place for consideration of how moral virtues may be integrated into the training of physicians. The virtues of a technician may be demanding but they are different from important moral qualities needed for the good doctor. Ladd's paper points us beyond a minimalistic ethics and also beyond a virtue ethics as classically conceived. I have emphasized the critical importance of the concepts of *moral community* and *persons in relation* for his view.

*University of Connecticut Health Center*
*School of Medicine*
*Farmington, Connecticut*

## BIBLIOGRAPHY

1. Macmurray, J.: 1957, *The Self as Agent*, Faber and Faber, Ltd., London.
2. Macmurray, J.: 1961, *Persons in Relation*, Faber and Faber, Ltd., London.

MYRON GENEL

# GOVERNMENT BY CASE ANECDOTE OR CASE ADVOCACY: A PEDIATRICIAN'S VIEW

I have decided to address the role of the pediatrician as an advocate for children in the development of health and social policy. Since this is a very broad charge, and there are many other pediatricians who have labored long and valiantly as advocates for children, it would be presumptuous and misleading for me to attempt to represent them all. Suffice it to say that the efforts of these pediatricians in a variety of areas ranging from calling attention to the horrors of a nuclear holocaust to the eradication of dread diseases such as smallpox and polio would have to rank as among some of the most noteworthy in the past and at present.

There is, in addition, a different form of advocacy performed on almost a daily basis by practicing pediatricians on behalf of their child patients – so called "case" advocacy. In his foreword to the report of the Ross Roundtable entitled "Child Advocacy in Pediatrics" Dr. Eli Newberger cites a North Carolina pediatrician, Dr. Oliver "Bo" Roddey of Charlotte, in this regard:

> On one end of the spectrum are the professional advocates and at the other end those in private practice, with academic pediatricians being somewhat in between. To get the job done we need all varieties, from those who would make "capital cases" out of the issues and are willing and prepared to do this on the most public scene, utilizing the media, etc., to those who, in a more behind-the-scenes manner, practice case advocacy with their patients and expand on this at times to benefit larger groups ([22], pp. 11–12).

In this essay I wish to share vignettes from a personal odyssey and to examine the policy implications of some recent cases which have achieved considerable notoriety. While there is an element of case advocacy in these stories, their far-reaching policy implications lend themselves more readily to what Knitzer has labelled "class" advocacy or advocacy on behalf of groups of "vulnerable" children [17]. I have chosen this approach, since it seemed that I should write from personal experience, which on reflection is quite limited, rather than regurgitate the activities and thoughts of others who could more deservedly be categorized as pediatrician-advocates.

The Babies Doe and Baby Fae are some of the more illustrative cases out of a recent series, all poignant and heart-wrenching, that have been presented for "national grand rounds" by the media during the past five years. Aside from the underlying medical conditions, the cases differ from each other in the extent to which publicity was sought by their parents and caretakers. Thus, in the landmark cases of babies John Doe [19, 24] and Jane Doe [6, 16] attention was not sought by the families and their caretakers, at least for the most part, but rather by advocacy groups who seized upon them as opportunities to establish public policy. With regard to the organ transplant children, quite the opposite has been true, beginning with the celebrated appearance of Charles Fiske before the national meeting of the American Academy of Pediatrics in October 1982 through the countless appeals for organs and funding that continue to the present. To a great extent all of this media attention and public curiosity have understandable origins, but without concerted thoughtful analysis of the issues presented by these cases, we risk descent into what I term "government by anecdote". It may well be that policy decisions which are appropriate for the individual case may not be as clear-cut when examined on the basis of competing claims and allocation of limited resources.

Some description of my entry into this field is probably appropriate. A pediatric endocrinologist, I became involved in child advocacy in the mid-1970s through stimulation of a New Haven chapter of the Juvenile Diabetes Foundation and later, with the assistance of the chapter's leadership, creation of a state-wide comprehensive program to provide multidisciplinary comprehensive care for children with diabetes [13]. This became possible through personal contacts with the Connecticut Senate majority leader at a time when there was a state fiscal surplus and was an object lesson in the practice of personal political advocacy. In 1982 I was awarded a Robert Wood Johnson Health Policy Fellowship at the Institute of Medicine and spent the year beginning September 1982 on Capitol Hill where I was assigned to the Subcommittee on Investigations and Oversight of the House Science and Technology Committee, headed by then Congressman Albert Gore Jr. from Tennessee.

After several weeks, the Congressman received a telegram requesting assistance in securing a liver for a child awaiting surgery in a Memphis hospital. Like most of us, congressional offices relish case advocacy and Congressman Gore's staff quickly contacted me, assuming that, as a pediatrician from academic medicine, I would surely know how to obtain a liver and the office could thereby share the credit for saving the child's life. As it turned out, I did know enough to recognize that the premier transplant center was at the University of Pittsburgh and called the organ procurement specialist there. When I inquired as to why people felt obliged to send telegrams to their congressional offices, I was informed that the problem was created by media sensationalism generated by anxious parents. Somehow this still didn't ring true and so, after I reported to the Congressman, we decided to launch a congressional inquiry since, in Gore's words: "it is inappropriate to ask families to make spectacles of themselves in order to secure the health of their child" [32].

The investigation that followed demonstrated shortcomings in the mechanism for obtaining and distributing solid organs for transplantation. In place was a loosely connected network of 110 procurement agencies, and little apparent rationale as to apportionment of organs between centers and patients. There was a helter-skelter system of reimbursement. Heart and liver transplantation was regarded as "experimental" by federal assessment agencies, principally structured to advise Medicare [14]; the lack of a federal imprimatur prevented many private insurers from coverage as well. Ironically, only a small number of patients were eligible for Medicare coverage, not by age, but by being placed on the disabled rolls.

At one of our early hearings, an army captain testified that his daughter, who had already undergone an unsuccessful palliative procedure, could not receive a liver transplant since the military dependent insurance program (CHAMPUS) would not pay for "experimental" procedures [32]. This child received her transplant, but only after Congressman Gore succeeded in attaching an amendment to the authorization legislation for the Department of Defense which required CHAMPUS to cover liver transplants [9]. Sadly, the transplant operation was not successful and the little girl died of postoperative complications. It is noteworthy, and a reflection of the timeliness of governmental process, that the final regulations implementing procedures for liver transplantation for CHAMPUS beneficiaries [27] were issued more than three years after the legislation was passed. Fortunately, Congress had the good sense to make the legislation retroactive to July 1, 1983.

It is also worth noting that the Reagan administration, while steadfastly insisting that liver transplants were not only experimental, but could not be afforded by the military health insurance program, was simultaneously offering the services of Air Force One to transport livers across the continent. The specific case on this occasion was Ashley Baily, a seven-month-old infant who was the subject of a personal radio appeal by the President. The appeal was unsuccessful in securing a liver for this very small child. It is also noteworthy that the White House maintained a full-time staff position for what was essentially a federal ombudsman who was very visible in procuring services and financing for families attracting his interest. At this same time the administration insisted that corrective legislation was not necessary and that the problems in organ procurement and transplant funding could be solved through the private sector.

Nonetheless, Congress, just before the end of the 98th session, passed the National Organ Transplant Act [21], legislation developed in response to problems uncovered by the congressional inquiry and Gore hearings. The National Organ Transplant Act established an Office of Organ Transplantation within HHS and a twenty-six member Task Force on Organ Transplantation which completed its report in April, 1986 [31]. The Task Force called for a much stronger and more assertive federal role in stimulating and organizing organ procurement and distribution, in financing of organ transplants including coverage of immunosuppressive medications, and in controlling the diffusion of these highly specialized and resource intensive procedures. To a great extent, all these issues had been uncovered by the congressional inquiry and hearings held during 1983 [32]. In actuality, all of this was stimulated by an exercise in congressional case advocacy.

This is not to suggest that the Task Force report has resolved these problems. To the contrary, there continue to be serious problems in supply of transplantable vital organs and tissues. To some extent there is an increasing demand generated by well-publicized successes resulting from improved surgical techniques and the advent of cyclosporine. Thus the federal Task Force estimated that organs were obtained from less than 20% of potentially suitable donors [31]. While these figures have improved with increasing publicity and public awareness, the gap between demand and supply, if anything, has increased. As the reluctance of many health professionals to solicit donation has been identified as a major impediment, many states have now passed so-called "required request" bills. In late 1986, Congress passed legislation [23] requiring institutions to have organ donation programs in place as a requirement for Medicare and Medicaid reimbursement. Implementing regulations were released several months later [30] to take effect at the beginning of the 1988 fiscal year.

The problems of scarce supply are well illustrated by the controversies that have surrounded provision of heart transplants to two infants, Baby Fae and Baby Jesse, both with the hypoplastic left heart syndrome, at Loma Linda Medical Center in California. Extension of heart transplant surgery to infants was adventuresome and by no means a generally accepted therapeutic approach, in part because of the technical difficulties, but more because of concerns regarding long-term dependence on very potent immunosuppressive agents such as cyclosporine. Moreover, there is an uncertain supply of available hearts from infant donors, so much so that the Loma Linda team performed its first heart transplant on Baby Fae in 1984 with a baboon heart, assuming that a human heart could not be obtained and, perhaps naively, confident of overcoming the immunologic barrier between species [4]. In the ensuing controversy, while questions were raised about the adequacy of consent and the experimental basis for performing the xenograft, professional and public opinion was generally assuaged by the dire prognosis and the limited availability of transplantable organs for infants. The Loma Linda group has subsequently completed several transplants using human infant donors with moderate success [5] and other centers are also performing infant heart transplants [11].

The issues of organ availability and distribution were highlighted even further by the case of Baby Jesse, born out of wedlock in circumstances that were initially judged to be poorly predictive of successful long-term care. Now it should be noted that, in a climate of organ scarcity, criteria for transplantation have generally included psychosocial factors thought essential in predicting compliance to a complicated regimen which requires constant monitoring of medication and close medical follow-up. Thus there was some basis for the initial denial by the heart transplant team. Some, however, seized on this case as discrimination and, after custody was awarded to the maternal grandparents, the team agreed to perform a transplant. A donor heart was secured from a brain-dead infant in Michigan while the parents were publicizing their appeal on the syndicated Phil Donahue program. Paradoxically, at the same time, another infant, Baby Calvin, was awaiting a heart transplant in nearby Louisville, Kentucky, but without benefit of the media attention. Although a donor was eventually found for Baby Calvin, the glaring weakness in our organ procurement and distribution system should be evident.

It is also evident that not much progress had been made between the appeal of Charlie Fiske in the fall of 1982 and the appeal of Baby Jesse's parents on the Donahue program in the spring of 1986, at least not as perceived by families grasping at any available straw. This is what I regard as "government by anecdote", a process by which policymakers become lulled into a sense of accomplishment by the success of individual case advocacy while avoiding the more difficult and complex problems of class advocacy.

A somewhat different picture emerges from examination of the issues engendered by the Baby Doe cases. These have become a more or less generic term to describe infants born with profound handicapping conditions who present equally profound moral, ethical, and medical decision-making dilemmas. In these situations, however, the opposite has occurred. Profound public policy decisions *have* been promulgated as a solution to issues engendered by individual cases, but without examination of the broad and far-reaching policy implications that would be based on true class advocacy.

The facts and circumstances surrounding the two infamous Baby Doe cases are well known. These cases and the attendant bioethical and public policy issues engendered by them have been the subjects of at least four recent books, not to mention countless essays and commentaries [12, 18, 19, 35]. The original Baby Doe, an infant with Down syndrome and associated esophageal atresia, was born in Bloomington, Indiana, in April 1982. The decision of the parents to withhold corrective surgery and the infant's subsequent death at 6 days of age unleashed a torrent of media attention and a cacophony of federal initiatives from presidential directives, a series of so-called "Baby Doe Regulations" and congressional legislation, to amendments to the Child Abuse Prevention and Treatment Act of 1984 [7]. While all of this was going on, the second case, Baby Jane Doe, was born in Long Island in October 1983 with a constellation of congenital deformities, including microcephaly and lumbar myelomeningocele. The parents, acting on professional advice, elected to defer surgery and pursue conservative medical management. This unleashed both a series of suits in the New York state courts, initiated by an unrelated and self-styled right-to-life advocate [34], and attempted federal intervention [33], in the guise of determining that the hospital was not in violation of section 504 of the 1973 Rehabilitation Act [26]. The federal regulations based on this act were recently overturned by the U.S. Supreme Court in a 5–3 decision [3].

The dilemmas attending the treatment decisions in severely imperiled newborns were not new; ten years earlier Ray Duff and Alex Campbell published their landmark article in the *New England Journal of Medicine* and called for a strong parental role in the decisions to treat or not to treat ambiguous cases [10]. This paper was published shortly after I joined the Yale faculty. Even then there was no unanimity regarding its conclusions and recommendations. Over the succeeding decade other articles, essays, and well-publicized cases appeared, but none attracted attention and galvanized public policy until the appearance of the Bloomington "Baby Doe." It is important to recognize that major advances in perinatal medicine and neonatalogy occurred during this short period of time so that by 1982 there were many more infants who could be described as "imperiled" and who yet had a chance for survival but with uncertain residual handicaps. By far the greatest number of these are low and very low birth weight babies.

I arrived in Washington while committee staffs on both sides of the Hill were drafting what they perceived to be "corrective" legislation. This represented the congressional parallel to the administration's response of promulgating regulations. As a pediatrician, sensitized to these questions from my Yale experience, I was struck by the over-simplified and somewhat doctrinaire approach that was being considered by congressional staff wholly unfamiliar with the subtleties and nuances of medical decisionmaking. It is important here to emphasize the crucial role played by congressional staff, especially on the various committees and subcommittees responsible for the shaping of legislation. This is especially so with regard to subjects that are fairly technical and on which congressional members may take broad positions with regard to overall policy while leaving specific details to staff.

In the spring of 1983 the staffs of the House Education and Labor and Senate Labor and Human Resources Committees were beginning work on reauthorization of the Child Abuse Prevention and Treatment Act. With the Bloomington Baby Doe case fresh in their minds, this legislation seemed an appropriate vehicle. I became involved in this process through the Academy of Pediatrics and its Washington government affairs office with which I became connected. By this time the Academy had taken leadership in opposition to the initial Baby Doe regulations, issued in response to the Bloomington, Indiana, case, successfully filing suit to invalidate the regulations on procedural grounds [2].

In the political process that followed, the House of Representatives overwhelmingly approved changes in the Child Abuse Prevention and Treatment Act, which provided rather strict criteria under which imperiled infants were to be treated, after first defeating a modulating amendment by a 231–182 margin [8]. Editorially, the *New York Times* labelled this "an abusive amendment" [1]. The final version, however, arose from a compromise developed by the staffs of six senators spanning the entire political spectrum [15]. This compromise was developed after months of tedious negotiations wih representatives of some twenty health care providers and interest groups and resulted in the often criticized Child Abuse Amendments of 1984. What is not appreciated by critics, including many of my colleagues, is the fragile nature of the compromise that was reached in the Senate version which prevailed as the final legislative language.

As enacted shortly before the close of the 98th Congress, the Child Abuse Amendments of 1984 [7] require states, as a condition of receiving child abuse prevention and treatment grants, to establish programs and procedures within the state's child protective service system to respond to reports of "medical neglect." This now includes the "withholding of medically indicated treatment" from infants with life-threatening conditions. The compromise language developed by the six senators carefully and explicitly defined "indicated treatment" as that which attending physicians, using reasonable medical judgment, would judge to be potentially effective in correcting life-threatening conditions except under three specific circumstances:

(A) the infant is chronically and irreversibly comatose;
(B) the provision of such treatment would
   (i) merely prolong dying,
   (ii) not be effective in ameliorating or correcting all of the infant's life-threatening conditions, or
   (iii) otherwise be futile in terms of the survival of the infant; or
(C) the provision of such treatment would be virtually futile in terms of the survival of the infant and the treatment itself under such circumstances would be inhumane.

The legislation also encouraged development of so-called "Infant Care Review Committees", a concept proposed by the President's Commission for the Study of Ethical Problems in Medicine and Biomedical and Behavioral Research and published in the Commission's milestone report, *Deciding to Forego Life-Sustaining Treatment. Ethical, Medical and Legal Issues in Treatment Decisions* [25]. The encouragement and stimulation of institutional mechanisms, including such so-called "Infant Bioethics Committees", may well be one of the most consequential outcomes of the protracted debate over the care of handicapped newborn infants.

Passage of the Child Abuse Amendments was not the final chapter in this continuing saga. To begin with, the interim regulations based on this legislation, published in the *Federal Register* of December 10, 1984 [29], went far beyond the spirit and wording of the compromise by expanding upon the definitions carefully crafted for the Senate compromise and by citing often inappropriate examples to illustrate the procedures. The final regulations, published in the April 15, 1985 *Federal Register* [28], were an improvement, in part by the device of leaving in an appendix the administration's "interpretive guidelines" as to the meaning of the actual regulations. The net effect of the 1984 legislation and the subsequent final regulations, it should be emphasized, was to reinforce the primacy of state agencies and state judicial systems in addressing instances of child abuse, alleged parental neglect or, in the expanded definition, provision of "medically indicated" care [20].

Finally the U.S. Supreme Court, in its decision of June 9, 1986 [3], declared invalid the earlier "Baby Doe" regulations which had been based on section 504 of the 1973 Rehabilitation Act [26]. The Supreme Court decision, by a small plurality of 5–3, was based on a narrow interpretation of intent of the Rehabilitation Act. Nonetheless, it reemphasizes, at least for the present, the primacy of medical decisions between parents and physicians.

In summary, I have provided a somewhat personal odyssey to dramatize the impact created by some of the more sensationalized pediatric cases of the recent few years. My point is that these represent examples of extreme case advocacy, not practiced by the pediatric practitioner, but rather by the national media and by policymakers and politicians. In the situation of organ transplants, I fear that solutions to individual cases have obfuscated the need for public policy to deal with very difficult resource allocation questions in an era in which cost containment has become everybody's theme. Somewhat paradoxically, with regard to treatment of severely handicapped newborns, we have seen attempts to evolve public policy from a few well-publicized cases. However, many pediatricians feel these "Baby Doe" cases are not typical of the decisionmaking dilemmas confronting parents and physicians in our modern tertiary care neonatal intensive care facilities. Both situations, in my view, represent misdirected case advocacy. We need more effective class advocacy and it is now more essential than ever that pediatricians emerge as its standard bearer.

*Yale University School of Medicine*
*New Haven, Connecticut*

## BIBLIOGRAPHY

1. 'An Abusive Amendment' (editorial), *New York Times*, February 9, 1984, p. A30.
2. *American Academy of Pediatrics v. Heckler*, 541 F. Supp. 395 (D.D.C., 1983).
3. *American Hospital Association v. Heckler, American Medical Association v. Heckler*, 585 F. Supp. 541 (S.D.N.Y. 1984), aff'd F.2d (2d Cir. Dec. 27, 1984), petition for cert. filed (Mar. 27, 1985), aff'd. (U.S. Supreme Court, June 9, 1986).
4. Bailey, L. L. *et al.*: 1985, 'Baboon-to-Human Cardiac Xenotransplantation in a Neonate', *JAMA* **254**, 3321–9.
5. Bailey, L. L. *et al.*: 1986, 'Cardiac Allotransplantation in Newborns as Therapy for Hypoplastic Left Heart Syndrome', *New England Journal of Medicine* **315**,

949–951.
6. Chambers, M.: 1983, 'Parents of "Baby Doe" Criticize "Intrusion" by U.S.', *New York Times* (November 6), p. 45.
7. Child Abuse Prevention and Treatment and Adoption Reform Act Amendments of 1984, Pub. L. 98–457, 42 U.S.C. 5101 *et seq.* (1984).
8. *Congressional Record*: 1984 (February 2), pp. H379–H402.
9. Department of Defense Authorization Bill, 1984, Pub. L. 96–354, 10 U.S.C. 1079–1086.
10. Duff, R. and Campbell, A.: 1973, 'Moral and Ethical Dilemmas in the Special-Care Nursery', *New England Journal of Medicine* **289**, 890–894.
11. Fricker, F. J. et al.: 1987, 'Experience with Heart Transplantation in Children', *Pediatrics* **79**, 138–146.
12. Frohock, F. M.: 1986, *Special Care. Medical Decisions at the Beginning of Life*, The University of Chicago Press, Chicago.
13. Giordano, B. *et al*.: 1977, 'Regional Services for Children and Youth with Diabetes', *Pediatrics* **60**, 492–498.
14. Iglehart, J. K.: 1983, 'Another Chance for Technology Assessment', *New England Journal of Medicine* **309**, 509–512.
15. Kerr, K.: 1985, 'Negotiating the Compromises', *Hastings Center Report* **15** (June), 6–7.
16. Kerr, K.: 1984, 'Reporting the Case of Baby Jane Doe', *Hastings Center Report* **14** (August) 7–9.
17. Knitzer, J.: 1978, 'Concepts of Advocacy, Definition and Levels of Professional Action' in E. H. Newberger (ed.), *Child Advocacy and Pediatrics*, Report of the Eighth Ross Roundtable on Critical Approaches in Common Pediatric Problems, pp. 13–20.
18. Kuhse, H. and Singer, P.: 1985, *Should the Baby Live? The Problem of Handicapped Infants*, Oxford University Press, New York.
19. Lyon, J.: 1985, *Playing God in the Nursery*, W. W. Norton & Co., New York.
20. Murray, T. H.: 1985, 'The Final, Anti-climactic Rule on Baby Doe', *Hastings Center Report* **15** (June), 5–9.
21. National Organ Transplant Act, Pub. L. 98–507, 42 U.S.C. 273 *et seq.* (1984).
22. Newberger, E. H.: 1978, 'Editor's Introduction', in E. H. Newberger (ed.), *Child Advocacy in Pediatrics*, Report of the Eight Ross Roundtable on Critical Approaches in Common Pediatric Problems, pp. 11–12.
23. Omnibus Budget Reconciliation Act of 1986, Pub. L. 99–509, 42 U.S.C. 1302, 1395x and 1395hh. (1986).
24. Pless, J. E.: 1983, 'The Story of Baby Doe' (letter), *New England Journal of Medicine* **309**, 664.
25. President's Commission for the Study of Ethical Problems in Medicine and Biomedical and Behavioral Research: 1983, *Deciding to Forego Life-Sustaining Treatment*, U.S. Government Printing Office, Washington, D.C.
26. Rehabilitation Act of 1973, Section 504, as amended, 29 U.S.C. 794 (1976).
27. U.S. Department of Defense: 1986, 'Civilian Health and Medical Program of the Uniformed Services (CHAMPUS); Liver Transplantation' (Amendment to Final Rule), *Federal Register* **51** (September 2), 31100–31103.
28. U.S. Department of Health and Human Services: 1985, 'Child Abuse and

Neglect Prevention and Treatment Program' (Final Rule), *Federal Register* **50** (April 15), 14878–14892.
29. U.S. Department of Health and Human Services: 1984, 'Child Abuse and Neglect Prevention and Treatment Program' (Proposed Rule), *Federal Register* **49** (December 10), 48160–48169.
30. U.S. Department of Health and Human Services: 1987, 'Medicare and Medicaid Programs, Organ Procurement Organizations and Organ Procurement Protocols' (Proposed Rule), *Federal Register* **52** (July 31), 28666–28677.
31. U.S. Department of Health and Human Services: 1986, *Organ Transplantation: Issues and Recommendations*, Report of the Task Force on Organ Transplantation.
32. U.S. House of Representatives: 1983, *Organ Transplants*, Hearings Before the Subcommittee on Investigations and Oversight of the Committee of Science and Technology.
33. *U.S. v. University Hospital of State University of New York*, 575 F. Supp. 607 (E.D.N.Y. 1983), aff'd, 729 F.2d 144 (2d Cir. 1984).
34. *Weber v. Stony Brook Hospital*, 60 N.Y. 2d 208, 211-213, 456 N.E. 2d 1186, 1187–1188.
35. Weir, R.: 1984, *Selective Non-treatment of Handicapped Newborns. Moral Dilemmas in Neonatal Medicine*, Oxford University Press, New York.

H. TRISTRAM ENGELHARDT, JR.

# ADVOCACY: SOME REFLECTIONS ON AN AMBIGUOUS TERM

Myron Genel's explorations bring the idea of advocacy into question: What is it to be an advocate? In Roman times, "advocatus" was the term for a legal assistant or a councillor. It was derived from the Latin "advocare", to call or to summon one to a place, especially for council or aid. It also meant to avail oneself of someone, such as an assistant, witness, or counselor, in some cause. The formal and forensic valence of these usages is an ancient one. In English it continues to suggest a conflictual circumstance. As is conveyed by Genel's essay, advocacy intimates the intrusion of an authority into decisions by parents regarding their children. Under such circumstances, it is unlikely that there will be a single understanding of proper advocacy or of the authority of an advocate.

To make sense of a notion of advocacy, one will at the very least need to ask:

(1) To whom does the advocate speak?
(2) On behalf of whom does the advocate speak?
(3) Whose values does the advocate defend?
(4) Whose rights does the advocate defend?
(5) By whose authority does the advocate speak?

To answer these questions, one will have to distinguish between advocacy focused on

(1) interfering with the claimed liberties of others, such as parents, to make decisions to decline treatment for their children, and

(2) securing funds from third parties for parents who would be pleased to use such for the benefit of their children.

The first sense of advocacy is advocacy on behalf of restricting the range of options open to parents. The second focuses on providing a material basis for a greater range of choices on the part of parents. Both of the senses are ambiguous because "best interests" are defined not by the individual to be treated, but by third parties.

Advocacy on behalf of additional funding for new and expensive treatment is morally problematic (1) insofar as there are difficulties in justly acquiring resources to empower parents to make choices beneficial to their children, and (2) insofar as such choices encourage a reliance on technology to increase marginally at great cost the quality years of life available to children. The problem of just acquisition involves the difficulty of drawing the proper line between public and private resources, a line that must in the end limit the authority of the state to tax private resources, even for the most important of public goals. Entwined with this question is the issue of the limits of state sovereignty and state authority over the lives of citizens. Insofar as there are limits on state authority or insofar as rights are not fully reducible to interests in goods and values, the matter is not simply whether there are good things to do for children, but whether one has the right to do them, all things considered. The additional problem with the allure of technology is that, with ever more funds, we can marginally always do a little bit more. However, if we do not want to spend all of our available resources on extending the lives of individuals, we will at some point need to decide not to invest resources (at least public resources), even though there will be some probability that they will do some good. We will need to support philosophy departments, even though, had the resources been invested in more neonatal intensive care units, more children could have been saved. At some juncture we will need to endorse policies that encourage parents to try to have another child, rather than to invest significant resources in saving the one just born.

The point is that it will not be enough simply to be an advocate on behalf of the patient. Responsible advocacy on the part of the physician should require more than seeking funds to provide treatment. As Genel suggests, sensationalizing particular technologies, failing to underscore the limits of treatment, and not acknowledging that costs can defeat obligations to treat may undermine the capacity of institutions to use new, emerging, and costly technologies effectively, and of parents to make responsible choices. Just as it is virtuous to give free and informed consent to a patient in order to empower the patient's choice, it is virtuous to aid institutions and society as a whole to make prudent decisions regarding the development of reimbursement policies for transplantation and other costly approaches to saving lives. In the passion to provide costly treatment that offers some chance of saving some lives, one may obscure the importance of the more routine endeavors of medicine. At the very least, one must balance the advocacy of the pediatrician on behalf of the child with an advocacy on behalf of citizens for the prudent use of resources put into the trust of the state or private insurance systems. If one radically alters the allocation of resources held by private or governmental insurance schemes in order to attempt to save lives at great cost and with little likelihood of success, one may have failed on a fiduciary responsibility to those who provided the resources. Thus, one is pressed also to endorse a counter-advocacy to bring the first advocacy into perspective.

The questions raised by advocacy for the provision of resources to save the lives of seriously handicapped or ill newborns and infants have some overlap with the questions raised by advocacy that leads to state intrusions that force the treatment of children over the protests of their parents. The overlap centers on how financial, psychological, and social costs to parents and to society defeat the reasonableness of providing care. It would be difficult if not impossible to argue successfully that life is so important that it should be saved at any cost. Such a course would lead to investing all resources in the incremental extension of life (presumably one could invest in the enjoyment of life only insofar as that helped indirectly to support the realization of this primary obsession). Insofar as individuals have rights to their resources for the enjoyment of their lives, and to the realization of goals other than the maximal extension of life, they should be able to set limits to such a single-minded commitment to saving life at all costs. The old distinction between ordinary and extraordinary care served the purpose of suggesting such limits as a line between treatments that are obligatory to provide and those that were not [1]. Likelihood of success, as well as financial and social and psychological costs, were recognized as defeating an obligation to provide treatment.

The advocacy involved in Baby Doe also raises the question of the limits of state authority. Only if one holds that state authority is full and unlimited, will one conclude that the state may in all cases intervene to protect children despite parents' wishes. After all, to protect a child is to protect it in terms of certain values and a particular vision of proper deportment. One must select one out of a range of views of best interests. The values and visions of individual families often diverge from those approved by the state. Parents and families may conclude that it is best for their child to die, not only because of the costs involved, but because the child's life would involve more harms than benefits. Particular families may wish to realize a view of proper treatment for the child which may lead to death and be contrary to the regnant ideology of the state. In fact, insofar as the state does not assume all of the costs of a child's treatment, the family may properly (if not legally) decide that a particular treatment is extraordinary (i.e., involves an undue burden) and morally non-obligatory, the wishes of the state to the contrary notwithstanding. Indeed, since a major portion of the costs are psychological, not simply financial, it will not be possible for the state ever to shoulder all the costs involved.

The bothersome question comes center-stage. When does the state have authority to impose a particular view of proper treatment for the child over, against, and despite the wishes of the parents. Until the Baby Doe regulations, the law allowed parents to choose among options endorsed by an accepted group of medical practitioners. Where there was uncertainty, parents were in authority to select among established medical authorities. This policy can be justified because claims of intrusive state authority require the state to show (1) that its policy is correct, (2) that it has the moral authority to impose it over the protests of those who do not consent, and (3) that the imposition will lead to more benefits than harms. When there is clear disagreement among experts regarding a proper line of treatment, such justification is fundamentally undermined. It is not just that this approach allowed parents to choose in areas of disagreement, it also placed the burden of proof on the state in the sense that a particular court order needed to be sought to show that a particular parental choice fell beyond the range of acceptable medical options. Families did not need in advance to meet a test to show that they were choosing correctly. They possessed a substantial exclave for their own choices.

Until Baby Doe, except for very egregious circumstances, pediatricians were advocates at most of a particular view of the best interests of a child to the parents within a parent/physician relationship. Pediatricians tended to function more as moderators among conflicting goals and possibilities. They served as counselors of the parents, not agents of a set of state regulations that disregard issues of cost and of quality of life (save for permanent loss of consciousness). The pediatrician as the physician for a family prior to the Baby Doe regulations helped the family to come to terms with painful possibilities and complicated medical information when selecting from a range of accepted options. In contrast, pediatricians are currently constrained to be advocates of a concrete state policy, which has significantly circumscribed the latitude for parental decision. The change is a fundamental one. Not only has decisionmaking by parents been bureaucratized and circumscribed, but the advocacy character of the 1985 legislation has shifted the burden of proof to parents, who must now show that they are acting within constraints set by the current Baby Doe regulations, even if a body of accepted medical opinion is in agreement with the parents. Particular concrete standards have been established in detail, removing the right to choose among established medical views of what is reasonable treatment. In addition, the privacy of the parent/physician relationship has been intruded upon not only by regulations, but also by infant care review committees established to oversee the implementation of the regulations. The notion of the pediatrician as advocate within the spirit of the Baby Doe regulations is thus at loggerheads with the notion of the family as having a moral integrity and vision of its own.

There is no one unambiguous sense of advocacy, as Myron Genel establishes through his series of vignettes. To be an advocate is to take a particular position for a particular person or group of persons over against other positions and other persons or groups of persons. The Baby Doe regulations make the pediatrician not simply the advocate of the child, but also generally the advocate of high-technology, high-cost interventions. The same can be said at least with regard to advocacy of funding for treatment when high-technology, high-cost solutions are embraced with little consideration of when high costs and low likelihood of success undermine the reasonableness of providing care.

It is beyond the scope of these brief reflections to determine how one may set limits to state authority or determine when obligations to care are defeated by costs. Such determinations cannot be provided piecemeal, but as a part of general reflections on the possibilities of resolving moral controversies in bioethics [2]. Still, even without invoking a fundamental theory, an analysis of the ambiguities of advocacy can show some of the pitfalls. As Genel underscores, national policy has developed out of emotional reactions to particular cases in a way that has failed to produce a justifiable policy for the usual range of medical choices. Just as one should fear the Greeks when they come bearing gifts, Genel gives us ample grounds to be suspicious of advocates. We thus return to where this analysis began. When confronted with well-meaning advocates of the best interests of children, we will need to ask hard questions regarding (1) the authority (correctness) of their visions of the best interests of children, and (2) their authority to impose a particular view of those best interests. Advocacy is not unambiguous, nor is it without its costs. Fashioning a sustainable concept of responsible advocacy in a high-technology society such as ours requires at the very least acknowledging the limits of the authority of advocates and of states, as well as the limits of our obligations to save lives at any cost. Often the role of moderator or counselor will be more appropriate than that of advocate.

*Baylor College of Medicine,*
*Houston, Texas, U.S.A.*

## BIBLIOGRAPHY

1. Cronin, D. A.: 1958, 'The Moral Law in Regard to the Ordinary and Extraordinary Means of Conserving Life', dissertation, Gregorian University, Rome.
2. Engelhardt, H. T., Jr.: 1986, *The Foundations of Bioethics*, Oxford, New York.

THOMAS G. IRONS

## LOVING THE CHRONICALLY ILL CHILD: A PEDIATRICIAN'S PERSPECTIVE

The doctor-patient relationship has historically been viewed as the most important aspect of the practice of medicine. Indeed, in the preantibiotic era, medical training and practice often had little else to offer. The enormous scientific advances of this century, especially in the last three decades, however, have shifted that focus substantially. Likewise, there are few who would deny that the relationship is seriously threatened and needs our attention.

The public's reaction to the dehumanization of medicine may be largely responsible for one of the profession's most visible problems, the malpractice crisis. To some, an ancient covenant seems broken; the patient no longer entrusts himself willingly into his doctor's hands, no longer is secure in the knowledge that his physician is his committed servant. He looks instead to a complex health care apparatus supervised by his doctors, who, if they push all the right buttons, will heal him. It is a silly twentieth-century, science-fiction fantasy, a uniquely American one, and the expected failures occur with discouraging regularity. The patient is left disappointed at best, and angry and vengeful at worst.

Beyond this anger, I believe, lies a deep yearning for the healing touch, the soothing voice and the listening ear. Nowhere is this more important than in the care of those children whose lives are qualitatively and quantitatively altered by chronic disease. Love is a word rarely used in medical textbooks, and that is exactly why it is in the title of this paper. Love sets the tone of the doctor-patient relationship. Indeed, it is where that relationship must begin, grow, and end. It is a posture of humility, matched with genuine commitment of the physician's physical, mental, and emotional resources to the child and family he serves.

If I were to paint a portrait of the ideal physician, I would draw considerably upon my long-term memory. I see this person at the sick child's bedside, perhaps auscultating the chest; the picture of patience, of kindness, of competence. The setting is not the hospital intensive care unit, but the child's bedroom. The faces of the child and parents reflect the comfort and relief brought on simply by the doctor's presence. In the physician's black bag are the few most useful diagnostic and therapeutic instruments of the profession. Before and after office hours each day, travelling with it from house to house, the doctor ministers to his or her patients.

I have quite purposefully placed my doctor in the child's bedroom. The house call is almost a dinosaur in the 1980s; my own children have only vague familiarity with the term. It is warmly remembered, though, by those of us who are a little older, and has come to symbolize for many the sense of personal commitment that sometimes seems to have disappeared from medicine. I have spent many impatient hours in the front seat of a 1953 Oldsmobile, waiting while my mother or father attended some sick infant, or child, or invalid grandmother. It was there that I first began to comprehend the business of chronic illness. The compromised child was usually kept at home. It was a safer environment, and the difficulty of transportation to the office or hospital was simply too great. Although thirty years have passed, I remember those times in vivid technicolor. It was much later when I realized how profoundly they had affected me.

There was an infant girl who had a brain tumor. Her death was perhaps days away, and it was cold and snowing. Transportation was limited. So my mother, her doctor, brought her to our home to spend her final days, her family tromping up the snow-covered steps to visit several times a day. These "reverse house calls" did not seem odd or uncomfortable at the time, nor did the sick baby's presence in our home. There were three children with cystic fibrosis, rural residents of a coastal county some distance from the eastern North Carolina community where my parents both practiced medicine. Our family's summer cottage was much closer, so our vacations were punctuated by their visits, or we might all drive over "just for a little while" to check on them. I feel as if I have a hundred such stories I need to tell, all of them reminders of the very personal relationship that existed between patients (especially the chronically ill) and their doctors thirty years ago. The relationship has changed significantly, and the care system dramatically, since that time, but the problem of chronic illness may even be greater than before. While we have made significant strides in children's cancer and cystic fibrosis, both of these remain very serious chronic health problems for children. Other diseases such as rheumatic fever, post-streptococcal glomerulonephritis, and tuberculosis are relatively rare now. It is good that they are gone, but they have been replaced by a flood of new problems and entities. Children who did not survive in the 50s, 60s, and even 70s are surviving today, but may continue to survive only through extraordinarily complex management regimens. The surviving very small premature infant is, indeed, fortunate if he can escape chronic cardiopulmonary problems or neurological impairment. No longer must the hematologist inform anxious parents that childhood leukemia is always fatal, but what lies ahead for the leukemic child and family in the eighties sometimes seems worse. The complex chemotherapy regimen that may save his life may also dramatically alter his appearance and make him at times absolutely miserable. Further, we are at an enormous distance from the time when we can guarantee the success of that regimen. Children with cystic fibrosis likewise survive longer, but require a coordinated program that consumes hours of parental time each day and requires the support of an entire team of health care workers.

Virtually all chronically ill children receive care from multiple sources, and communication among caregivers is often suboptimal. The cystic fibrosis team or the leukemia team, for example, may only be available at referral centers, and offer care to only a small number of those affected. Further, because of the severe financial constraints imposed on the families of these children, most are forced to depend on government programs for support. Gaining access to these programs is never easy, and multiple programs may serve a single child. It is ironic that one vital member of the chronic health care team, the social worker, who might be particularly sensitive to the human needs of the child and family, must spend a great percentage of time assuring patient access to these programs. To further complicate matters, several physicians are often involved, and communication may be limited. It is interesting to ask chronically ill children, "Who is your doctor?" There is almost always an answer, and only one answer, but that physician might be at any level in the health care system. It is pointed testimony for the child's need to identify and depend on one special person as his physician, in what may seem to be a hopelessly fragmented health care system.

Few of us would return to the time when the healing touch was all we had to offer, but neither would we sacrifice it to high-tech medicine. Still, our efforts toward its preservation are often misdirected. In a recent editorial in the *American Journal of Diseases of Children*, William Weil cites another major threat to the child and his family. "The medical profession," he says, "has failed children and their families by defining their needs for them, ... by deciding where they are to get their care, what treatments they are to carry out (whether they have the resources, personal and concrete to do so or not), and labels them as noncompliant when they do not do what the doctor orders" ([1], p. 268).

Paternalism is not new to the medical profession, of course. But it is a particular problem in chronic illness, when the child and his family desperately need some sense of autonomy, some control over the exercise of their therapeutic options, a sense of sharing the decisions with the physician. Physicians are sometimes reluctant to admit to families that they do not have all the answers to their patients' problems. Such an admission implies vulnerability, something that many of us find uncomfortable.

Paternalism may be viewed as one of a whole group of defenses, reactions to the vulnerability we feel when we acknowledge our own humanity, when we submit ourselves to the inevitable pain we must suffer if we are to share the pain of our patients. There are many others. A highly visible one is the practice of defensive medicine. Physicians may abdicate their responsibility for making difficult care decisions by allowing fear of malpractice suits to make the decisions for them. The physical and mental expense of such practice is borne by the health care system and our patients. Sometimes we isolate and protect ourselves by retreating into science, by focusing on the illness rather than on the patient. We busy ourselves making fine adjustments in antibiotic dosage, or chemotherapy protocols, or seizure medication programs, all the while ignoring the deeper personal needs of the patient and family. One of our favorite ploys, when things are not going well, is to call in one or more consultants, shifting responsibility away from ourselves, and leaving the patient and family with a sense of abandonment. These may be crucial parts of the child's care program, but they simply must never be set between the physician and the patient he serves.

The doctor-patient relationship is, then threatened, not only by the complexity of the health care system itself, but by the dangerous tendency of modern physicians to isolate themselves from the humanity of the people they serve. What are we to do? My view is that the solution is quite simple, at least in principle. Given that the relationship between a chronically ill child, his family, and his physician is worth preserving and nurturing, we must attend to that goal as a first priority. We must turn our attention away from ourselves and toward our patients. On the one hand, the growing complexity of the treatment of chronic illness has certainly been more beneficial than harmful. The team-coordinated approach has been particularly useful, both in enhancing our perception of the needs of children and families, and in meeting those needs. On the other hand, the fragmentation of health care programs is deplorable, and probably irremediable at the moment. No matter how complex the individual child's situation may be, the physician continually must look through the technical and bureaucratic maze to see that the system has not swallowed up the patient. The child must be assured that he is not abandoned to his care program, that the program has not replaced his doctor.

There is usually flexibility in assumption of the primary physician role. Some seriously ill children can totally identify with the subspecialist, while others identify with the primary physician and see the specialist as a consultant. There is even room for sharing of the role, but sharing requires close communication between the physicians involved. When conflict occurs, the child's interest is paramount and should determine the resolution. Often a physician will find that a child's health care program requires rather heavy-handed control, especially when many sources are involved. Here paternalism is a particularly dangerous trap, and must be carefully avoided. It is imperative that the child and family be involved in decisionmaking on a continual basis.

Ideally, the physician-patient relationship is an intensely personal business. It is at the personal level, therefore, where we should concentrate our efforts to support and strengthen that relationship. The key word in this paper is "loving" so I will conclude by discussing the implications of that imperative. It implies first that we open ourselves, our *persons* rather than our intellects, to the child and his family. We must take the time to listen, and be willing to hear. Listening requires time, but hearing requires honesty and commitment. Likewise, the physician must be willing to speak openly and honestly; not to protect himself, but to assure the child and family of his sincere advocacy.

Honesty need not be mistaken for gloominess, however. A young child was recently heard to say to his mother, "I like it when you love me even though you are mad." Honesty in the context of love is never harmful; sometimes painful for both physician and patient, but never harmful. We cannot, and must not, try to avoid that pain. Communication is, of course, not only verbal. The touch carries special significance in medicine. It conveys to the child, in a way that words cannot, a soothing confidence and security. Even the youngest infant responds to it, and needs it desperately. It is too easy to stand at the bedside and speculate, or prognosticate, or forget to take the child's hand before examining his abdomen. The family must be touched and loved as well, not bypassed on the evening rounds because we know they will ask too many questions, or maybe have the effrontery to disagree with us. It is especially difficult to communicate lovingly with families who are angry, yet a simple touch and smile often dissipate anger instantly.

It is unrealistic, especially in chronic illness, to expect the physician-patient relationship to be always positive. Friction is inevitable. The doctor, though, is duty-bound to apply all his skills to the relief of that friction for the benefit of his patient. Such a feat can only be accomplished through the assumption of the posture of service, service given freely and willingly in love. I will close with a story. It is about an encounter with a child I once knew, and loved. ...

Thirteen years have passed since that San Francisco spring evening. Standing in the front of a crowded end-of-the-day elevator, I felt her calling. Debbie had been one of my first patients. A fresh and uncertain pediatric intern, I had cared for her for three weeks in July. I had become close to her and her mother, and I had visited her often on subsequent admissions. Now it was March, and she was in again. Cystic fibrosis had all but destroyed her lungs, and word had passed among the house staff that this admission would be her last. She was thirteen and feisty, in almost constant conflict with her caregivers. I had been imprisoned in the neonatal intensive care unit for six weeks. Concentrating on my own survival, I had neglected my visiting for a time.

I stepped off the elevator, glanced briefly out at the parking lot, went to a phone, and called home. "I'll be late," I said. She understood. "I love you," she said. "I love you too," I said back. I took the stairs, up six flights. Detouring briefly to the supply closet for a scrub shirt, I walked down the long corridor and paused at her door ... "NO SMOKING – OXYGEN IN USE" shouted the sign. I imagined it otherwise ... "ATTENTION – DYING PERSON INSIDE".

The shades were pulled and the room dark. In a rocking chair near the window was Debbie's mother. She smiled a greeting as I touched her arm. "Surprised to see me?" I thought but did not say. "She's been unconscious all day," she said. Debbie was almost hidden from view in a cloud of humified oxygen. Standing at the bedside, I took off my shirt and tie and pulled on the scrub shirt. Turning back the curtain of the mist test and slipping my upper body inside, I took her hand. It was deeply cyanotic and cold. Debbie was gasping, throwing her head back with each breath. "You are here," she said, her brown eyes wide open and clear. "I want to hug you." She had never hugged me. I bent over so that she could get her arms around my neck. It was an uncomfortable position. After a few minutes, her mother brought over a chair. I couldn't sit, but climbed on it on my knees and relaxed a little. "I love you," she whispered. "I love you, too," I said. There were no more words. Perhaps a half hour passed before the breathing stopped and her arms fell away.

Her mother and I turned off the monitors before they had time to make their obnoxious noises. Then we sat together and talked for a while. "I haven't seen you for days," she said. "What made you come tonight?" "Debbie called me," I replied. There was a long silence before either of us could say anything. "I'm glad I was listening," I said, and we parted.

*East Carolina University School of Medicine*
*Greenville, North Carolina*

## BIBLIOGRAPHY

1. Weil, W. B.: 1986, Review of 'Neglect of Chronically Ill Children', *American Journal of the Diseases of Children* **140**, 628.

JOHN C. MOSKOP

## LOVE AND THE PHYSICIAN:
## A REPLY TO THOMAS IRONS

In 'Loving the Chronically Ill Child: A Pediatrician's Perspective' [4], Thomas Irons exhorts his fellow physicians to rededicate themselves to the traditional but endangered values of service and love for their patients. These values are symbolized by and embodied in the house call, the healing touch, the soothing voice, and the listening ear, activities and skills central to the practice of an earlier generation of physicians. Irons identifies a number of major obstacles preventing today's physicians from achieving the ideals of service and love, including fragmentation in health care, limited access to care, paternalism, defensive medicine, retreat into science, and over-reliance on consultation. The latter four of these – paternalism, defensive medicine, retreat into science, and over-reliance on consultation – are described as attempts by the physician to avoid the vulnerability and pain which accompany identification with the patient and his suffering. I believe that physicians would do well to dedicate themselves to the ideals Irons has presented, and I agree that the obstacles he discusses seriously threaten the provision of compassionate care for chronically ill children. I do, however, also believe that describing the physician-patient relationship as a relationship of love requires further elaboration if it is not to be misunderstood, and this is what I will try to provide.

  Love is, of course, a central concept in Western thought; the term has been used to describe human activities as different as sexual passion and the beatific vision. How, then, should physicians or medical students interpret the call to love their patients? Irons describes this love as "genuine commitment of the physician's physical, mental, and emotional resources to the child and family he serves ([4], p. 323)." Clearly, then, Irons has in mind neither sexual passion nor the beatific vision. Rather, he speaks of love as a commitment of one's physical, mental and emotional resources to another. The physician should, in other words, dedicate his actions, thoughts, and feelings to the good of the persons he serves. Thus described, Irons' notion of love has a clear affinity with the Christian concept of love of neighbor, not in the sense of a generalized feeling of benevolence, but in the sense of a personal and active commitment to fostering the good of other persons. This commitment, moreover, is not just one of thought and action, but also of emotions; as the

*Loretta M. Kopelman and John C. Moskop (eds.),*
*Children and Health Care: Moral and Social Issues,* 331–335.
© *1989 by Kluwer Academic Publishers.*

story of Debbie suggests, Irons expects physicians to express feelings of warmth and affection to their patients.

No one, I think, would deny that physicians should be genuinely committed to their patients; the difficulty lies in determining what such a commitment entails. Note that Irons does not describe this commitment as total or unconditional. Clearly, it cannot be, since physicians have many obligations: to different patients, to family, to co-workers, to the community, and to themselves. If any one of these obligations, such as the obligation to serve one's patients, were total or unconditional, satisfying that obligation could make it impossible to fulfill any of the others. For example, the needs (and demands) of just one severely, chronically ill child or her family for care, guidance, emotional support, etc., can be enormous [5]. How far, then, should the physician's commitment to ministering to those needs extend? This is a crucial question, for without a sense of the limitations of commitment to patients, a physician may feel guilty or inadequate whenever he or she does not completely satisfy a patient's needs.

Irons has focused on the danger of physicians who love too little and on the mechanisms such physicians employ to distance themselves from their patients. In the rest of this commentary, I will consider the other side of the coin, namely, situations in which Irons' ideal of love, unless carefully interpreted and clearly understood, may be either counter-productive or unrealistic. Such situations include overidentification with patients, caring for difficult or hateful patients, and confronting scarcity of health care resources.

Irons criticizes the tendency of some physicians to distance themselves from their patients by means of defense mechanisms like paternalism and immersion in scientific questions. There is also, in the medical literature, a longstanding recognition of the dangers of over-identification with the patient and his suffering and a call for an attitude of "detached concern" on the part of the physician. Borgenicht summarizes this view as follows:

Many have argued that a good, well-adjusted physician needs both a conscious and unconscious sense of detached concern to function effectively as a clinician. For a physician to perform certain procedures, for example, a certain degree of emotional distance from the patient is necessary. Overinvolvement with the patient may prevent the physician from functioning objectively in crises ([1], p. 923).

More needs to be said here about what constitutes this "objective functioning" and why it is of value. Let me offer an example to illustrate what I think is meant here. Consider the great suffering and anxiety experienced by the loving parents of a young child with a life-threatening illness or injury. This child's physicians cannot experience the same emotions as the parents and still care effectively for the child, let alone all of their other patients. Thus, the physician's love for his or her patients should not involve the same emotional commitment as that of parental love.

Of course, the skills most needed by a neonatologist, or a surgeon, or an obstetrician may differ in important ways from those needed by a general pediatrician, internist, or family physician caring for a chronically or terminally ill patient. It will, for latter the group, probably be much more important to forge a strong personal relationship with the patient and his or her family. But even here, overidentification with the patient may interfere with good judgment and effective care. Stephens expresses this view as follows:

> I believe that I could have been a better intern and a better young physician, and that I would have learned more and suffered less if someone could have told me explicitly, repeatedly, and patiently that the dying at hand was not my own, that the patient whose death I attended was not, in fact, myself, nor was it my wife, nor my child, nor my parents, nor, fortunately, was it often my friend. And most important, I needed to be told and taught that the dying which I was attending did not, in itself, increase my vulnerability nor the vulnerability of those for whom I cared most deeply. The confusion involved in the sympathetic relationship, wherein identities merge and blur – this is what is intolerable and excruciating and blinding ([7], p. 323).

An example from the satirical novel *The House of God* [6] makes this same point in a humorous way. A senior resident in the novel develops a set of rules for the interns on his service. The first rule for interns upon entering the room of a patient who has arrested is to take *their own* pulse, lest they identify too greatly with the patient and faint dead away. Some defense mechanisms against identification with the patient, like this one, may not be all bad.

Thus far, I have suggested some reasons why physicians should, for their own sake and that of their patients, avoid too great an emotional attachment to their patients. Other situations raise the question whether physicians *can* form strong emotional commitments to all of their patients. (Remember that Irons describes love as a commitment of emotional as well as physical and mental resources.) I am thinking here of what are sometimes referred to as hateful patients [3], patients who are so angry, dependent, abusive, or self-destructive that they evoke strong negative reactions in the physicians who try to care for them. At least some children and adolescents with serious chronic or terminal illnesses will fit into one of these categories, and it may be difficult to find any way to forge a good relationship with them.

Physicians should not, of course, abandon these patients, and should strive to work with them. It may, however, be a tactical error to insist that physicians should *love* them. "How can I love this patient," a physician might respond, "when I can't stand him?" Perhaps it would be better here to speak not of love, but of a responsibility to provide service as respectfully and effectively as one can.

Finally, how does the commitment to love patients square with the painful prospect of shortages in health care? Irons points out that services may not be available for all children with cystic fibrosis or cancer who need them. If that is the case, physicians may sometimes have to deny their patients care which would benefit them and which is available to others.

Physicians might be able to refuse to provide certain kinds of care on the grounds that the resources saved could provide greater benefits for other patients. This is, according to Daniels [2], a possible justification for denial of health care in a closed system like that of the British National Health Service, but not in the United States, where denying care to one patient does not mean that the resources saved thereby will be used for other patients. Note that even in a closed system, one's commitment to individual patients cannot be total, since it is tempered by claims of justice in the distribution of care among all the patients served by the system. If the concrete expression of the physician's love for his patients is the provision of needed care, then scarce resources limit the physician's ability to express that love.

For American physicians, Daniels points out, the situation may be much more problematic from a moral point of view, since prospective payment systems and HMO's may attempt to control costs by offering physicians a financial incentive for limiting the care of their patients in certain ways. Such incentives may represent a strong temptation for physicians to compromise their commitment to their patients. Scarcity, then, may limit the physician's commitment to individual patients, and current attempts at cost containment may be driving physician and patient interests even further apart. Thus, resource limitations may come to be as serious an obstacle to the physician's ability to serve his patients as the defense mechanisms Irons describes.

In summary, then, I have tried to clarify Irons' application of the language of love to the physician-patient relationship by illustrating some of its potential dangers and limitations. Though I prefer Irons' term 'service' to 'love', I suspect that our images of the ideal pediatrician are very much alike.

*East Carolina University School of Medicine*
*Greenville, North Carolina*

## BIBLIOGRAPHY

1. Borgenicht, L.: 1984, 'Richard Selzer and the Problem of Detached Concern', *Annals of Internal Medicine* **100**, 923–928.
2. Daniels, N.: 1986, 'Why Saying No to Patients in the United States is so Hard', *New England Journal of Medicine* **314**, 1380–1383.
3. Groves, J. E.: 1978, 'Taking Care of the Hateful Patient', *New England Journal of Medicine* **298**, 883–887.
4. Irons, T.: 1989, 'Loving the Chronically Ill Child: A Pediatrician's Perspective', in this volume, pp. 323–329.
5. Perrin, J. M.: 1985, 'Clinical Ethics and Resource Allocation: The Problem of Chronic Illness in Childhood', in J. C. Moskop and L. Kopelman (eds.), *Ethics and Critical Care Medicine*, Kluwer Academic Publishers, Dordrecht, Holland, pp. 105–116.
6. Shem, S.: 1979, *The House of God*, Dell, New York.
7. Werner, E. R. and Korsch, B. M.: 1976, 'The Vulnerability of the Medical Student: Posthumous Presentation of L. L. Stephens' Ideas', *Pediatrics* **57**, 321–328.

# NOTES ON CONTRIBUTORS

Dan W. Brock, Ph.D., Professor and Chair, Department of Philosophy, Brown University, Providence, Rhode Island.

H. Tristram Engelhardt, Jr., M.D., Ph.D., Professor, Center for Ethics, Medicine and Public Issues, Baylor College of Medicine, Houston, Texas.

Peter C. English, M.D., Ph.D., Associate Professor of Pediatrics and Associate Professor of History, Duke University, Durham, North Carolina.

Myron Genel, M.D., Professor of Pediatrics, Associate Dean and Director, Office of Government and Community Affairs, Yale University School of Medicine, New Haven, Connecticut.

Angela R. Holder, L.L.M., Counsel for Medicolegal Affairs, Yale-New Haven Hospital and Yale University School of Medicine, Clinical Professor of Pediatrics (Law), Yale University School of Medicine, New Haven, Connecticut.

Robert L. Holmes, Ph.D., Professor, Department of Philosophy, The University of Rochester, Rochester, New York.

Thomas G. Irons, M.D., Associate Professor, Department of Pediatrics, School of Medicine, East Carolina University, Greenville, North Carolina.

Loretta M. Kopelman, Ph.D., Professor and Chair, Department of Medical Humanities, School of Medicine, East Carolina University, Greenville, North Carolina.

John Ladd, Ph.D., Professor, Department of Philosophy, Brown University, Providence, Rhode Island.

Rosalind Ekman Ladd, Ph.D., Professor, Department of Philosophy, Wheaton College, Norton, Amherst, Massachusetts.

Robert J. Levine, M.D., Professor, Department of Internal Medicine, Yale University School of Medicine, New Haven, Connecticut.

Gareth B. Matthews, Ph.D., Professor, Department of Philosophy, University of Massachusetts at Amherst, Amherst, Massachusetts.

John C. Moskop, Ph.D., Associate Professor, Department of Medical Humanities, School of Medicine, East Carolina University, Greenville, North Carolina.

William Ruddick, Ph.D., Professor, Department of Philosophy, New York University, New York, New York.

*Loretta M. Kopelman and John C. Moskop (eds.)*
*Children and Health Care: Moral and Social Issues*, 337–338.
© 1989 *by Kluwer Academic Publishers.*

Todd L. Savitt, Ph.D., Professor, Department of Medical Humanities, School of Medicine, East Carolina University, Greenville, North Carolina.

Stuart F. Spicker, Ph.D., Professor of Community Medicine and Health Care (Philosophy), School of Medicine, University of Connecticut Health Center, Farmington, Connecticut.

Barbara F. Starfield, M.D., M.P.H., Professor and Head, Division of Health Policy, School of Hygiene and Public Health, The Johns Hopkins University, Baltimore, Maryland.

Ann L. Wilson, Ph.D., Professor, Department of Pediatrics and Adolescent Medicine, School of Medicine, University of South Dakota, Sioux Falls, South Dakota.

# INDEX

Abbott, Grace, 44, 49
Abortion, 222
  and minors, 169, 170
    availability of, 13
    parental consent for, 169
    right to refuse, 164
  debate over, 223
Ackerman, T.F., 75
Adams, Samuel S., 266
Addams, Jane, 30
Adolescents
  and informed consent, 162–165
  autonomy vs. parental rights, 161–162
  choices in circumstances of death, 115
  frequency of births to, 9
  heightened concern with physical appearance in, 187
  literature, 118
  suicide in, 8
Advocacy, 244, 305–322, 327
  by case, 305, 313
  by class, 309–310, 313
  of pediatrician, 305-313, 318, 320–321
AFDC Program, 7, 16
Afterlife, 135, 136
Alcohol, use by high school students, 8
American Academy of Pediatrics, 46–49, 50, 51, 56, 59, 60, 61, 153, 231–232, 248, 306, 311
American Medical Association (AMA), 28, 46–49, 50, 51, 61, 69, 70, 231–232, 248, 253, 263, 265
  Section on Diseases of Children, 265
*American Medical News,* 57–58
American Pediatric Society, 253
  establishment of, 28
American Public Health Association, 47
Amniocentesis, 222

Anemia, 11, 14, 17
*Apology, The*, 108, 114
Appendicitis, 17
Arena, Jay M., 251
Aristotle, 286, 291, 292, 298, 303
Assent,
  capacity to, 79
  definition of, 80
  in research 78–80
Asthma, 9
Authority
  decisionmaking, 213, 218–219
  of the state 33, 46, 253
  of U.S. federal government, 27, 56
  parental, 234
Autonomy. *See also* Self-determination
  and children, 75–76
  respect for children's, 75

Babies Hospital, New York, 258
Baby Calvin, 309
Baby Doe cases, 303–306, 309, 313
Baby Doe Regulations, 67, 70, 232, 235–247, 244, 282–284, 310–313, 319–321
  and Handicapped Infant Hotline, 59
Baby Fae, 244, 308, 309
Baby Jesse, 244, 308, 309
Bailey, Joseph, 32, 33, 38
Baily, Ashley, 308
Baker, Josephine, 30, 39
Barnett, Henry, 261
Barnhart, Henry, 35
Bayard, Thomas, 46
Beecher, H.K., 98
Behrman, Richard, 261
Bellevue Hospital
  Outdoor Bureau of Relief, 249, 251
Benevolence, 332
Bentsen, Lloyd, 60
Best interest standard, 221–229, 321–237

340 INDEX

Biomedical research. *See* Research.
Births, registration of, 35, 46, 47
Birthweight, influences on, 14
Bluebond-Langner, Myra, 105, 107, 121, 128, 142–145, 147–148, 149, 150
Borah, William, 33
Borgenicht, Louis, 332
Bradley, William N., 262
Brazelton, T. Berry, xii
Brennemann, Joseph, 265
Brent, Sandor B., 135, 147
British Medical Association, 263
Brock, Dan, 148, 181–212, 213, 216, 218
Brothers, A., 251
Bureau of Child Hygiene, 30

Campbell, Alexander, 236, 310
Cancer, 122, 164
Capacity, legal, 73–85, 89–98
Capron, Alexander, 206–207
Caraway, T.H., 48
Carey, Susan, 133–136
Care
  comparison of ordinary and extraordinary, 232–234
  extraordinary, 222, 319
Carnegie, Andrew, 67
Case advocacy, 243–244, 305–315
Casebeer, J.B., 254
Catholic Church,
  moral theology of, 232
  opposition to Sheppard-Towner Act, 46
Centers for Disease Control, 11
Certified Milk movement, 29–30
CHAMPUS, 307
Chapin, Henry Dwight, 242, 255, 259, 261
*Charlotte's Web*, 103–139
Child abuse, 173, 174, 217, 235, 252
Child Abuse Prevention and Treatment Act of 1984, 310, 311, 312, 317
Child development, 185, 253, 255, 259–261
  measured by X-rays, 260, 265

Child labor, 28, 29, 33, 34
Child neglect, 277–278
Children
  advocacy for, 27–65
  and consent, 78
  benefit of research to, 78, 84, 85
  best interests of, 226
  constitutional issues in health care, 174
  decisionmaking in terminal illness, 105
  diseases of, 248–252
  effects of poverty on health, 58
  health care needs, xiii
  moral issues in treatment of, 173
  obligations to, 231–237
  organ transplants in, 306–309
  pretense and the terminally ill, 149
  protection from harm, 75
  research on, xiii, 3, 73–86
  respect for autonomy of, 75
  right to custody, 75
  right to health care, 56
  suicide in, 8
  trust and the terminally ill, 149
  U.S. federal role in health care, 27
  well-being of, 27–65, 216, 241–242, 244
Children's Bureau, 4, 5, 27–65, 67–69
  educational role of, 37
Children's Defense Fund, 58
Children's health
  social aspects of, 255
Children's health policy, 7–21
Children's hospitals
  history of, 28, 242, 257–259, 278
Children's literature, 121–131
  as therapeutic, 102–131
Chloramphenicol, 77
Christopher, W.S., 255
Choice
  children's capacity for, 194
Chronically ill children, 323–328
  demands on physician, 332
Class advocacy, 309–310, 313
Cocaine, 7
Coit, Dr. Henry L., 29–30
Commitment
  and due process, 167

INDEX 341

of adults, 166
of minors, 175
Committee on Economic Security, 50
Common law
  view of children as chattel, 161
Communication
  and terminally ill children, 103, 105, 110, 133–146, 150–151, 327
  in physician-patient relationship, 245
Communism, 46, 48
Community
  as moral entity, 303–304
  family as, 234
  importance of, 243
  moral, 74
Compassion, of physicians, 295
Competence
  and informed consent, 189
  and legal policy, 189–212, 205–209
  capacities needed for, 183–185
  children's in health care decision-making, 75, 181, 185–189
  concept of, 182–184
  implications for medical practice, 205–209
  in consent for research, 90
  standards of, 181–219
  values at stake in determining, 190–198
*Conceptual Change in Childhood*, 133
Congenital metabolic conditions
  screening for, 14
*Congressional Record*, 47, 62
Consent
  deception studies, 92
  informed. *See* Informed consent.
  parental or guardian proxy, 73–85
Constitutional issues, in health care of children, 174
Contraception
  and the minor, 168–169, 175
  parental responsibility and use in minors, 175
  right to privacy in use of, 175
Copeland, Royal, 47
Crippled Children's program, 52, 54
*Crito, The*, 108, 114
Cystic fibrosis, 325

Daniels, Norman, 334
Darwin, Charles, 67
Davis, James, 48
Death
  and literature, 108, 121–131
  and self-determination for children, 115, 149, 181–209
  as annihilation, 136
  as natural, 114, 121–122
  children's conceptions and developmental psychology, 104, 133–146
  children's understanding of, 105, 107–120, 128–129, 133–147
  communication, 102, 103, 105, 110, 122, 133–146, 148–151
  in children, 107–120, 128–129, 319
  registration of, 47
Decisional authority, of children in health care. *See* Competence
Decisionmaking
  and best interest of children, 226
  capacities, 148, 185–189, 213–219
  child-parent role, 199, 209, 221–229, 232
  cultural differences in, 221
  discontinuing life-saving treatment, 231–232, 319
  parent-physician, 313
  physician's role in, 235–247
  risk/benefit ratios in, 225
  values at stake, 181–212
Declaration of Helsinki, 73
Defense mechanisms, 128, 245, 326, 332, 333
Defensive medicine, 245
Dehumanization, as threat to doctor-patient relationship, 244
Demilt Dispensary, 248–249
Department of Health and Human Services (DHHS), 54, 55, 56, 57, 59, 60, 79, 81–83, 91, 98
  'squeal rule', 168
Detached concern, 332
Developmental Psychology, and children's concept of death, 104, 133–146
Diabetes, 17
Dignity, of dying child, 116

342 INDEX

Diphtheria anti-toxin, 259, 264
Disclosure of terminal illness, 105, 137, 140
Divorce, effect on children, 8, 12
Doctor, Good (virtuous) 242–243, 281–302, 323, 331–334
Doctor-patient relationship. *See* Physician-patient relationship
Drug trials, 73–85
Duff, Raymond, 236, 294, 303, 310
Duty and moral goodness, 284–286

Eaton, Charles, 51
Education
  of mothers and children, 43, 46
  role of Children's Bureau in, 37
Elderly, 4, 8
Eliot, Martha, 52, 55
Emancipated minor. *See* Minor.
Embryology, 253
Emergency Maternity and Infant Care (EMIC) program, 53
Emergency, treatment of children in, 161
Emotions, in physician care, 332
Engelhardt, H. Tristram, Jr., 231–237, 244, 317–322
English, Peter, 241, 242, 247–273, 275–278
Environmental toxins, risk to children, 12
Ethics, *passim*
  minimalistic, 288–291, 303
  occupational, 222
  quandary, 282
Euthanasia, infant, 222
Evil, problem of, 104, 121, 125–127
Evolution, theory of, 67

Fairness
  in allocation of resources, 3
  in children's health policy, 7–12
Family, 103, 197, 231–237
  interests of, 221–229
  as community, 234
  cost of treatment as burden to, 231–237
  government involvement in, 70
  of World War II servicemen, 53

  relationship with physician, 245
Federal government, U.S. *See* U.S. government.
Fetal research, 76
Fiske, Charles, 306, 309
Food and Drug Act (FDA), 92
Food support programs, 4
Foundling homes, 252–253, 257, 278
France, Joseph, 43, 62
Freedman, Benjamin, 75
Frelinghuysen, Joseph, 43
*Frothingham v. Mellon et al*, 44

Gallinger, Senator Jacob, 32
Genel, Myron, 243, 306–307, 311, 317–321
General Federation of Women's Clubs, 36
General practitioner, 254–258
Gerstemberger, Henry J., 261
Gesell, Arnold, 260
Gesell, Judge, and Handicapped Infant Hotline, 59
Ginott, Haim, xii
*Good Housekeeping*, 39, 44, 46, 47
Goodness and virtue, 281–302
Good, conception of, 193–194, 216–217
Gore, Albert, Jr. 306, 307
Gould, Jay, 67
Grandparent responsibility statute and adolescent contraception, 169
Great Depression, and social security for children, 49–53
Griffith, J.P. Crozer, 256–257, 261, 265
Grisso, T., 185–186, 188

Handicapped Infant Hotline, 59
Handicapped newborns, 317–321
Harriet Lane Home, 251–252
Harris, R.R., 252
Hatfield, Henry, 48
Healing touch, 323, 325, 327–328
Health care
  federal involvement in children's, 7–21, 27–62
  fragmentation of, 325–327
  right of children to, 27, 56
Health care professional, 293

assessment of capacity to consent, 186
assessment of children's best interests, *passim*
  roles and relationships, 281–302, 323–338
  virtues of, 281–302, 323–338
Health Insurance Organizations, 17
Health insurance
  Americans without, 23
  compulsory, 69
  in Europe, 69
Health Maintenance Organizations (HMO's)
Health status, and plight of children in U.S., 23
*Healthy People*, 57
Healy, Edwin, 233–234
Heckler, Margaret, 57
Hegel, G.F., 24
Henderson, L.H., 293
Henry Street Settlement, establishment of, 30
Hepatitis, Willowbrook studies, 91
Hippocratic principle, 222
Hitchcock, Gilbert, 33
Holder, Angela R., 161–172, 174, 213, 218
Holmes, Robert, 149, 173–179, 213–219
Holt, Luther Emmett, 258, 260–261, 264
House calls, 323–324
*House of God, The*, 333
House Science and Technology Committee, 244
Howland, John, 258, 261
Hull House, foundation of, 30
Hunger, eradication of, 18
Hygiene
  bureaus for children's, 38
  in infants and children, 253, 255, 260–261, 264
Hypothyroidism, 18

Identification, of physician with patient, 245, 332
Illness, terminal. *See* Death.
Immigrants, 68

Immortality. *See also* Afterlife.
  hope for, 104
  suggestion of in *Charlotte's Web*, 111
Immunizations, 14, 16, 18
Incompetence, of minors, 213–219. *See also* Competence.
Infant Care Review Committees, 236, 312, 320
*Infant Care*, 37
Infant mortality, 5, 13, 29, 35–37, 44, 47, 49, 54, 57, 114, 264
  black, 58
  downward trend in, 49
  in the 1980's, 56–60
  international, 58, 62
  neonatal, 13
  postneonatal, 13, 14, 58
  white, 58
Infants
  feeding of 259–261
  impaired, 22
  low birthweight in, 58
  premature, 22
  rights of handicapped, 59
Informed consent, 73–85, 89–98
  adults' right to waiver, 182
  and compliance, 193
  and determining standards of competence, 189
  and law, 161–177
  assessment in health care professionals, 186
  by parental proxy, 181–212
  capacities needed for, 213–219
  in research, 79–80
  of minors, 78, 182, 205
  third-party, 92
  voluntary, 188–189, 214
Institute of Medicine, 58
Institutional Review Board (IRB) 78–83, 91, 92, 97
Intimacy, value of, 197
Irons, Thomas, 242, 244, 323–329, 332, 333, 334

Jacobi, Abraham, 241, 252–254, 256, 259–261, 266, 275, 276
Jacobi, Mary Corinna Putnam, 253

Janofsky, B.A., 97–98
Jenkins, Thomas, 52
Job, Book of, 123
Johns Hopkins Hospital, 251–252
Johnson, Lyndon B., 55
Justice
　distributive, 317–319
　for research subjects, 78

Kant, Immanuel, 286, 297
Kelley, Florence, 31, 37, 40, 62
Kelley, Samuel, 256
Kennedy, John F., 54
Kenyon, Senator, 41
Ketoacidosis, 11, 17
Keyes, Frances, 44
Kopelman, Loretta, 6, 89–99, 112–113, 118–119, 121–131
Krugman, Saul, 91
Kubler-Ross, Elizabeth, 118

Ladd, John, 242, 281–304
Ladd, Rosalind Ekman, 103, 104, 107–131
*Ladies Home Companion*, 39
*Ladies Home Journal*, 39
*Laissez-faire*, 28, 67–69
Lathrop, Julia, 30, 35, 38, 44
Law, and consent, 161–177
Layton, Caleb, 42, 51, 62
Lead poisoning, 8, 9, 11
Lea, Clarence, 43
Legal policy
　and competence, 205–209
　comparison of minors and adults, 181
Lenroot, Katharine, 52
Lesser, A.J., 28
Leukemia, 105, 142, 325
Levine, Robert J., 6, 73–87, 89, 90, 94, 97
Life expectancy, 8
Lindsay, John D., 34
*Literary Digest*, 41
Literature. *See* Children's literature.
*Little Women*, 107
Locus of control, 186, 188, 216
Loma Linda Medical Center, 309
London, Arthur, 251, 263

Love, 112, 245, 323–329, 331–334
Low birthweight, 9, 12
Low-income children. *See also* Poverty.
　frequency of low birthweight in, 9
　immunizations in, 9

MacMurray, John, 303–304
Malaria, 18
Malpractice, crisis in, 323
Marijuana, 7
*Massachusetts v. Mellon*, 44
Maternal and Child Health Care Block Grant, 56, 57
Maternity and infancy, bill for protection of, 38
Maternity care, 54
Maternity centers, establishment of, 46
Matthews, Gareth, 104, 105, 133–146, 147, 148, 151
Mayo, Charles, 49
McClanahan, Harry, 263
McCormick, R.A., 74–75
McKay, R. James, 261
Media, 243, 305–315, 309, 313
Medicaid, 5, 7, 13, 14, 16, 17, 24, 55, 56, 57, 165
Medical advances
　affecting mothers and children, 54
Medical education, 281–282
　pediatrics, 256–257, 261–263, 267–270
Medical neglect, withholding of medically indicated treatment, 312
Medical profession
　authority in decisionmaking, 221–229
　dehumanization of, 323
　influence on public policy, 67
　legalism in, 284
　moral goodness in, 284
　opposition to Sheppard-Towner Act, 5
　optimism in decisionmaking, 221–229
　roles and relationships, 331–334
　specialization, 242, 275–278
　values compared to parental, 222

Medical reductionism, 292
Medical roles and relationships, 281–301
Medication, for children, 116, 128
Meissner, Friedrich Ludwig, 276
Meningitis, 17
Mental health system, and the minor, 11, 165–168
Mental retardation, 54
Midwife, 241
Military spending, cost compared to health care, 4, 18, 24
Minimal risk, 76, 78, 81–85, 89–98
   critique of, 95–99
   definition of, 92, 95–99
   minor increments above, 82–85, 89–99
Minor
   and abortion, 169
   and contraception, 168–169
   and sexual activity, 170, 175
   and the mental health system, 165–168
   commitment of, 167–168
   emancipated, 162, 206
   law in relation to, 162–163, 165–166, 206
   mature, 161, 163, 165, 206
   rights of, 161–179
Mitchell, A. Greene, 261
Moral community, 74
Moral theology, Catholic, 232
Morgan, J.P., 67
Morse, John Lovett, 257, 262
Mortality and morbidity, childhood, 7–12, 29
Mortality rates, 36–37, 44, 47, 49
   in foundling homes, 252
   maternal, 54
Mortality, infant. *See* Infant Mortality.
Moskop, John, 105, 147–152, 331–338, 245

Nagel, Thomas, 117
Nagy, M.H., 133
National Association for Retarded Children, 54
National Association Opposed to Woman Suffrage, 41

National Child Research Act, 54
National Commission for the Protection of Human Subjects of Biomedical and Behavioral Research, 6, 73–86
National Consumer's League, 31
National Health Service, British, 334
National Health Service Corps, 18
National Institute of Child Health and Human Development, 54
National Institutes of Health (NIH), 92
National Maternal and Child Health Resource Center, 56
National Organ Transplant Act, 308
*Nation*, 40
Natural
   as contrasted with the supernatural, 123
   as good, 103, 109, 118, 122–126
   meaning of, 104, 123
Nelson, Waldo E., 261
Neonatologists, 224
Newberger, Eli, 305
Newborns, severely handicapped, 309, 317–321
Newsholme, A., 29, 57
Nineteenth Amendment, 40
Noeggerath, Emil, 256
Northrop, William P., 258
Nuremberg Code, 73, 85
Nurse
   as pediatric professional, 241
   as 'social visitor,' 261
Nutrition
   in infants and children, 254, 259

Ober, William B., 115
Obligations, to children, 321–237
Obstetrician
   and prenatal care, 224
   as expert in care of mother, 241
Office of Maternal and Child Health Care, 57
Omnibus Reconciliation Act of 1981, 56
Opportunity, equality of, 3, 4, 18
Organ transplants in children, 244, 306–309
Orphan drugs, 77

Orphanages. *See* Foundling homes.
Osgood, Kenneth, 60
*Outlook, The*, 37
Outpatient department, 265
O'Dwyer, Joseph, 258

Pain medication, for children, 128
Palliative treatment, 128
Parental consent
  emergency exception to, 161
  for abortion in minor, 169
  in treatment of adolescents, 164
Parent-physician decisionmaking, 313
Parent-physician relationship, privacy of, 320
Parents
  and protection of children's welfare, 149
  and terminally ill child, 150
  as clients, 22
  authority in research, 81
  decisionmaking authority regarding children, 175–178, 196–197, 221–229, 234, 244
  interests of, 191, 196–197, 209, 217–219, 227–229, 317–320
  responsibilities of, 217–220
  role in decisionmaking, 221–229, 232, 310
  suffering of, 333
  values of, 222–225
Park, Edwards A., 258
Pasteurization, 264
Paternalism, 105, 138, 221–229
  and parental concern for children, 174
  and protection of children, 174, 199
  and the terminally ill child, 146
  as defense mechanism of physician, 245
  as problem in chronic illness, 325
Patient
  commitment of physician's resources to, 245
  rights of, 283
  suffering of, 295
Pediatrician
  as advocate, 242–243, 305–313, 318, 320–321
  characteristics of, 247
  relationship with parents, 222, 224, 283
  role of, 256
Pediatrics, xiii, 28, 241
  American textbooks on, 260–261
  history of 247, 275–278
  scope of, 253–259
  social, 261–263
Peters, Andrew, 33
*Phaedo, The*, 108, 114–116, 122–125, 129
Phenylketonuria, 18
Phocomelia, 77
Physician
  as professional, 243, 282
  as technologist, 282
  as craftsman, 281–301
  attitude in care of sick child, 245, 333, 342
  commitment to patients, 332
  compassion of, 295
  minimalistic model of, 291–301
  moral requirements of, 283
  role in decisionmaking, 235–237
  virtues of, 323–329, 331–334
  visits, 16
Physician-patient relationship, 245, 292–294, 323–328, 331–334
Piaget, Jean, 186
Pincoffs, Edmund, 282
Pirquet, Clemens F. von, 257
Pittman, Key, 43
Planned Parenthood, 168
Plato, 101, 103, 104, 107, 116, 118, 122, 125, 126–127, 174
Pope Pius XII, 233–234
Prenatal care, 222
  and obstetrician, 224
Poverty
  as contributor to disease, 254
  effect on children, 3, 4, 7, 8, 18, 19, 38
Preferred Provider Organizations, 17
*Prenatal Care*, 37
President's Commission for the Study of Ethical Problems in Medicine and Biomedical and Behavioral Research, 148, 312

Primary care-giver role, 323–328
Privacy
   constitutional right to, 168–170, 175
   of parent-physician relationship, 320
   right to in abortion decision, 196
*Private Worlds of Dying Children, The*, 107
Professionalism
   model of physician, 292–301
   requirements of, 288
Progressive Era, 60, 68
   and children's health care, 4, 27–44
Progressive Movement, 33, 56, 70
Prospective payment systems, 334
Psychology. *See* Developmental psychology.
Public Health Service (PHS), 92
   Office of Maternal and Child Health Care, 57
Public Health Service Act of 1970, Title X, 168
Public policy
   influence of medical profession on, 67
   neglect of children in, 3, 4
   on research, 78–85
   relation to children's health care, 241

Quality of life, 231–232

Ramsey, Paul, 6, 73–75
Rankin, Jeanette, 38
Reagan administration, 5, 56, 57, 59, 60, 67, 70, 307–308, 310–311
Reagan, Ronald, 56, 60
Reed, James, 34, 40, 41, 42
Rehabilitation Act of 1973, 59, section 504, 310, 313
*Republic, The*, 107
Research, 6, 73–86, 89–99
   and fetuses, 76
   and pregnant women, 76
   and public policy, 78–85
   animal studies, 79
   assent and consent in, 78–80, 97
   benefit to children of, 73, 84, 85, 91
   children in, 73–85, 89–98, xiii
   child's 'deliberate objection' to, 80
   classification of studies by
      risk/benefit ratio, 91
   confidentiality in, 97
   informed consent in, 97
   labeling in, 97
   minimal risk in, 78, 81–85
   non-therapeutic studies prohibited in, 90
   risk categories, 91–92
   risk in, 73–85, 89–98
   stigmatization in, 97
   therapeutic, 73–85, 89–98
*Research Involving Children*, 73
Research subjects,
   children as, 73–85, 89–98
   fetuses as, 76
   non-consenting, 73–85, 89–98
   pregnant women as, 76
Resource allocation, 245, 307–308
Respect for persons, as ethical principle, 73–85, 89–98
Respect, for autonomy of children, 75–76
Rheumatic fever, 9, 14
Rights
   of minors
      and commitment, 168
      in giving birth, 170
      to health care, 4
      to refuse treatment, 163
   of parents in child-raising, 197
   of privacy in abortion, 169
Risk
   assessment in research, 89–99
   in research, 73–85, 89–98
   adolescent consideration of, 163
   minimal, 76, 78, 81–85, 89–98, 121–131
   minor increase over minimal, 92–99
   no discernible, 74–76
   physical, 96–99
   psychosocial, 97–99
   routine, 96–99
Risk/benefit ratios, in decisionmaking, 225
Rockefeller, John D., 67
Roddey, Oliver, 305
Roosevelt, Franklin, 5, 50, 52, 67, 70
Roosevelt, Theodore, 4, 29, 31, 32, 40, 62, 67

Rosenthal, Edwin, 259
Rotch, Thomas Morgan, 255–256, 259–260
Rothstein, William, 276
Royster, Lawrence T., 256
Ruddick, William, 221–229, 231–232, 235–237
Rudolph, Abraham, 261

Sauthoff, Harry, 51
Savitt, Todd L., 67–71, 275–279
School lunch programs. *See* Nutrition
Schools, racial segregation in, 54
Self-determination
   child's capacity for, 78
   of children in health care decision-making, 107–131, 149, 162–170, 193–196, 198, 199, 325
Self-interest, individual, 67
Self-sacrifice, 227–229
Service, of physician, 327, 334
Shaw, Henry, 263
Sheppard Towner Act, 5, 61, 68–70
Sheppard, Morris, 38
Shirkey, H.C., 76, 90
Smith, Henry F., 173
Smith, Job Lewis, 258, 260
Social Darwinism, 5, 68
Social responsibility, 75
Social Security Board, 51
Social Security Act of 1935, 5, 56
   Title V, 51, 56, 70
Social security, for children, 49–53
Social worker, 325. *See* also Health care professional.
Socrates, 104, 114, 116–117, 136
Specialties, formation of, 276–278
Speece, Mark W., 135, 147
Speece-Brent survey, 135
Spicker, Stuart, 4, 23–26, 243, 303–304
Spock, Benjamin, xii
Sprago, J.S., 28, 29
Squeal rule, DHHS regulations regarding contraception, 168
Standard Developmental Account
   and Bluebond-Langner, 144
   and disclosure, 140–141
   and paternalism, 138
   and view of immortality, 136–137
Stanford, Leland, 67
Starfield, Barbara, 3, 7–21, 23, 70, 97
State's rights, 27–50, 70
State
   authority in maternal and child health care, 46
   involvement in family affairs, 70
Stevens, Supreme Court Justice, 168
Stillman, William O., 34
Stoicism, 121
Straus, Nathan, 30, 34
Suffering
   of children, 128–129
   of patients, 295
   of sick child's parents, 333
Suicide, 8
Supererogation, 242, 288–291
Surgeon General's Report on Health Promotion and Disease Prevention, 57
Survival-of-the-fittest, philosophy of, 68–69

Team care, 326
Terminal illness. *See* Death.
Thalidomide, 77
Therapeutic orphans, children as, 6, 76–78
Thomas, Charles, 40
Thomas, Dylan, 118
*Times*, 41
Title V of Social Security Act, 51, 52, 53, 54, 55
Touching, wrongful, 74
Towner, Horace, 38
Transplantation
   in children, 244, 306–309
   required request, 308
   Task Force on, 308
Truman, Harry S., 54
Trust
   and the capacity to consent, 215–216
   between patient and physician, 216
Truthfulness, 141, 146, 327
Twentieth Annual Report of the Chief of the Children's Bureau to the Secretary of Labor, 49

Understanding. *See* Competence.
U.S. Census Bureau, mortality rates, 36
U.S. Census of 1870, child labor statistics, 28
U.S. Congress, 27–65, 67, 306, 311
U.S. Constitution, 27, 33
U.S. government
    authority of, 56, 60
    funding through block grants, 56
    role in social security, 53
U.S. Public Health Service, 55
U.S. Supreme Court, 49, 310, 313
    and abortion, 169
    and contraception, 168
    and rights of minors regarding commitment, 167

Vaccines, effectiveness of, 14
Values, parental compared to professional, 222
Vaughn, Victor C., 261
Veeder, Borden, 265–266
Venereal disease, and the treatment of minors, 162
Vierling, L., 185–186, 188
Virtue and goodness, 281–302
Virtues
    as character traits, 286–288
    as dispositions, 298
    as intrinsic
    as praiseworthy, 287–288
    as supererogatory, 288–291
    concept of, 296–300
    craftsman model, 303
    defined, 286–288
    listed 286
Voluntariness and capacity, 181–212

Wald, Lillian, 30
Walker, Bailus, 57
Ward, Robert, 91
Weil, William, 325
White House Conference on Children and Youth, The, (1960) 54, 57
White, E.B., 103, 104
Wiles, Ira S., 261
Willowbrook hepatitis studies, 91
Wilson, Ann, 4, 27–65, 67, 69, 70
Witte, Edwin, 50
*Woman Patriot*, 46
Women, pregnancy and drug trials, 76
Women, Infants and Children Program (WIC), and nutrition for children, 14
Women's suffrage, xiii, 40, 44
World Health Organization, annual budget cost, 18
World War II, 56
    health care for wives and children of military, 5

*The Philosophy and Medicine Book Series*

*Editors*

H. Tristram Engelhardt, Jr. and Stuart F. Spicker

1. **Evaluation and Explanation in the Biomedical Sciences**
   1975, vi + 240 pp.   ISBN 90–277–0553–4
2. **Philosophical Dimensions of the Neuro-Medical Sciences**
   1976, vi + 274 pp.   ISBN 90–277–0672–7
3. **Philosophical Medical Ethics: Its Nature and Significance**
   1977, vi + 252 pp.   ISBN 90–277–0772–3
4. **Mental Health: Philosophical Perspectives**
   1978, xxii + 302 pp.   ISBN 90–277–0828–2
5. **Mental Illness: Law and Public Policy**
   1980, xvii + 254 pp.   ISBN 90–277–1057–0
6. **Clinical Judgment: A Critical Appraisal**
   1979, xxvi + 278 pp.   ISBN 90–277–0952–1
7. **Organism, Medicine, and Metaphysics**
   *Essays in Honor of Hans Jonas on his 75th Birthday, May 10, 1978*
   1978, xxvii + 330 pp.   ISBN 90–277–0823–1
8. **Justice and Health Care**
   1981, xiv + 238 pp.   ISBN 90–277–1207–7(HB)   90–277–1251–4(PB)
9. **The Law-Medicine Relation: A Philosophical Exploration**
   1981, xxx + 292 pp.   ISBN 90–277–1217–4
10. **New Knowledge in the Biomedical Sciences**
    1982, xviii + 244 pp.   ISBN 90–277–1319–7
11. **Beneficence and Health Care**
    1982, xvi + 264 pp.   ISBN 90–277–1377–4
12. **Responsibility in Health Care**
    1982, xxiii + 285 pp.   ISBN 90–277–1417–7
13. **Abortion and the Status of the Fetus**
    1983, xxxii + 349 pp.   ISBN 90–277–1493–2
14. **The Clinical Encounter**
    1983, xvi + 309 pp.   ISBN 90–277–1593–9
15. **Ethics and Mental Retardation**
    1984, xvi + 254 pp.   ISBN 90–277–1630–7
16. **Health, Disease, and Causal Explanations in Medicine**
    1984, xxx + 250 pp.   ISBN 90–277–1660–9
17. **Virtue and Medicine**
    *Explorations in the Character of Medicine*
    1985, xx + 363 pp.   ISBN 90–277–1808–3
18. **Medical Ethics in Antiquity**
    *Philosophical Perspectives on Abortion and Euthanasia*
    1985, xxvi + 242 pp.   ISBN 90–277–1825–3(HB)   90–277–1915–2(PB)
19. **Ethics and Critical Care Medicine**
    1985, xxii + 236 pp.   ISBN 90–277–1820–2

20. **Theology and Bioethics**
    *Exploring the Foundations and Frontiers*
    1985, xxiv + 314 pp.   ISBN 90–277–1857–1
21. **The Price of Health**
    1986, xxx + 280 pp.   ISBN 90–277–2285–4
22. **Sexuality and Medicine**
    *Volume I: Conceptual Roots*
    1987, xxxii + 271 pp.   ISBN 90–277–2290–0(HB)   90–277–2386–9(PB)
23. **Sexuality and Medicine**
    *Volume II: Ethical Viewpoints in Transition*
    1987, xxxii + 279 pp.   ISBN 1–55608–013–1(HB)   1–55608–016–6(PB)
24. **Euthanasia and the Newborn**
    *Conflicts Regarding Saving Lives*
    1987, xxx + 313 pp.   ISBN 90–277–3299–4(HB)   1–55608–039–5(PB)
25. **Ethical Dimensions of Geriatric Care Value Conflicts for the 21st Century**
    1987, xxxiv + 298 pp.   ISBN 1–55608–027–1
26. **On the Nature of Health**
    *An Action-Theoretic Approach*
    1987, xviii + 204 pp.   ISBN 1–55608–032–8
27. **The Contraceptive Ethos**
    *Reproductive Rights and Responsibilities*
    1987, xxiv + 254 pp.   ISBN 1–55608–035–2
28. **The Use of Human Beings in Research**
    *With Special Reference to Clinical Trials*
    1988, xxii + 292 pp.   ISBN 1–55608–043–3
29. **The Physician as Captain of the Ship**
    *A Critical Appraisal*
    1988, xvi + 254 pp.   ISBN 1–55608–044–1
30. **Health Care Systems**
    *Moral Conflicts in European and American Public Policy*
    1988, xx + 368 pp.   ISBN 1–55608–045X
31. **Death: Beyond Whole-Brain Criteria**
    1988, x + 276 pp.   ISBN 1–55608–053–0
32. **Moral Theory and Moral Judgments in Medical Ethics**
    1988, vi + 232 pp.   ISBN 1–55608–060–3
33. **Children en Health Care: Moral and Social Issues**
    1989, xxiv + 349 pp.   ISBN 1–55608–078–6